Progress in Optical Science and Photonics

Volume 21

The purpose of the series Progress in Optical Science and Photonics is to provide a forum to disseminate the latest research findings in various areas of Optics and its applications. The intended audience are physicists, electrical and electronic engineers, applied mathematicians, biomedical engineers, and advanced graduate students.

More information about this series at https://link.springer.com/bookseries/10091

Cheng Liu · Shouyu Wang · Suhas P. Veetil

Computational Optical Phase Imaging

 Springer

Cheng Liu
Jiangnan University
Wuxi, China

Shouyu Wang
Jiangnan University
Wuxi, China

Suhas P. Veetil
Higher Colleges of Technology
Dubai, United Arab Emirates

ISSN 2363-5096   ISSN 2363-510X   (electronic)
Progress in Optical Science and Photonics
ISBN 978-981-19-1640-3   ISBN 978-981-19-1641-0   (eBook)
https://doi.org/10.1007/978-981-19-1641-0

This Springer imprint is published by the registered company Springer Nature Singapore Pte Ltd.
The registered company address is: 152 Beach Road, #21-01/04 Gateway East, Singapore 189721,
Singapore

Preface

Over the last several decades, contrast enhancement techniques based on spatial phase distribution have provided a new paradigm for advanced visualization in scientific and clinical applications involving samples that are transparent or have uniform transmittance amplitudes. Recent advances in phase retrieval algorithms and numerical techniques have led to the development of new visualization modes that have evolved from being a stunning visual tool to providing a quantitative interpretation of all physical parameters based on phase imaging. With computational phase imaging, many physical limitations of conventional imaging can be overcome, resulting in several advances in image magnification, spatial resolution, and contrast. By combining the advantages of microscopy, interferometry, holography, and iterative numerical computations, several techniques have been developed to record and restore phase information, including ptychographic iteration machines, Fourier ptychographic microscopy, and coherent modulation imaging. Artificial intelligence (AI), usually in the form of deep learning, has also been successfully applied to phase imaging, achieving accuracy comparable to that of classical interferometry. Many new applications are now possible in the field of optical phase imaging with the help of advanced numerical computational algorithms.

The purpose of this book is to provide a comprehensive and self-contained introduction to computational phase imaging. The book is divided into six chapters. In the first chapter, we provide a theoretical understanding of optical imaging, spatial resolution, magnification, and optical phase imaging. In the second part, we discuss classical qualitative phase contrast techniques, including phase contrast and differential interference contrast microscopy. Our third chapter discusses the most popular interferometric techniques. There is a detailed discussion of non-interferometric methods in chapter four, including Shark-Hartmann sensing, the transport intensity equation, and phase retrieval techniques that iteratively calculate the light phase via a mathematical approach of derivative descent, including the G-S algorithm, the Fienup algorithm, and coherent modulation mapping. Using MATLAB® as a simulation tool, this book shows the reader how phase estimation works for different computational phase imaging modalities, enabling them to gain a deeper understanding of the physical and mathematical principles behind phase estimation. This fifth part presents some of the typical applications of the above techniques, including the measurement of deformations, observation of biological samples, and diagnosis of laser beam quality. The last part of this book discusses some recent developments in computational phase imaging methods.

The book provides researchers and professionals with a comprehensive and rigorous overview of the theoretical and applied aspects of computational phase imaging. In recent decades, both computation and instrumentation have substantially improved in the field of computational phase imaging. It was challenging to cover so many developments in six chapters, however, the authors did their best to cover key concepts, methods, applications, as well as recent trends in almost 300,000 words. The book developed largely from papers published by the authors and lectures and computer programs used in photo-electronic information science and engineering graduate courses at Jiangnan University, Wuxi, China. Students who have completed a course in modern optics or a similar one, such as optoelectronics, optical engineering, photonics, biophotonics, or applied physics may find this book helpful. By publishing the book, we hope to encourage researchers interested in optical imaging to explore new possibilities.

Wuxi, China   Cheng Liu
Wuxi, China   Shouyu Wang
Dubai, United Arab Emirates   Suhas P. Veetil
January 2022

Contents

Chapter 1
Introduction to Computational Phase Imaging

Smooth surfaces of metal, glass, and water can all create a virtual image by changing the direction of light according to the principle of geometric optics [1]. Thus, mirrors became the oldest imaging devices and are widely used in daily life, although the image produced cannot be recorded directly and can only be observed by the naked eye. A tiny pinhole is the simplest device to produce a recordable image, which is the geometric projection of the object being observed, and its spatial resolution is inversely proportional to the size of the pinhole [2]. To obtain images with sufficiently high resolution, the diameter of the pinhole must be very small, which greatly reduces the energy of the received light. Pinhole methods are therefore only suitable for imaging extremely bright objects such as candle flames and electronic sparks, and have limited applications in both scientific research and industry. Our understanding of how we see things around us and our ability to make good lenses led to the development of optical imaging technology [3].

The human eye is a sophisticated optical imaging system, as shown in Fig. 1.1. When we look at something, our eye lens transmits the intensity distribution of light on its surface to the retina, generates a bioelectronic signal proportional to the intensity, and transmits it to our brain to create the sensation of "seeing" [4]. Practical optical imaging techniques essentially mimic this biological process. Glass and other transparent materials are used to create lenses that capture light and produce accurate reproductions of the intensity distribution of objects [5]. A light-sensitive film or electronic device is used to record the intensity distribution [6]. All imaging systems, such as giant astronomical telescopes, optical microscopes, transmission electron microscopes, and household cameras, operate on the same principle. The only difference is the spatial resolution, which is determined by the wavelength and the numerical aperture of the optical systems [7].

© The Author(s), under exclusive license to Springer Nature Singapore Pte Ltd. 2022
C. Liu et al., *Computational Optical Phase Imaging*, Progress in Optical Science
and Photonics 21, https://doi.org/10.1007/978-981-19-1641-0_1

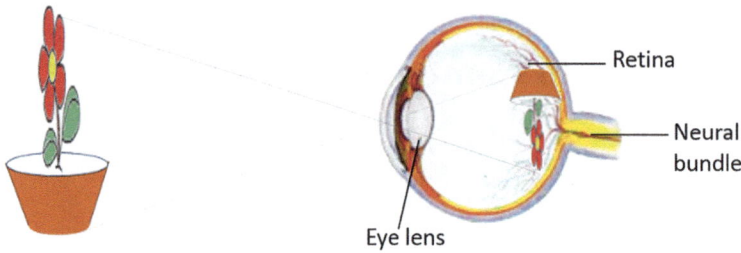

Fig. 1.1 Imaging principle of human eye

1.1 Fundamentals of Optical Imaging

Image formation by a convex lens of focal length f is shown in Fig. 1.2. The object O is at a distance of S_0 from the lens and has a height of h_0. The axial and vertical positions of the image I are determined by the well-known Gaussian formulas as [8]

$$\frac{1}{S_I} + \frac{1}{S_0} = \frac{1}{f}, \quad \frac{h_I}{S_I} = \frac{h_0}{S_0} \tag{1.1}$$

These positions are determined by the principle of geometry optics. The transverse amplification of such an imaging system can be defined as $M = \frac{S_I}{S_0}$ which gives an axial amplification of M^2 [8]. Thus for $M > 1$, the image obtained will be "fat" in axial direction as shown in Fig. 1.2.

Since light is electric–magnetic, the point object O radiates divergent spherical wave fronts which are shown in Fig. 1.2. The function of a lens is to transfer the divergent spherical wave into that of a converging one. The divergent light wave reaching the lens plane can be written as

$$U(y) = \frac{1}{\sqrt{S_0^2 + (y - h_0)^2}} e^{i2\pi k \sqrt{S_0^2 + (y - h_0)^2}} \tag{1.2}$$

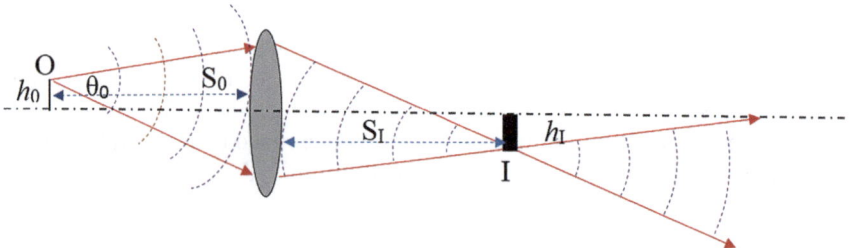

Fig. 1.2 Basic principle of imaging with a lens

Under paraxial approximation $U(y)$ can be written as

$$U(y) = \frac{1}{S_0} e^{i2\pi k\left[S_0 + \frac{(y-h_0)^2}{2S_0}\right]} \tag{1.3}$$

By neglecting the third and higher orders of Taylor series of $\sqrt{S_0^2 + (y - H_0)^2}$, the convergent wave leaving lens plane can be simplified as

$$U'(y) = \frac{1}{S_0} e^{-i2\pi k\left[S_I + \frac{(y+H_I)^2}{2S_I}\right]} \tag{1.4}$$

The transmission function of lens is calculated as

$$t(y) = \frac{U'(y)}{U(y)} = e^{-i2\pi k\left[S_I + \frac{(y+h_I)^2}{2S_I} + S_0 + \frac{(y-h_0)^2}{2S_0}\right]} \tag{1.5}$$

By neglecting the constant phase terms and using $\frac{h_I}{S_I} = \frac{h_0}{S_0}$,

$$t(y) = e^{-i2\pi \frac{ky^2}{2}\left[\frac{1}{S_I} + \frac{1}{S_0}\right]} = e^{-i\pi \frac{k}{f} y^2} \tag{1.6}$$

The light leaving a point object O radiates at an angle of 4π in space and only a small amount of light falling within a cone of angle θ_0 is collected by the lens. It is then transformed into a convergent light that falls in a cone of angle $\frac{S_0}{S_I}\theta_0$. For simplicity without losing generality, for a given point object on the optical axis, the complex amplitude of light on the image plane can be expressed as

$$h_c(y) = \int_{-k\sin\left(\frac{S_0}{2S_I}\theta_0\right)}^{k\sin\left(\frac{S_0}{2S_I}\theta_0\right)} e^{i2\pi yk\sin\alpha} d(k\sin\alpha) \tag{1.7}$$

Equation (1.7) is essentially the Fourier transform of rectangular function $\mathrm{rect}\left[\dfrac{k_y}{2k\sin\left(\frac{S_0}{2S_I}\theta_0\right)}\right]$, and thus

$$h_c(y) = 2k\sin\left(\frac{S_0}{2S_I}\theta_0\right)\sin c\left[2k\sin\left(\frac{S_0}{2S_I}\theta_0\right)y\right] \tag{1.8}$$

$h_c(y)$ is plotted as a solid curve in Fig. 1.3a, whose first zero emerges at $y = \left[2k\sin\left(\frac{S_0}{2S_I}\theta_0\right)\right]^{-1}$. Thus the transverse size of the image of a point object becomes $\left[k\sin\left(\frac{S_0}{2S_I}\theta_0\right)\right]^{-1}$. Hence any two point objects that are separated by a distance smaller than $\left[2Mk\sin\left(\frac{S_0}{2S_I}\theta_0\right)\right]^{-1}$ on object plane cannot be distinguished.

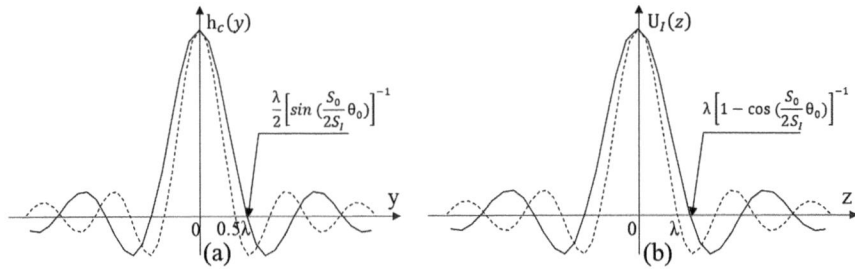

Fig. 1.3 Spatial resolutions in **a** transverse and **b** axial directions

$\left[2Mk\sin\left(\frac{S_0}{2S_I}\theta_0\right)\right]^{-1}$ essentially corresponds to the so-called spatial resolution of the optical imaging system. An increase in spatial resolution requires a higher value for $2k\sin\left(\frac{S_0}{2S_I}\theta_0\right)$ which can be achieved by using a shorter wavelength to obtain a larger k and using a bigger lens to increase θ_0. The maximum value of $\sin\left(\frac{S_0}{2S_I}\theta_0\right)$ is 1.0, and it is shown as a dotted curve in Fig. 1.3a where the first zero emerges at $y = 0.5\lambda$. This explains why the highest transverse resolution of an optical imaging system is half of the wavelength used.

$h_c(y)$ in Eq. (1.8) is the coherent point spread function of the imaging system. For an extended object $O(y)$, each point on the object at a vertical distance of y_0 from optical axis will generate an image with amplitude of $h_c(y - My_0)$, and the overall amplitude on the image plane becomes

$$U(y) = \int O(y_0)h_c(y - My_0)dy_0 = \frac{1}{M}O\left(\frac{y}{M}\right) \otimes h_c(y) \tag{1.9}$$

Equation (1.9) means that the obtained image amplitude is the convolution of ideal geometrical image with the point spread function of the optical system.

With the same method, the light field along the optical axis can be written as

$$U_I(z) = 2 \int_{k\cos\left(\frac{S_0}{2S_I}\theta_0\right)}^{k} e^{i2\pi zk\cos\alpha}d(k\cos\alpha) \tag{1.10}$$

Equation (1.10) is the Fourier transform of $\text{Rect}\left[\frac{k_y - 0.5k\left(1+\cos\left(\frac{S_0}{2S_I}\theta_0\right)\right)}{k\left(1-\cos\left(\frac{S_0}{2S_I}\theta_0\right)\right)}\right]$. So $U_I(z)$ can be rewritten as $2e^{i\pi kz\left(1+\cos\left(\frac{S_0}{2S_I}\theta_0\right)\right)}\sin c\left[kz\left(1 - \cos\left(\frac{S_0}{2S_I}\theta_0\right)\right)\right]$. The amplitude of $U_I(z)$ is shown in Fig. 1.3b as a solid curve, whose first zero emerges at $z = \lambda\left[1 - \cos\left(\frac{S_0}{2S_I}\theta_0\right)\right]^{-1}$ where $\lambda\left[1 - \cos\left(\frac{S_0}{2S_I}\theta_0\right)\right]^{-1}$ is the axial resolution of the imaging system. Since light should propagate along the optical axis, $\cos\left(\frac{S_0}{2S_I}\theta_0\right)$

cannot take negative values. When $\cos\left(\frac{S_0}{2S_I}\theta_0\right)$ is equal to zero, the first zero of $U_I(z)$ emerges at $z = \lambda$ as shown in Fig. 1.3b with a dotted line and thus the highest axial resolution reachable for any imaging system is lower than the wavelength used.

From the above, it is clear that the position of the image of a point object is determined by the geometric parameters of the optical alignment used, and the spatial resolution of the optical system depends on the wavelength and the acquisition angle of the lens. The ultimate goal of most imaging techniques is to achieve higher spatial resolution and gain.

1.2 Limitations of Common Intensity Imaging

Textbooks often treat optical imaging as if the sample were 2D; however, in reality, all practical samples have finite thickness. In other words, all samples are 3D in practice. The imaging of a 3D object with a convex lens is shown schematically in Fig. 1.4. The outermost layers of a 3D object, a giraffe and a flower, are represented in two discrete planes with a separation of d_0 between them with respective transmission functions $t_1(x, y)$ and $t_2(x, y)$.

With parallel illumination, the transmitted light through the giraffe layer $t_1(x, y)$ forms an illumination of $U_{\text{illu}}(x, y)$ on the flower layer which is the Fresnel propagation of $t_1(x, y)$ by a distance d_0, that is, $U_{\text{illu}}(x, y) = \mathcal{F}_{\text{res}}[t_1(x, y), d_0]$. The transmitted light leaving this layer is $U_{\text{trans}}(x, y) = \mathcal{F}_{\text{res}}[t_1(x, y), d_0]t_2(x, y)$. When d_0 is much smaller than S_0, the amplification of both layers can be approximated to be the same and can be assumed to be 1.0. If a detector is placed at the image plane of the flower, it records an intensity $|U_{\text{trans}}(x, y)| = |\mathcal{F}_{\text{res}}[t_1(x, y), d_0]t_2(x, y)|$, which is intensity image of flower layer $t_2(x, y)$ multiplied by diffraction pattern of giraffe $\mathcal{F}_{\text{res}}[t_1(x, y), d_0]$. If the detector is moved further by Δd from the image plane of flower, the recorded intensity becomes $\mathcal{F}_{\text{res}}[U_{\text{trans}}(x, y), \Delta d]$. So, the recorded intensity is always a mixture of two layers which makes it difficult to identify the structural information contained in each layer separately. A 3D sample consists of several such layers and the structural information of the whole sample becomes inseparable along the axial direction. This is the reason why we need thin layered samples in biological research and disease diagnosis.

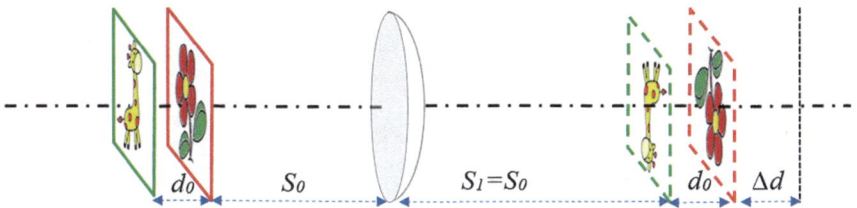

Fig. 1.4 Optical imaging of thick sample

Diffraction within such a thin sample is negligible and the light beam passes directly through the sample without changing its direction. The transmission function of the sample under uniform illumination U_0 can be approximated as follows [9]

$$t(x, y) = U_0 e^{-u(x,y)d} \tag{1.11}$$

Generally, $u(x, y)$ has complex value and thus can be written as $u(x, y) = u_r(x, y) + iu_i(x, y)$

$$t(x, y) = U_0 e^{-u_r(x,y)d - iu_i(x,y)d} \tag{1.12}$$

d is the thickness of sample. $u_r(x, y)$ is the 2D absorbing coefficient, with its value varying between 10 mm^{-1} and 100 mm^{-1}, depending on light wavelength and nature of the sample. If d is only a few micrometers, the range of $u_r(x, y)d$ is much smaller than 0.1, so the sample becomes transparent, resulting in poor contrast of the transmitted light intensity $|t(x, y)|$, thereby suppression of relevant information about the sample. Several dying methods were developed as an alternative to enhance the contrast and a huge number of dying chemicals were invented to enhance the variation of $u_r(x, y)$. However, dying the sample changes its characteristics or functionality.

Theoretically, optical imaging systems cannot provide useful information about the structure of purely transparent samples because their absorption coefficients $u_r(x, y)$ are uniform. Optical imaging systems suffer from the diffraction effect, and the intensity drops off at points where the term $u_i(x, y)d$ changes abruptly. For some samples, such as binary phase plates and cultured cells, their shapes or edges appear in the intensity and phase images, as in Fig. 1.5. The acquired image clearly shows the boundaries between regions with zero phase and π-phase.

The phenomenon in Fig. 1.5 is explained by the convolution effect of imaging with optical lens, that is, the formed image is essentially the convolution between then real object $U_0(y)$ and the point spread function $h_c(y)$, that is $U_I(y') = \int_{-\infty}^{+\infty} h_c(y' - y)U_o(y)dy$ [14]. If the object is a pure phase grating with phase $\varphi(y) = \pi \text{comb}(y - 2n) \otimes \text{rect}(2y)$ as shown in Fig. 1.6a, the complex amplitude of the image formed will be $U_I(y') = \int_{-\infty}^{+\infty} h_c(y' - y)e^{i\varphi(y)}dy$. For an ideal diffraction limited imaging system with pupil of a in diameter, $h_c(y)$ is equal to $\text{sinc}\left(\frac{ay}{\lambda d_i}\right)$,

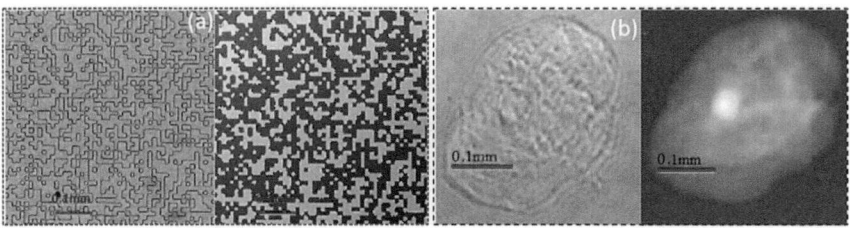

Fig. 1.5 Wide field intensity and phase images of **a** binary pure phase plate and **b** cultured oral cell

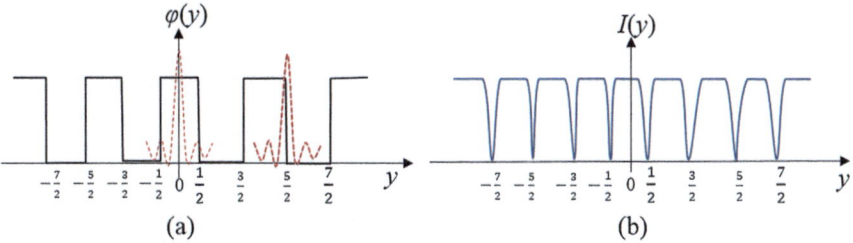

Fig. 1.6 **a** Convolution effect in imaging of a pure phase grating and **b** corresponding intensity image

where λ is the wavelength used, and d_i is the distance of image plane to lens. Then $U_I(y')$ is $\int_{-\infty}^{+\infty} \sin c\left[\frac{a(y'-y)}{\lambda d_i}\right] e^{i\varphi(y)} dy$. Since the value of $\sin c\left(\frac{ay}{\lambda d_i}\right)$ decreases quickly with increasing $|y|$, the value of $U_I(y')$ is dominated by the integration in the range of the main lobe of $\sin c\left[\frac{a(y'-y)}{\lambda d_i}\right]$. That is, $U_I(y') \approx \int_{y'-\frac{\lambda d_i}{a}}^{y'+\frac{\lambda d_i}{a}} \sin c\left[\frac{a(y'-y)}{\lambda d_i}\right] e^{i\varphi(y)} dy$.

For $y' = 0$, $\sin c\left(\frac{ay}{\lambda d_i}\right)$ is at the origin of y-axis, $U_I(0) = \int_{-\frac{\lambda d_i}{a}}^{\frac{\lambda d_i}{a}} \sin c\left[\frac{ay}{\lambda d_i}\right] e^{i\pi} dy$, and

$|U_I(0)|^2 = \left|\int_{-\frac{\lambda d_i}{a}}^{\frac{\lambda d_i}{a}} \sin c\left[\frac{ay}{\lambda d_i}\right] dy\right|^2$, since $\sin c\left[\frac{ay}{\lambda d_i}\right]$ is positive in the range $\left[-\frac{\lambda d_i}{a}, \frac{\lambda d_i}{a}\right]$,

$|U_I(0)|^2$ is definitely larger than 0. For $y' = \frac{5}{2}$, $\sin c\left(\frac{ay}{\lambda d_i}\right)$ is at the edge of π phase and

zero phase. $\left|U_I\left(\frac{5}{2}\right)\right|^2 = \left|\int_0^{\frac{5}{2}+\frac{\lambda d_i}{a}} \sin c\left[\frac{a\left(\frac{5}{2}-y\right)}{\lambda d_i}\right] dy - \int_{\frac{5}{2}-\frac{\lambda d_i}{a}}^{0} \sin c\left[\frac{a\left(\frac{5}{2}-y\right)}{\lambda d_i}\right] dy\right|^2 \approx 0$.

The intensity image formed is shown in Fig. 1.6b, which has zero intensity at $y = n + \frac{1}{2}$ and outlines the edges between the regions of zero phase and π-phase. This is the reason why many snaky segments appear in Fig. 1.5a. In natural transparent samples, there is no remarkable phase jump in most cases, and since $U_I(y')$ is never zero, the contrast of the images taken with a classical bright microscope is quite poor and insufficient to see many details of the sample details as shown in Fig. 1.5(b).

1.3 Intensity Imaging to Phase Imaging

When light passes through a sample, it experiences a phase delay of $\varphi(x, y) = \frac{2\pi d}{\lambda} n(x, y)$ where $n(x, y)$ is the refractive index and its value depends on the physical and chemical structure of the sample under study. For most biological samples $n(x, y)$ varies around 1.3, λ is around 0.5 μm for visible light. When d is only 10 microns, the phase delay can reach a value of 1.2, even with a small variation of 0.01 in $n(x, y)$. This shows that phase measurement can be an alternative solution for imaging problems with transparent samples. Since light is an electromagnetic wave, it has also a temporal component $2\pi vt$ in addition to the spatial phase $(x, y) = \frac{2\pi d}{\lambda} n(x, y)$. Therefore, the phase of the transmitted light can be written in the form $\varphi(x, y, t) = 2\pi\left(\frac{d}{\lambda}n(x, y) + vt\right)$. The electric field of the light in the x–y plane can then be expressed as $E(x, y, t) = |E(x, y)|\cos[\varphi(x, y) + \omega t]$, where $|E(x, y)|$ is the amplitude of light and its square $|E(x, y)|^2$ can be measured by CCD or other electronic devices, $\varphi(x, y) + \omega t$ is its phase at the coordinate (x, y) of time t when its strength oscillates at the angular frequency of ω. When the detector has a faster response rate than the angular frequency of ω, the recorded light changes periodically between $-|E(x, y)|$ and $+|E(x, y)|$ with the frequency of ω, where the negative sign "$-$" means electric field changes direction and the spatial phase can be determined as $\varphi(x, y) = \cos^{-1}\left[\frac{E(x,y,t)}{|E(x,y,t)|_{max}}\right] - \omega t$. However, detectors can only record the time average of the square of the electronic field according to Eq. (1.13), because the frequency of light is of the order of 10^{17} Hz for visible light, much higher than the response rate of any available electronic device. The spatial phase information of the light $\frac{2\pi d}{\lambda}n(x, y)$ which is able to highlight the structural difference, is lost during the recording and the detector records only the intensity as

$$\overline{I(x, y) = |E(x, y, t)|^2} = |E(x, y)|^2\overline{\cos^2[\varphi(x, y) + \omega t]} = 0.5|E(x, y)|^2 \quad (1.13)$$

1.4 Basic Principles of Computational Phase Imaging

Since it is not possible for physical detectors to capture the optical phase information directly, only indirect decoding methods can recover the lost phase from the recordable intensity. Phase contrast microscopy and differential interference contrast microscopy are two such classical methods of phase imaging based on the fact that the generated intensity $I(x, y)$ is proportional to the phase term $\varphi(x, y)$ and its spatial derivatives [10, 11] and therefore $I(x, y)$ itself is a direct indication of the phase distribution $\varphi(x, y)$ obtained without additional calculations. However, this is not the case for thick samples with large values of $\varphi(x, y)$ where the intensity $I(x, y)$ is not strictly proportional to $\varphi(x, y)$ or its derivatives. Therefore, these two methods have always been used for biological sample observations and cannot be used for quantitative measurements. In other phase imaging techniques such as interferometry, digital holography, transport intensity equation (TIE), and coherent diffraction imaging (CDI) etc., the distribution of recorded $I(x, y)$ is significantly different from that of $\varphi(x, y)$, and various algorithms have been developed to compute $\varphi(x, y)$ from $I(x, y)$.

The main difficulty in measuring the phase term $\varphi(x, y)$ of light is that $E(x, y, t)$ changes very rapidly and no electronic detectors can operate on the frequency scale of light. Therefore, to realize phase measurements, we should generate a slowly changing signal to replace $E(x, y, t)$ and use it to calculate the phase of light $\varphi(x, y)$. If two light fields with frequencies ω_1 and ω_2 reach the same coordinate (x, y), a beat signal with frequency $\omega_1 - \omega_2$ can be generated, which is resolvable for a common detector. Combining two such light fields $E_1(x, y, t) = |E_1(x, y)|e^{i[\varphi_1(x,y)+\omega_1 t]}$ and

$E_2(x, y, t) = |E_2(x, y)|e^{i[\varphi_2(x,y)+\omega_2 t]}$, we obtain the resultant intensity as

$$I(x, y, t) = |E_1(x, y)|^2 + |E_2(x, y)|^2$$
$$+ 2|E_1(x, y)||E_2(x, y)|$$
$$\times \cos[\varphi_1(x, y) - \varphi_2(x, y) + (\omega_1 - \omega_2)t] \quad (1.14)$$

If $E_2(x, y, t)$ is a uniform parallel light beam, which can be written as $E_2(x, y) = E_0$ and $\varphi_2(x, y) = 0$, $I(x, y, t)$ will become

$$I(x, y, t) = |E_1(x, y)|^2 + |E_0|^2 + 2|E_1(x, y)||E_0| \cos[\varphi_1(x, y) + (\omega_1 - \omega_2)t] \quad (1.15)$$

If $(\omega_1 - \omega_2)$ is much lower than the response rate of detector, a time sequential signal in the form of cosine function can be recorded at coordinate (x, y), and then $\varphi_1(x, y)$ can be calculated as $\varphi_1(x, y) = \cos^{-1}\left[\frac{I(x,y,t)-|E_1(x,y)|^2-|E_0|^2}{2|E_1(x,y)||E_0|}\right] - (\omega_1 - \omega_2)t$, realizing phase imaging.

If $E_1(x, y, t)$ and $E_2(x, y, t)$ in Eq. (1.14) are from the same laser source, their frequencies become $\omega_1 = \omega_2 = \omega$, then

$$I(x, y) = |E_1(x, y)|^2 + |E_0|^2 + 2|E_1(x, y)||E_0| \cos[\varphi_1(x, y)] \quad (1.16)$$

The intensity of $I(x, y, t)$ in Eq. (1.16) becomes static and is recordable for common detectors, and the phase $\varphi_1(x, y)$ can be calculated as $\varphi_1(x, y) = \cos^{-1}\left[\frac{I(x,y,t)-|E_1(x,y)|^2-|E_0|^2}{2|E_1(x,y)||E_0|}\right]$. Since $\cos^{-1}\left[\frac{I(x,y,t)-|E_1(x,y)|^2-|E_0|^2}{2|E_1(x,y)||E_0|}\right]$ cannot distinguish $\varphi_1(x, y)$ from $-\varphi_1(x, y)$, additional techniques like phase shifting and off-axis illuminating reference were developed and widely applied to decide the sign of $\varphi_1(x, y)$ [12, 13].

The principle for indirect phase measurement shown in Eq. (1.16) was called interferometry. Various techniques and instruments have been developed based on this idea, such as the Twyman-Green interferometer, Fizeau interferometer, and ordinary off-axis holography, in which a separate parallel or spherical light beam $E_2(x, y, t)$ was used as a reference light to obtain a static $I(x, y)$. Interferometry can also be realized via an approach called self-interference. Shearing interferometers, for example, detect light $E(x, y, t)$ interferences with its shifted replica $E(x + \Delta x, y, t)$, and the phase deviation $\varphi'(x, y)$ can then be deduced from static $I(x, y)$, and finally obtain $\varphi(x, y)$ as an integral of $\varphi'(x, y)$ along the x-axis.

Phase imaging can also be realized by computationally retrieving the phase from the recorded intensity pattern. According to the principle of Fresnel diffraction, when a light field $|E(x_0, y_0)|e^{i\varphi(x_0, y_0)}$ propagates a distance of z from x_0-y_0 plane to x-y plane, its complex light amplitude $E(x, y)$ on x-y plane can be written as [14]

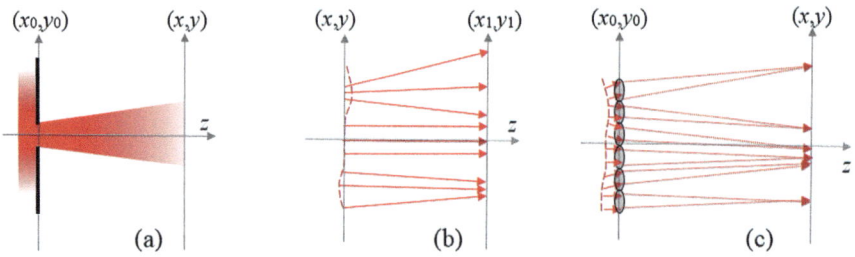

Fig. 1.7 Imaging methods of **a** coherent diffraction imaging, **b** transport intensity equation, and **c** Shark-Hartmann sensor

$$E(x, y) = \frac{1}{j\lambda z}e^{j\frac{k}{2z}(x^2+y^2)} \iint |E(x_0, y_0)|e^{i\varphi(x_0, y_0)}e^{j\frac{k}{2z}(x_0^2+y_0^2)}e^{-j\frac{2\pi}{\lambda z}(x_0xy+y_0y)}dy_0dx_0$$

$$(1.17)$$

Though the phase distribution $\varphi(x_0, y_0)$ is encoded into the spatial distribution of $|E(x, y)|^2$, it is lost during data acquisition and cannot be calculated analytically in most of the cases. However, by adding additional constraints on $|E(x_0, y_0)|e^{i\varphi(x_0, y_0)}$, its phase $\varphi(x_0, y_0)$ can be iteratively retrieved using various computational algorithms.

Figure 1.7a shows such an example, where $|E(x_0, y_0)|e^{i\varphi(x_0, y_0)}$ is confined by a tiny hole and a detector at some distance records a diffraction pattern $|E(x, y)|^2$. Then, $|E(x_0, y_0)|$ and $\varphi(x_0, y_0)$ can be iteratively retrieved using Error-Reduction (ER) or Hybrid-Input-Out (HIO) algorithm developed by Fienup [15]. These types of imaging modalities [16, 17] have been evolved into a sub-branch of phase imaging, coherent diffraction imaging (CDI). In contrast to iterative methods, the transport of intensity equation (TIE) is another category of phase extraction method that uses propagation to extract the phase directly rather than iteratively. The principle of this method is shown in Fig. 1.7b. An intensity image $|E(x, y)|^2$ is recorded in the xy plane and a defocused image $|E(x_1, y_1)|^2$ is recorded at a slightly defocused x_1y_1 plane, and the light phase $\varphi(x, y)$ in xy plane is calculated by comparing the intensities $|E(x, y)|^2$ and $|E(x_1, y_1)|^2$. The dotted curve indicates the curvature of the wave-front $\varphi(x, y)$ of a light beam at xy plane with a uniform intensity $|E_0|^2$. Due to waves having different curvatures at different parts of a wavefront, the transmitted light is reorganized either convergently or divergently, resulting in an intensity redistribution. Since the upper part of $\varphi(x, y)$ has a convex wave-front, light rays are diverging and the intensity $|E(x_1, y_1)|^2$ formed by these rays is smaller than $|E_0|^2$. The middle part of $\varphi(x, y)$ is planar where the light rays travel parallel to the optical axis and leave the intensity $|E(x_1, y_1)|^2$ roughly equal to $|E_0|^2$. The lower part of $\varphi(x_0, y_0)$ has a concave wavefront, causing light rays to converge to redistribute the intensity $|E(x, y)|^2$ to be stronger than $|E_0|^2$. The phase $\varphi(x, y)$ can be computed by comparing intensities in both planes, and this is the basic idea behind Transport Intensity Equation (TIE) [18–20], which is widely studied and applied to biological sample observations. The Shack-Hartmann wavefront sensor is another phase measurement technique works on the principle of geometric optics. The $|E(x_0, y_0)|e^{i\varphi(x_0,y_0)}$ in Eq. (1.16) is tilted by a small angle θ with respect to the optical axis, it causes $E(x, y)$ to shift transversely and the phase ramp of $|E(x_0, y_0)|e^{i\varphi(x_0,y_0)}$ can be computed from the shift in $E(x, y)$. As shown in Fig. 1.7c, by splitting $|E(x_0, y_0)|e^{i\varphi(x_0,y_0)}$ into many sub-apertures and focusing it using micro-lens arrays, the phase ramp of light in each sub-aperture can be computed from the transverse shift of each focus from its central positions. This is the basic principle of Shark-Hartmann sensor which has found several applications for astronomical observations.

Technologies for imaging and measuring phase can be broadly divided into two basic groups based on their ability to recover phase: a qualitative method for phase imaging and a quantitative method for phase measurement. The first group typically includes phase contrast microscopy, differential interference contrast microscopy, and the schlieren method. The second group includes interferometry, holography, coherent diffraction imaging, Hartmann sensing, transport intensity equation, and artificial intelligence-based methods, etc. The category of these methods is shown in Fig. 1.8. We try to present all these techniques and their applications in this book systematically according to their characteristics and history. The second chapter discusses qualitative phase representation and imaging. Typical interferometric methods of quantitative phase imaging are discussed in Chap. 3 and methods of phase representation without interferometry are discussed in Chap. 4. Several typical

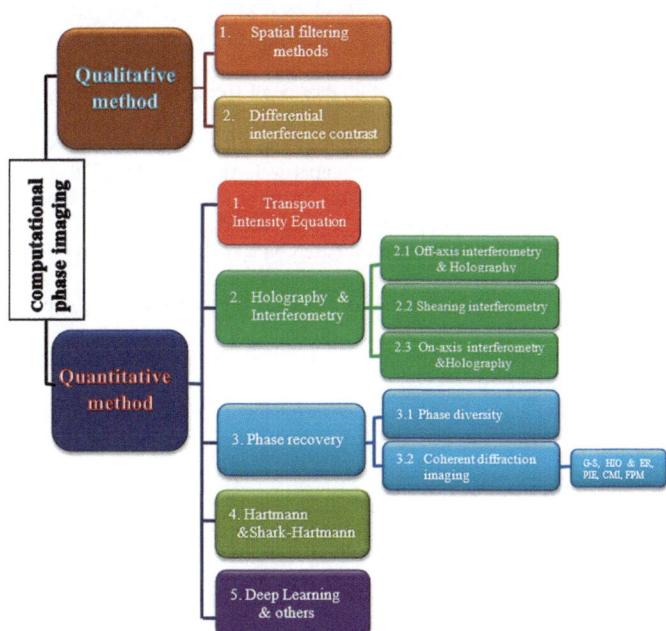

Fig. 1.8 Category of computational phase imaging method

applications of these methods are discussed in Chap. 5, and the trend and future of phase representation is briefly discussed in Chap. 6.

References

1. Romano, A., Cavaliere, R.: Geometric Optics. Springer, Berlin (2016)
2. Renner, E.: Pinhole Photography, 4th edn. Focal Press, Oxford (2008)
3. Sun, H.: Lens Design: A Practical Guide (Optical Sciences and Applications of Light), 1st edn. CRC Press, Boca Raton (2016)
4. Atchison, D., Smith, G.: Optics of the Human Eye. Butterworth- Heinemann, Oxford (2000)
5. Malacara-Hernández, D., Malacara-Hernández, Z.: Handbook of Optical Design, 3rd edn. CRC Press, Boca Raton (2013)
6. Schwartz, K.: The Physics of Optical Recording, 1st edn. Springer-Verlag, Berlin (1993)
7. Sakai, K., Hirayama, N., Tamura, R. (eds.): Novel Optical Resolution Technologies. Springer-Verlag, Berlin (2007)
8. Rowlands, A.: Physics of Digital Photography. Institute of Physics Publishing, Philadelphia (2017)
9. Mazda, F. (ed.): Telecommunications Engineer's Reference Book. Butterworth-Heinemann, Oxford (1993)
10. Benford, J.R., Richard, L.S.: Phase contrast microscopy for opaque specimens. J. Opt. Soc. Am. **40**, 314–316 (1950)
11. Holmes, T.J.: Signal-processing characteristics of differential-interference-contrast microscopy 2: noise considerations in signal recovery. Appl. Opt. **27**, 1302–1309 (1988)
12. Ishiguro, K.: The phase-shift measurement of thin films and its amplification. J. Opt. Soc. Am. **40**, 789–790 (1950)
13. Hernandez, G.: Analytical description of a fabry-perot spectrometer 3: off-axis behavior and interference filters. Appl. Opt. **13**, 2654–2661 (1974)
14. Steward, E.G.: Fourier Optics: An Introduction, 2nd edn. Dover Publications, New York (2011)
15. Fienup, J.R.: Phase retrieval algorithms: a comparison. Appl. Opt. **21**, 2758–2769 (1982)
16. Faulkner, H.M.L., Rodenburg, J.M.: Movable aperture lensless transmission microscopy: a novel phase retrieval algorithm. Phys. Rev. Lett. **93**, 023903 (2004)
17. Zhang, F., Rodenburg, J.M.: Phase retrieval based on wave-front relay and modulation. Phys. Rev. B **82**, 121104 (2010)
18. Teague, M.R.: Deterministic phase retrieval: a Green's function solution. J. Opt. Soc. Am. **73**, 1434–1441 (1983)
19. Roddier, F.: Curvature sensing and compensation: a new concept in adaptive optics. Appl. Opt. **27**, 1223–1225 (1988)
20. Roddier, F.: Wavefront sensing and the irradiance transport equation. Appl. Opt. **29**, 1402–1403 (1990)

Chapter 2
Qualitative Phase Imaging

The traditional imaging methods depend on the specimen acting on the incident illumination to create an observable image. Light absorbed by different parts of the object produces a brightness contrast or amplitude contrast that gives a visual impression of the object. Many biological and industrial specimens differ more in refractive indices than in light absorption, so a contrast in intensity does not reveal all of their details. Furthermore, such methods do not produce a satisfactory contrast when observing transparent or thick samples. Staining and labeling are widely used to improve contrast in images, but they can damage biological specimens. Besides biomedical applications, some plasma assisted applications also require optical diagnosis of phenomena occurring in transparent media. Consequently, label-free solutions are preferred for obtaining high-contrast images. If a microscopic specimen has structural details that differ in optical path, it will alter the phase of light passing through the specimen, so every part of the wavefront will have its own phase. Phase change can be characterized as a change in brightness, allowing us to identify microscopic areas causing the change. This can be achieved effectively with phase contrast imaging and provides an alternative to intensity contrast imaging.

Several label-free high-contrast imaging techniques for phase objects have been developed since Zernike introduced them in the 1930s. Hoffman contrast microscopy and differential interference contrast (DIC) microscopy are excellent tools for studying cells and tissues. In plasma applications, Schlieren photography is another technique used to observe air turbulence. The basic principle behind all of these methods is the transformation of phase into intensity, thereby improving the contrast of the image. But each method does not rely on a quantifiable technique to measure phase. Since modulated intensity and specimen phase are mixed in the images, it is not possible to reconstruct phase directly. Thus, these methods are considered qualitative phase imaging methods.

A mathematical description of qualitative phase imaging techniques is presented in this chapter, along with numerical simulations using MATLAB®. Furthermore, the chapter discusses recent developments in phase contrast imaging and the capabilities and limitations of each technique.

C. Liu et al., *Computational Optical Phase Imaging*, Progress in Optical Science and Photonics 21, https://doi.org/10.1007/978-981-19-1641-0_2

2.1 Phase Contrast Microscopy

Described first by Zernike [1–3], phase contrast microscopy enhances the contrast of an image of a transparent sample using the phase difference between direct and scattered light beams from the sample. When these beams are combined, they create a contrast either by strengthening or cancelling the beams based on the phase difference. We present here both theoretical and numerical models of phase contrast microscopy. Figure 2.1a represents the principle of phase contrast imaging. For a transparent specimen in the object plane, its transmission distribution t can be described as

$$t = e^{i\varphi(x,y)} \tag{2.1}$$

where $\varphi(x, y)$ indicates the specimen phase distribution. The complex wavefront distribution Ψ at the back focal plane of the first lens (the front focal plane of the second lens) can be computed through the Fourier transform F as

$$\Psi = \mathcal{F}(t) \tag{2.2}$$

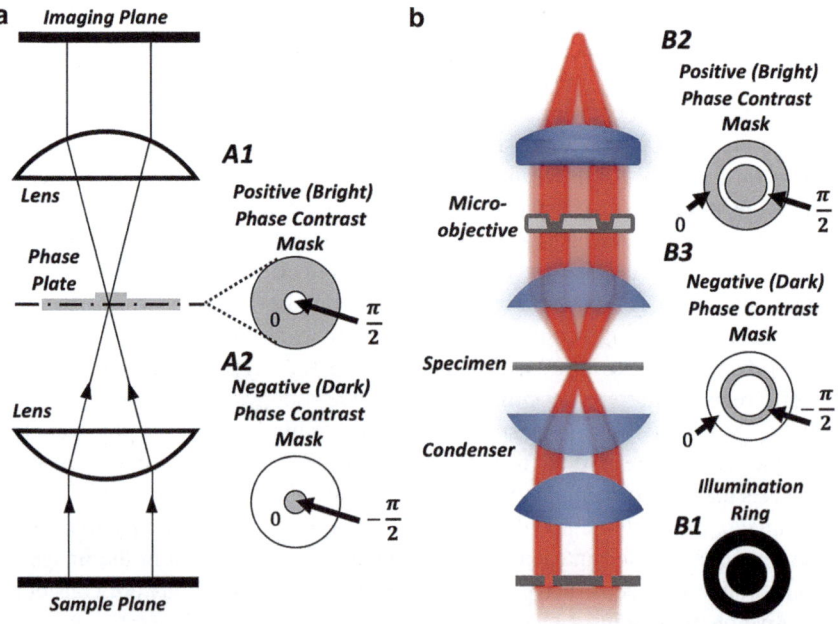

Fig. 2.1 Phase contrast microscopy. **a** Phase contrast imaging principle; (A1) positive phase contrast mask; (A2) negative phase contrast mask; **b** phase contrast imaging in commercial microscope; (B1) illumination ring; (B2) positive phase contrast mask; (B3) negative phase contrast mask

When the specimen phase is small, Eq. (2.1) can be simplified as Eq. (2.3).

$$t = e^{i\varphi(x,y)} \approx 1 + i\varphi(x, y) \tag{2.3}$$

Substituting Eq. (2.3) into Eq. (2.2), the complex wavefront distribution Ψ becomes

$$\Psi = \delta + i\mathcal{F}[\varphi(x, y)] \tag{2.4}$$

The first term in Eq. (2.4) is zeroth order specimen spectrum and the second term represents non-zeroth order specimen spectrum. A phase contrast mask as shown in Fig. 2.1(A1) is introduced at the back focal plane of the first lens, which has the function to delay the phase of the zeroth order spectrum by $\pi/2$. The modified complex wavefront distribution Ψ' is

$$\Psi' = i\delta(\xi, \eta) + i\mathcal{F}[\varphi(x, y)] \tag{2.5}$$

The complex wavefront distribution Φ' at the back focal plane of the second lens is

$$\Phi' = \mathcal{F}(\Psi') = i + i\varphi(x, y) \tag{2.6}$$

The image I' can be computed as

$$I' = |\Phi'|^2 = |i + i\varphi(x, y)|^2 \approx 1 + 2\varphi(x, y) \tag{2.7}$$

where the square term of $\varphi(x, y)$ is ignored since the specimen phase is small.

Alternatively, using a phase contrast mask as shown in Fig. 2.1(A2) would delay the phase of the zeroth order spectrum by $3\pi/2$, equivalent to advancing the phase by $\pi/2$. Re-writing Eqs. (2.5)–(2.7),

$$\Psi' = -i\delta(\xi, \eta) + i\mathcal{F}[\varphi(x, y)] \tag{2.8}$$

$$\Phi' = \mathcal{F}(\Psi') = -i + i\varphi(x, y) \tag{2.9}$$

$$I' = |\Phi'|^2 = |-i + i\varphi(x, y)|^2 \approx 1 - 2\varphi(x, y) \tag{2.10}$$

Equations (2.7) and (2.10) describe two ways of phase contrast-positive (bright) phase contrast microscopy and negative (dark) phase contrast microscopy. However, both tactics improve the imaging contrast by transforming phase into intensity.

Commercial microscopes often implement phase contrast microscopy in a configuration similar to Fig. 2.1b, which is very similar to dark field microscopy. The light passing through the illumination ring shown in Fig. 2.1(B1) is modulated by the positive or negative phase contrast mask shown in Fig. 2.1(B2) and 1(B3). Zeroth order light is within the ring region, which is different from Fig. 2.1a, but the principle of phase contrast imaging remains the same: the phase contrast image is caused by the interference between directly transmitted light within the phase contrast mask ring region and phase shifted light containing the specimen information from other portions of the phase contrast mask. The phase contrast mask and lens are often integrated in a phase contrast micro-objective.

Phase contrast microscopy described in Fig. 2.1a is numerically implemented using the following MATLAB code. Figure 2.2 shows the results of numerical simulation using samples of red blood cells. Parameters used were obtained statistically following massive measurements [4]. The phase and amplitude of these red blood cells are shown in Fig. 2.2a. Simulations disregard absorption and scattering effects. The bright field image obtained in Fig. 2.2b does not show any specimens due to the unique amplitude distribution shown in Fig. 2.2a. The images shown in Fig. 2.2c, d are obtained with positive and negative phase contrast masks. Simulation results show a significant improvement in contrast. Readers may find more images of various samples in phase contrast microscopy image galleries [5, 6].

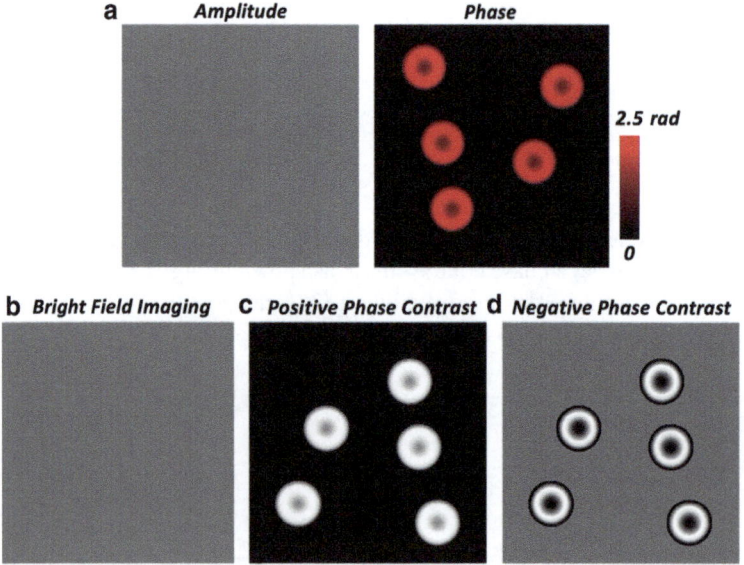

Fig. 2.2 Numerical simulation of phase contrast microscopy. **a** Amplitude and phase of red blood cells; **b** bright field image; **c** positive phase contrast image; **d** negative phase contrast image

Matlab Code for Phase Contrast Imaging

```
1   % Computational Optical Phase Imaging
2   % Phase Contrast Microscopy
3
4   %% CCC
5   clear all;
6   close all;
7   clc;
8
9   %% System Parameters
10  PixelNumber = 2^8; % Image recorder size
11  PixelSize = 6*10^(-6); % Image recorder pixel size, unit: m
12  MagnificationRatio = 40; % Magnification ratio of optical
    system, 40x
13  RIwater = 1.33; % Refractive index of water
14  RIcell = 1.40; % Refractive index of red blood cell
15  lambda = 532*10^(-9); % Wavelength
16
17  %% Red Blood Cell Model [4] in Chapter 2
18  Radius = 7.65*10^(-6)/2;
19  Size = floor(2*Radius/PixelSize*MagnificationRatio);
20  Xaxis = linspace(-Radius, Radius, Size);
21  Yaxis = linspace(-Radius, Radius, Size);
22  [Xmatrix, Ymatrix] = meshgrid(Xaxis, Yaxis);
23  Rho = (Xmatrix.^2+Ymatrix.^2).^0.5;
24  RhoNormalized = Rho/Radius;
25  Mask = double(RhoNormalized <= 1);
26  RhoNormalized = Mask.*RhoNormalized;
27  RBC = ((1-
    RhoNormalized.^2).^(0.5)).*(0.72+4.152*RhoNormalized.^2-
    3.426*RhoNormalized.^4);
28  RBC = 2*Mask.*RBC*10^(-6);
29
30  %% Specimen
31  Specimen = zeros(PixelNumber,PixelNumber);
32  Specimen(20:20+Size-1,30:30+Size-1) = RBC;
33  Specimen(120:120+Size-1,150:150+Size-1) = RBC;
34  Specimen(100:100+Size-1,50:50+Size-1) = RBC;
35  Specimen(40:40+Size-1,180:180+Size-1) = RBC;
36  Specimen(170:170+Size-1,60:60+Size-1) = RBC;
37  SpecimenPhase = 2*pi*Specimen*(RIcell-RIwater)/lambda;
38  SpecimenAmplitude = ones(PixelNumber,PixelNumber); % Almost no
    absorption
39
40  %% Bright Field Imaging & Phase Contrast Imaging
41  Wavefront = SpecimenAmplitude.*exp(1i*SpecimenPhase);
42  BFimaging = Wavefront.*conj(Wavefront); % Bright field imaging
43  Spectrum = fftshift(fft2(ifftshift(Wavefront)));
44  PositiveFilter = ones(PixelNumber,PixelNumber);
45  PositiveFilter(PixelNumber/2+1,PixelNumber/2+1) = exp(1i*pi/2);
46  NegativeFilter = ones(PixelNumber,PixelNumber);
```

```
47  NegativeFilter(PixelNumber/2+1,PixelNumber/2+1) =
    exp(1i*3*pi/2);
48  WavefrontPositive =
    fftshift(fft2(ifftshift(Spectrum.*PositiveFilter)));
49  WavefrontNegative =
    fftshift(fft2(ifftshift(Spectrum.*NegativeFilter)));
50  PCimagingPositive =
    WavefrontPositive.*conj(WavefrontPositive); % Positive phase
    contrast imaging
51  PCimagingNegative =
    WavefrontNegative.*conj(WavefrontNegative); % Negative phase
    contrast imaging
52  PCimagingPositive =
    PCimagingPositive/max(max(PCimagingPositive)); % Normalization
53  PCimagingNegative =
    PCimagingNegative/max(max(PCimagingNegative)); % Normalization
54
55  %% Plot
56  figure
57  subplot(1,3,1); imagesc(BFimaging); axis image; axis off;
    colormap(gray); title('Bright Field Imaging');
58  subplot(1,3,2); imagesc(PCimagingPositive); axis image; axis
    off; colormap(gray); title('Positive Phase Contrast Imaging');
59  subplot(1,3,3); imagesc(PCimagingNegative); axis image; axis
    off; colormap(gray); title('Negative Phase Contrast Imaging');
```

By combining phase contrast and phase shifting interference, improved techniques such as spatial light interference microscopy can provide quantitative phase distributions of specimens [7, 8]. We will discuss these quantitative phase contrast techniques in the next chapter. Phase contrast microscopy is also affected by the halo artefact, which is caused by higher-order light transmitting specimen information passing through the phase mask. Despite these limitations, phase contrast microscopy remains a convenient and cost-effective method of imaging phase objects that are frequently studied by biologists.

2.2 Differential Interference Contrast (DIC) Microscopy

Differential interference contrast (DIC) microscopy, designed by Nomarski in 1952, can produce relief-like images and eliminate halos [9–11]. The use of DIC microscopy in biological research is therefore more widespread than phase contrast microscopy.

Figure 2.3a illustrates the principle of DIC microscopy, which relies on light polarizations. Illumination beam passes through a linear polarizer and is subsequently split into two orthogonal components by a Wollaston prism as illustrated in Fig. 2.3b. Linearly polarized wavefronts with polarization angles 0° and 90° are represented by

$$\Psi_e = A(x + \Delta x, y + \Delta y)e^{-i[\varphi_0(x+\Delta x, y+\Delta y)+\varphi_\Delta]} \tag{2.11}$$

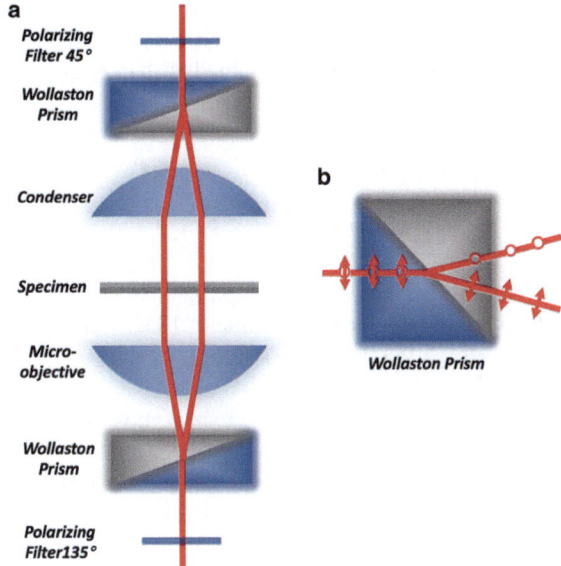

Fig. 2.3 DIC microscopy. **a** DIC microscopy principle; **b** Wollaston prism

a

Polarizing
Filter 45°

Wollaston
Prism

b

Condenser

Specimen

Micro-
objective

Wollaston Prism

Wollaston
Prism

Polarizing
Filter135°

$$\Psi_o = A(x - \Delta x, y - \Delta y)e^{-i[\varphi_0(x-\Delta x, y-\Delta y)]} \tag{2.12}$$

A represents the amplitude, φ_0 denotes the initial phase of the illumination light, $(2\Delta x, 2\Delta y)$ is the shearing, and φ_Δ is the bias retardation occurring while shifting phase between ordinary and extraordinary light. The shearing is determined by both the Wollaston prism (separation angle) and the condenser lens (axial position). Because ordinary and extraordinary light have different refractive indices, the bias is determined solely by the Wollaston prism. The specimen has a transmission distribution t as described in Eq. (2.1) and treated as a transparent object. It should also be considered polarization insensitive in the DIC microscopy mode. After passing through the specimen, the two transmitted wavefronts become Ψ_e' and Ψ_o' as described by Eqs. (2.13) and (2.14).

$$\Psi_e' = A(x + \Delta x, y + \Delta y)e^{-i[\varphi_0(x+\Delta x, y+\Delta y)+\varphi_\Delta]} \cdot e^{i\varphi(x,y)} \tag{2.13}$$

$$\Psi_o' = A(x - \Delta x, y - \Delta y)e^{-i[\varphi_0(x-\Delta x, y-\Delta y)]} \cdot e^{i\varphi(x,y)} \tag{2.14}$$

Using a combination of micro-objective and another Wallaston prism, two more new wavefronts Ψ_e'' and Ψ_o'' are formed according to the Wallaston prism principle described in Eqs. (2.15) and (2.16).

$$\Psi_e'' = A(x, y)e^{-i[\varphi_0(x,y)+2\varphi_\Delta]} \cdot e^{i\varphi(x-\Delta x, y-\Delta y)} \tag{2.15}$$

$$\Psi_o'' = A(x, y)e^{-i[\varphi_0(x,y)]} \cdot e^{i\varphi(x+\Delta x, y+\Delta y)} \tag{2.16}$$

Finally, the DIC image is obtained as an interference between these two wavefronts as

$$I = \frac{1}{2}\left|\Psi_e'' + \Psi_o''\right|^2 = A^2$$
$$+ A^2 \cos[\varphi(x + \Delta x, y + \Delta y) - \varphi(x - \Delta x, y - \Delta y) - 2\varphi_\Delta] \tag{2.17}$$

The term $\frac{1}{2}$ denotes that the light intensity is lost while it is passes through another polarizer with a polarization angle of 135°. With $\Delta\varphi$ as the phase gradient along the shearing direction, Eq. (2.17) becomes

$$I = A^2 + A^2 \cos(\Delta\varphi - 2\varphi_\Delta) \tag{2.18}$$

The DIC image is also equivalent to interference between the two sheared specimen transmission distributions $t(x + \Delta x, y + \Delta y)$ and $t(x - \Delta x, y - \Delta y)$ as described in Eq. (2.19).

$$I = \left|t(x + \Delta x, y + \Delta y) + t(x - \Delta x, y - \Delta y)\Delta e^{-i2\varphi_\Delta}\right|^2 = 2 + 2\cos(\Delta\varphi - 2\varphi_\Delta) \tag{2.19}$$

Similar to phase contrast microscopy, DIC microscopy also improves the imaging contrast by introducing phase gradient into the intensity. The following Matlab code implements the DIC microscopy. Figure 2.4 lists the numerical simulation results

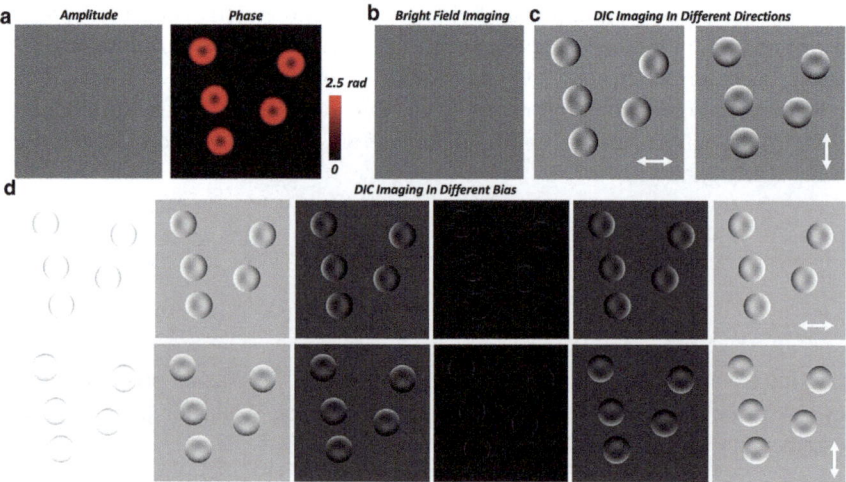

Fig. 2.4 Numerical simulation of DIC microscopy. **a** Amplitude and phase of red blood cells; **b** bright field image; **c** DIC images in different shearing directions; **d** DIC images in different bias

using the standard red blood cell models [4]. Figure 2.4a describes both the amplitude and phase of these red blood cells. The bright-field image shows no imaging contrast as shown in Fig. 2.4b. Figure 2.4c reveals the DIC images in different shearing directions and Fig. 2.4d, in different bias. Since the shearing is always fixed in commercial DIC microscopes, DIC images in high contrast and quality are obtained by adjusting bias to different values. More images from various samples can be seen in image galleries for DIC Microscopy [12, 13].

Matlab Code for DIC Imaging

```
1   % Computational Optical Phase Imaging
2   % Differential Interference Contrast Microscopy
3
4   %% CCC
5   clear all;
6   close all;
7   clc;
8
9   %% System Parameters
10  PixelNumber = 2^8; % Image recorder size
11  PixelSize = 6*10^(-6); % Image recorder pixel size, unit: m
12  MagnificationRatio = 40; % Magnification ratio of optical
    system, 40x
13  RIwater = 1.33; % Refractive index of water
14  RIcell = 1.40; % Refractive index of red blood cell
15  lambda = 532*10^(-9); % Wavelength
16
17  %% Red Blood Cell Model [4] in Chapter 2
18  Radius = 7.65*10^(-6)/2;
19  Size = floor(2*Radius/PixelSize*MagnificationRatio);
20  Xaxis = linspace(-Radius, Radius, Size);
21  Yaxis = linspace(-Radius, Radius, Size);
22  [Xmatrix, Ymatrix] = meshgrid(Xaxis, Yaxis);
23  Rho = (Xmatrix.^2+Ymatrix.^2).^0.5;
24  RhoNormalized = Rho/Radius;
25  Mask = double(RhoNormalized <= 1);
26  RhoNormalized = Mask.*RhoNormalized;
27  RBC = ((1-
    RhoNormalized.^2).^(0.5)).*(0.72+4.152*RhoNormalized.^2-
    3.426*RhoNormalized.^4);
28  RBC = 2*Mask.*RBC*10^(-6);
29
30  %% Specimen
31  Specimen = zeros(PixelNumber,PixelNumber);
32  Specimen(20:20+Size-1,30:30+Size-1) = RBC;
33  Specimen(120:120+Size-1,150:150+Size-1) = RBC;
34  Specimen(100:100+Size-1,50:50+Size-1) = RBC;
35  Specimen(40:40+Size-1,180:180+Size-1) = RBC;
36  Specimen(170:170+Size-1,60:60+Size-1) = RBC;
37  SpecimenPhase = 2*pi*Specimen*(RIcell-RIwater)/lambda;
38  SpecimenAmplitude = ones(PixelNumber,PixelNumber); % Almost no
    absorption
```

```
39
40  %% Bright Field Imaging & Phase Contrast Imaging
41  Wavefront = SpecimenAmplitude.*exp(1i*SpecimenPhase);
42  BFimaging = Wavefront.*conj(Wavefront); % Bright field imaging
43  PhaseBias = 0.2*pi;
44  ShearingPixel = 2;
45  LeftShearingWavefront = Wavefront(:,1:PixelNumber-
    ShearingPixel)*exp(1i*PhaseBias);
46  RightShearingWavefront =
    Wavefront(:,1+ShearingPixel:PixelNumber)*exp(-1i*PhaseBias);
47  UpShearingWavefront = Wavefront(1:PixelNumber-
    ShearingPixel,:)*exp(1i*PhaseBias);
48  DownShearingWavefront =
    Wavefront(1+ShearingPixel:PixelNumber,:)*exp(-1i*PhaseBias);
49  DICHorizontal =
    (LeftShearingWavefront+RightShearingWavefront).*
    conj(LeftShearingWavefront+RightShearingWavefront); % DIC image
    with horizontal shearing
50  DICVertical = (UpShearingWavefront+DownShearingWavefront).*
    conj(UpShearingWavefront+DownShearingWavefront); % DIC image
    with vertical shearing
51
52  %% Plot
53  figure
54  subplot(2,3,1); imagesc(BFimaging); axis image; axis off;
    colormap(gray); title('Bright Field Imaging');
55  subplot(2,3,2); imagesc(DICHorizontal); axis image; axis off;
    colormap(gray); title('DIC Imaging with Horizontal Shearing');
56  subplot(2,3,3); imagesc(DICVertical); axis image; axis off;
    colormap(gray); title('DIC Imaging with Vertical Imaging');
57  for num = 1:1:100;
58  PhaseBiasStep = pi/100;
59  LeftShearingWavefront = Wavefront(:,1:PixelNumber-
    ShearingPixel)*exp(1i*num*PhaseBiasStep);
60  RightShearingWavefront =
    Wavefront(:,1+ShearingPixel:PixelNumber)*exp(-
    1i*num*PhaseBiasStep);
61  UpShearingWavefront = Wavefront(1:PixelNumber-
    ShearingPixel,:)*exp(1i*num*PhaseBiasStep);
62  DownShearingWavefront =
    Wavefront(1+ShearingPixel:PixelNumber,:)*exp(-
    1i*num*PhaseBiasStep);
63  DICHorizontal =
    (LeftShearingWavefront+RightShearingWavefront).*
    conj(LeftShearingWavefront+RightShearingWavefront);
    with horizontal shearing
64  DICVertical = (UpShearingWavefront+DownShearingWavefront).*
    conj(UpShearingWavefront+DownShearingWavefront); % DIC image
    with vertical shearing
65  subplot(2,3,5); imagesc(DICHorizontal); axis image; axis off;
    colormap(gray); title('Dyanmic DIC Imaging');
66  subplot(2,3,6); imagesc(DICVertical); axis image; axis off;
    colormap(gray); title('Dyanmic DIC Imaging');
67  pause(0.1)
68  end
```

When compared to phase contrast microscopy, DIC microscopy completely eliminates the problem of halos; therefore, DIC microscopy is more commonly used in biological applications, especially in live cell imaging. Note that the specimen should be polarization-insensitive, and that plastic cannot be used in DIC microscopy because many polymers have a depolarizing effect on light. Although classical DIC microscopy cannot provide quantitative phase distributions of specimens, improved DIC methods can obtain quantitative phase imaging also by phase shifting [14–16]. In the following chapter, we will introduce these quantitative DIC methods.

2.3 Spectrum Modulation Contrast Imaging

Spectrum modulation is another widely used technique to enhance image contrast. Most of the techniques share a similar optical system as shown in Fig. 2.5 which consisting of a 4-f system that is used for optical spatial filtering. Based on the filter used in the spectrum plane, there are several approaches such as Laplace field imaging [17], gradient field imaging [18–21] and spiral phase contrast imaging [22, 23]. The general principle of spectrum modulation contrast imaging is discussed in this section.

The wavefront under observation at the object plane is represented by $u_o(x, y)$. Its amplitude and phase are shown in Fig. 2.6a. The wavefront obtained by its Fourier transform $U(\xi, \eta)$ is allowed to pass through a filter $F(\xi, \eta)$ set at the spectrum plane. The collected wavefront $u_i(x, y)$ at the image plane is illustrated by Eq. (2.20) with its intensity given in Eq. (2.21).

$$u_i(x, y) = \mathcal{F}\{F(\xi, \eta) \cdot \mathcal{F}[u_o(x, y)]\} \tag{2.20}$$

$$I = |u_i(x, y)|^2 = |\mathcal{F}[F(\xi, \eta) \cdot U(\xi, \eta)]|^2 = |\mathcal{F}\{F(\xi, \eta) \cdot \mathcal{F}[u_o(x, y)]\}|^2 \tag{2.21}$$

Figure 2.6(C1) shows the filter used in Laplace field imaging [17], represented by

$$F(\xi, \eta) = \xi^2 + \eta^2 \tag{2.22}$$

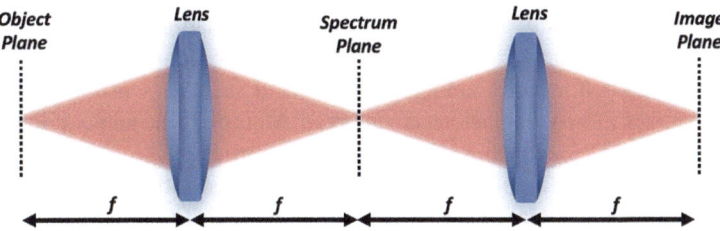

Fig. 2.5 Spectrum modulation contrast imaging

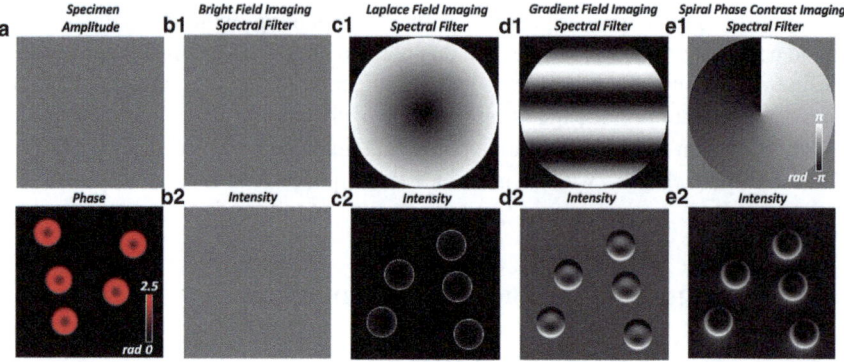

Fig. 2.6 Numerical simulation of spectrum modulation contrast imaging. **a** Amplitude and phase of red blood cells; (**b1**) spectral filter and (**b2**) intensity distribution of bright field image; (**c1**) spectral filter and (**c2**) intensity distribution of Laplace field imaging; (**d1**) spectral filter and (**d2**) intensity distribution of gradient field imaging; (**e1**) spectral filter and (**e2**) intensity distribution of spiral phase field imaging

With the above filter function, the wavefront passing through the filter at the spectrum plane becomes

$$F(\xi, \eta) \cdot U(\xi, \eta) = \left(\xi^2 + \eta^2\right) \cdot U(\xi, \eta) \tag{2.23}$$

The collected wavefront $u_i(x, y)$ in Eq. (2.20) becomes

$$u_i(x, y) = \mathcal{F}\left\{\left(\xi^2 + \eta^2\right) \cdot U(\xi, \eta)\right\} = \nabla^2 u_o(x, y) \tag{2.24}$$

According to Eq. (2.21), the corresponding intensity can be derived as Eq. (2.25).

$$I = \left[\nabla^2 \varphi(x, y)\right]^2 + \left\{\left[\frac{\partial \varphi(x, y)}{\partial x}\right]^2 + \left[\frac{\partial \varphi(x, y)}{\partial y}\right]^2\right\}^2 \tag{2.25}$$

Equation (2.25) shows how Laplace field imaging is able to transform phase information to an intensity distribution. Numerical simulation result in Fig. 2.6(C2) shows a significant enhancement in the imaging contrast compared to the bright field image in Fig. 2.6b. It is worth noting that such tactic is rather sensitive to the edges since derivatives in two dimensions are adopted. Moreover, the name of the method comes from Eq. (2.23), which is the Laplace transform [17].

Besides Laplace field imaging, Popescu group also proposed gradient field imaging [18–20] and commercialized this technique [21]. The filter of gradient field imaging is shown in Fig. 2.6(D1) and described by Eq. (2.26), which is actually a sinusoidal grating.

$$F(\xi, \eta) = 1 + \sin a\xi \tag{2.26}$$

Similarly, substituting Eq. (2.26) into Eq. (2.20), the wavefront passing through the filter at the spectrum plane can be derived as Eq. (2.27), and the collected wavefront at the image plane can be computed as Eq. (2.28).

$$F(\xi, \eta) \cdot U(\xi, \eta) = (1 + \sin a\xi) \cdot U(\xi, \eta) \tag{2.27}$$

$$u_i(x, y) = \mathcal{F}\{(1 + \sin a\xi) \cdot U(\xi, \eta)\} = u_o(x, y) + ia\frac{\partial u_o(x, y)}{\partial x} \tag{2.28}$$

According to Eq. (2.21), the corresponding intensity can be derived as Eq. (2.29), ignoring high order components.

$$I = \left| e^{i\varphi(x,y)} - a\frac{\partial \varphi(x, y)}{\partial x} e^{i\varphi(x,y)} \right|^2 \approx 1 - 2a\frac{\partial \varphi(x, y)}{\partial x} \tag{2.29}$$

Figure 2.6(D2) also lists the simulated intensity distribution using the following Matlab code for gradient field imaging. According to Eq. (2.29) and the simulation result in Fig. 2.6d, the gradient field imaging introduces a phase derivative in the intensity, thus significantly enhances the imaging contrast. In contrast to Laplace field imaging which uses two-dimensional phase derivative, the gradient field imaging only considers one-dimensional phase derivative, therefore it is rather similar to DIC imaging. Moreover, the name of the gradient comes from Eq. (2.29), indicating that the intensity is proportional to the phase gradient. In addition, Popescu group also updated the gradient field imaging into a quantitative method [24–26].

Ritsch-Marte group proposed spiral phase contrast imaging [22, 23]. Its filter is shown in Fig. 2.6(E1) and it is described in Eqs. (2.30) and (2.31). Unlike an amplitude filter used in Laplace field imaging and gradient field imaging, a phase vortex is used as a phase filter in spiral phase contrast imaging.

$$F(\xi, \eta) = e^{i\theta(\xi,\eta)} \tag{2.30}$$

$$\theta(\xi, \eta) = \tan^{-1}\frac{\eta}{\xi} \tag{2.31}$$

Substituting Eq. (2.30) into Eq. (2.20), the wavefront passing through the filter at the spectrum plane can be derived as Eq. (2.32), and the collected wavefront at the image plane can be computed as Eq. (2.33), in which H indicates two-dimensional Hilbert transform.

$$F(\xi, \eta) \cdot U(\xi, \eta) = e^{i\theta(\xi,\eta)} \cdot U(\xi, \eta) \tag{2.32}$$

$$u_i(x, y) = \mathcal{F}\{e^{i\theta(\xi,\eta)} \cdot U(\xi, \eta)\} = \mathcal{H}[u_o(x, y)] \tag{2.33}$$

According to Eq. (2.21), the corresponding intensity can be derived as Eq. (2.34).

$$I = \left| \mathcal{H}\left(e^{i\varphi(x,y)}\right) \right|^2 \tag{2.34}$$

Figure 2.6(E1) lists the simulated intensity distribution obtained with the Matlab code for spiral phase contrast imaging. According to Eq. (2.34) and the simulation result in Fig. 2.6(E1), the spiral phase contrast imaging significantly enhances the specimen edges according to the principle of Hilbert transform. Moreover, the name of the spiral comes from Eq. (2.30) since the optical filter is a spiral phase. In addition, Ritsch-Marte group also updated the spiral phase contrast microscopy into a quantitative method [27].

Matlab Code for Different Types of Spectrum Modulation Contrast Methods

```
1    % Computational Optical Phase Imaging
2    % Spectrum Modulation Contrast Imaging
3
4    %% CCC
5    clear all;
6    close all;
7    clc;
8
9    %% System Parameters
10   PixelNumber = 2^8; % Image recorder size
11   PixelSize = 6*10^(-6); % Image recorder pixel size, unit: m
12   MagnificationRatio = 40; % Magnification ratio of optical
     system, 40x
13   RIwater = 1.33; % Refractive index of water
14   RIcell = 1.40; % Refractive index of red blood cell
15   lambda = 532*10^(-9); % Wavelength
16
17   %% Red Blood Cell Model [4] in Chapter 2
18   Radius = 7.65*10^(-6)/2;
19   Size = floor(2*Radius/PixelSize*MagnificationRatio);
20   Xaxis = linspace(-Radius, Radius, Size);
21   Yaxis = linspace(-Radius, Radius, Size);
22   [Xmatrix, Ymatrix] = meshgrid(Xaxis, Yaxis);
23   Rho = (Xmatrix.^2+Ymatrix.^2).^0.5;
24   RhoNormalized = Rho/Radius;
25   Mask = double(RhoNormalized <= 1);
26   RhoNormalized = Mask.*RhoNormalized;
27   RBC = ((1-
     RhoNormalized.^2).^(0.5)).*(0.72+4.152*RhoNormalized.^2-
     3.426*RhoNormalized.^4);
28   RBC = 2*Mask.*RBC*10^(-6);
29
30   %% Specimen
31   Specimen = zeros(PixelNumber,PixelNumber);
32   Specimen(20:20+Size-1,30:30+Size-1) = RBC;
33   Specimen(120:120+Size-1,150:150+Size-1) = RBC;
34   Specimen(100:100+Size-1,50:50+Size-1) = RBC;
35   Specimen(40:40+Size-1,180:180+Size-1) = RBC;
```

```
36   Specimen(170:170+Size-1,60:60+Size-1) = RBC;
37   SpecimenPhase = 2*pi*Specimen*(RIcell-RIwater)/lambda;
38   SpecimenAmplitude = ones(PixelNumber,PixelNumber); % Almost no
     absorption
39
40   %% Bright Field Imaging
41   Wavefront = SpecimenAmplitude.*exp(1i*SpecimenPhase);
42   BFImaging = Wavefront.*conj(Wavefront); % Bright field imaging
43   BFFilter = ones(PixelNumber,PixelNumber); % Filter of bright
     field imaging
44
45   %% Specimen Spectrum and Spectral Coordinate
46   Spectrum = fftshift(fft2(ifftshift(Wavefront))); % Specimen
     spectrum
47   Freqency = 1/PixelSize;  % Spectral coordinate
48   Fxvector = linspace(-Freqency/2,Freqency/2,PixelNumber);
49   Fyvector = linspace(-Freqency/2,Freqency/2,PixelNumber);
50   [FxMat, FyMat] = meshgrid(Fxvector,Fyvector);
51   Mask = zeros(PixelNumber,PixelNumber);
52   Mask((FxMat.^2+FyMat.^2).^0.5<Freqency/2) = 1;
53
54   %% Laplace Field Imaging [21] in Chapter 2
55   LaplaceFilter =
     ((FxMat.^2+FyMat.^2).^0.5/(Freqency/2)^0.5).*Mask; % Spectral
     Filter of Laplace field imaging
56   LaplaceWavefront =
     fftshift(fft2(ifftshift(Spectrum.*LaplaceFilter))); % Spectral
     filtering
57   LaplaceImaging = LaplaceWavefront.*conj(LaplaceWavefront); %
     Laplace field imaging
58
59   %% Gradient Field Imaging [22-24] in Chapter 2
60   GradientFilter = (1+sin(10^(-4)*FyMat)).*Mask; % Spectral
     Filter of gradient field imaging
61   GradientWavefront =
     fftshift(fft2(ifftshift(Spectrum.*GradientFilter))); % Spectral
     filtering
62   GradientImaging = GradientWavefront.*conj(GradientWavefront); %
     Gradient field imaging
63
64   %% Spiral Phase Contrast Imaging [26,27] in Chapter 2
65   SpiralFilter = exp(1i*(atan2(FxMat,FyMat))).*Mask; % Spectral
     Filter of spiral phase field imaging
66   SpiralWavefront =
     fftshift(fft2(ifftshift(Spectrum.*SpiralFilter))); % Spectral
     filtering
67   SpiralImaging = SpiralWavefront.*conj(SpiralWavefront); %
     Spiral phase field imaging
68
```

```
69  %% Plot
70  figure
71  subplot(2,4,1); imagesc(BFImaging); axis image; axis off;
    colormap(gray); title('Bright Field Imaging');
72  subplot(2,4,5); imagesc(BFFilter); axis image; axis off;
    colormap(gray); title('Spectral Filter');
73  subplot(2,4,2); imagesc(LaplaceImaging); axis image; axis off;
    colormap(gray); title('Laplace Field Imaging');
74  subplot(2,4,6); imagesc(LaplaceFilter); axis image; axis off;
    colormap(gray); title('Spectral Filter');
75  subplot(2,4,3); imagesc(GradientImaging); axis image; axis off;
    colormap(gray); title('Gradient Field Imaging');
76  subplot(2,4,7); imagesc(GradientFilter); axis image; axis off;
    colormap(gray); title('Spectral Filter');
77  subplot(2,4,4); imagesc(SpiralImaging); axis image; axis off;
    colormap(gray); title('Spiral Phase Contrast Imaging');
78  subplot(2,4,8);
    imagesc(atan2(imag(SpiralFilter),real(SpiralFilter))); axis
    image; axis off; colormap(gray); title('Spectral Filter');
```

In contrast to phase contrast microscopy, DIC microscopy and Hoffman modulation contrast microscopy require modifications to the illumination and sometimes even the micro-objective, while spectrum modulation contrast microscopy can be accomplished by setting simple systems outside the C-mount of the commercial microscopes. Additionally, different filters can be used to achieve varied contrasts, making it suitable for different applications.

2.4 Hoffman Modulation Contrast Microscopy

Although DIC microscopy can provide high contrast images, it has several limitations. There are some downsides, including that it is not suitable for samples that alter polarized light and that plastic culture vessels cannot be used. It also requires a pair of Wollaston prisms, which are relatively expensive. Hoffman designed a modulation contrast microscope in 1975 [28, 29] based on cost-effective elements that work well even when polarizing materials such as plastics are introduced into the optical path. As with DIC microscopy, Hoffman modulation contrast microscopy also produces relief-like images.

Figure 2.7 illustrates the principle of the Hoffman modulation contrast microscopy. In the classical microscopic system shown in Fig. 2.7a, a slit is located at the front focal plane of a condenser and its image through condenser and micro-objective system is generated at the conjugate plane, which is the back focal plane of the micro-objective. Figure 2.7b, c explain how a specimen with a phase gradient shifts the generated slit image by the deflection of light. An optical amplitude spatial filter (termed by modulator by Hoffman in [28, 29]) is set at the back focal plane of the objective, and it is composed of three regions: (1) dark region D, with a transmittance less than 1%, (2) grey region G, with a transmittance of ~15%, and (3) bright region B, with a transmittance of nearly 100%. When the light is not deflected

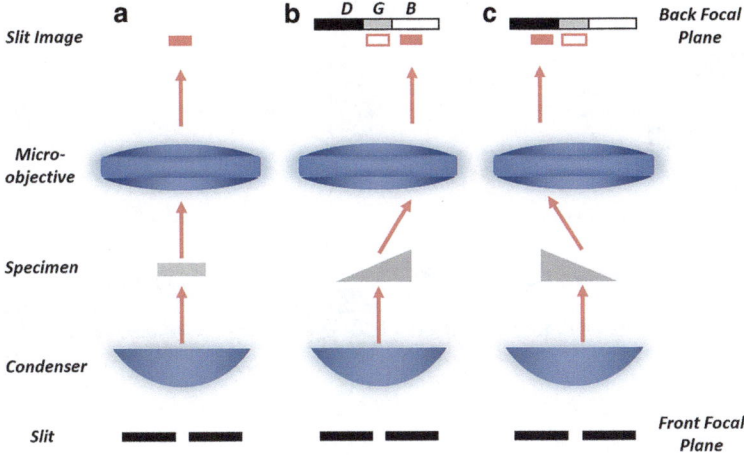

Fig. 2.7 Principle of Hoffman modulation contrast microscopy. **a–c** show image formation while using specimens with different phase gradient

by the specimen, transmitted light intensity is attenuated by 85%; when the light is deflected by the specimen rightwards as in Fig. 2.7b, it is almost fully transmitted and it is fully blocked when the light is deflected by the specimen to the left as in Fig. 2.7c. The specimen phase gradient thereby modulates the intensity distribution. The image of a real specimen is essentially generated by the interference of specimen induced deflected waves passing that are passing through bright and grey regions.

In practical Hoffman modulation contrast microscopy, the slit width set at the front focal plane of the condenser can be adjusted via a pair of polarizers revealed in Fig. 2.8a. When two polarizers are crossed, the slit width is the narrowest, often providing high contrast. However, in some special conditions, such as thick samples with large differences in refractive index, a wider slit offers a higher quality image even when the contrast is reduced [30]. Besides slit adjustment, specimen rotation may dramatically improve or degrade the imaging contrast. Additionally, there are symmetric and asymmetric slit-modulator designs in Hoffman modulation contrast microscopy as demonstrated in Fig. 2.8b, c. It should be noted that asymmetric designs have nearly twice the resolution of symmetric designs based on Abbe's theory. The asymmetric design of Hoffman modulation contrast microscopy is often preferred because it makes full use of the numerical aperture of the micro-objective and typically produces a good resolution [30].

Different from phase modulation in phase contrast microscopy, Hoffman modulation contrast microscopy adopts Hoffman modulator to modulate the amplitudes in different deflection directions and provides high-contrast images due to the presence of phase gradient in the specimen. More images of various samples observed by Hoffman modulation contrast microscopy can be seen in image galleries [31]. Compared to phase contrast microscopy, Hoffman modulation contrast microscopy does not suffer from halos, and also does not require expensive phase modulator.

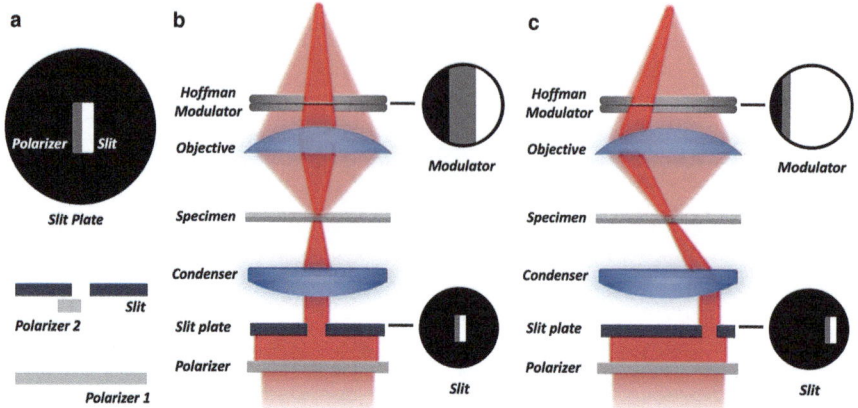

Fig. 2.8 Design considerations in Hoffman modulation contrast microscopy. **a** Slit design; **b** symmetric Hoffman modulation contrast microscopy; **c** asymmetric Hoffman modulation contrast microscopy

Compared to DIC microscopy, it does not rely on expensive optical elements such as Wollaston prisms. It can also be used for polarization sensitive specimen imaging. Therefore, Hoffman modulation contrast microscopy provides a cost-effective way for relief-like high-contrast imaging. However, it should be noted that Hoffman modulation contrast microscopy is most sensitive to specimen phase gradients perpendicular to the length of the slit and therefore orientation of the specimen should be adjusted to obtain high-quality relief-like images. Additionally, much of the incident illumination is blocked by Hoffman modulator and hence making it less popular among biological applications in contrast to DIC microscopy.

2.5   Schlieren Photography

Schlieren photography is an older technique in comparison to other qualitative phase imaging techniques. Though Schlieren was first observed by Hooke in early in 1665, it was Foucault and Toepler who proposed classical Schlieren systems in 1859 [32]. Considering the simple optical system and high-contrast imaging performance, Schlieren photography has been widely used in observing phenomena occurring in transparent media.

Schlieren photography is illustrated in Fig. 2.9. An image of a point source is projected on a screen behind the lens. The divergent light emitted from the point source is collected by the lens and it passes through point source image, leaving the intensity on the screen to almost unity as shown in Fig. 2.9a. If specimens with a gradient of refractive index occur within the optical system, the localized light will be bent as shown in Fig. 2.9b. In spite of the fact that light bending has an effect

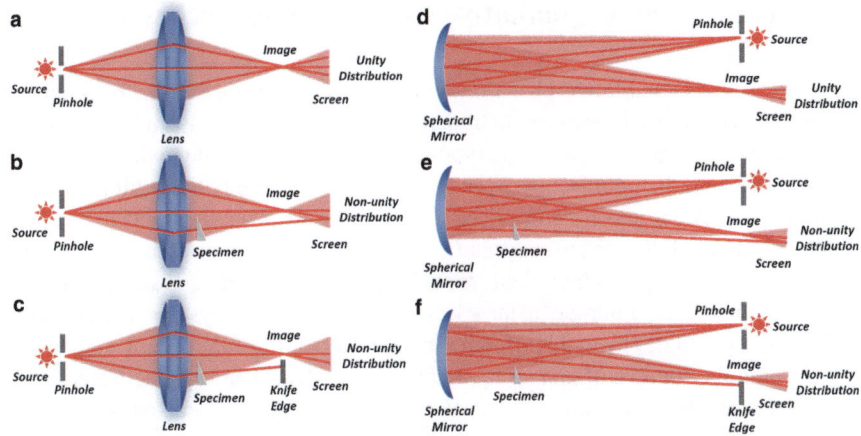

Fig. 2.9 Schlieren photography. **a–c** Transmission mode; **a** without sample; **b** with sample and without knife edge; **c** with sample and with knife edge; **d–f** reflection mode; **d** without sample; **e** with sample and without knife edge; **f** with sample and with knife edge

on the screen's intensity distribution, it is still difficult to distinguish because of the minimal change in brightness caused by the refractive index gradient. As a means of enhancing the contrast, an obstruction (often as a knife edge) is placed next to the point source image. According to theory, the obstruction does not affect the non-deflected light, but avoids partial deflection of light due to specimen refractive index gradient, as shown in Fig. 2.9c. In other words, the bent light is removed from the screen, thereby enhancing the image contrast. This principle also applies to the reflection model in Fig. 2.9d–f. In addition to the classical Schlieren photography systems in both transmittance and reflection modes, many other optical system designs have been proposed, as outlined in [33]. Moreover, Schlieren photography can be further explored using mathematical models, images, and applications [34, 35].

Schlieren photography is very similar to Hoffman modulation contrast microscopy: the non-unity relief-like intensity distributions are induced by specimen refractive index gradients. Hoffman modulation contrast microscopy, however, mostly reduces the non-deflected light while either blocking or allowing the deflected light according to deflection direction using a specially designed Hoffman modulator, as illustrated in Fig. 2.7. Schlieren photography, on the other hand, blocks deflected light while allowing the non-deflected light by using a knife edge, as revealed in Fig. 2.9.

2.6 Comparison of Qualitative Phase Imaging Techniques

There are several qualitative phase imaging techniques discussed in this chapter, which all work toward the same goal: transforming phase information into a measureable intensity distribution. Two basic types of techniques are employed: interference and modulation. DIC microscopy works on the basis that the collected image is the interference between the wavefront passing through the specimen and its sheared wavefront. Thus, DIC microscopy can be regarded as an interference technique. Laplace field imaging, gradient field imaging, and spiral phase contrast imaging are realized with different filters set at the spectrum plane in the 4-f system. The concepts of Schlieren photography and Hoffman modulation contrast microscopy are similar in enhancing the contrast of images by modulating deflected and non-deflected light. They both fall under modulation-based techniques. Furthermore, phase contrast microscopy works by an interference between the phase shifted directly transmitted light as well as the deflected light from the specimen, so it is both an interference and modulation combined technique. Phase contrast microscopy and spiral phase contrast use phase modulation, while Laplace field imaging, gradient field imaging, Hoffman modulation contrast microscopy, and Schlieren photography use amplitude modulation. All of these methods can effectively increase the contrast of an image, thereby enabling label-free specimen imaging.

Quantitative phase imaging can be used in a wide variety of applications, but their success depends on selecting the method that yields the best results. Table 2.1 summarizes qualitative phase imaging techniques, their specific advantages and disadvantages, as well as their applications. In biological live cell imaging, DIC provides high-quality, relief-like images without halos and often with high illumination efficiency, but it can be polarization sensitive and uses expensive Wollaston prisms. Schlieren photographs make it possible to observe air and heat turbulence.

We discussed a variety of qualitative phase imaging techniques based on interference and spectrum modulation, respectively. By combining numerical simulations with theoretical descriptions, researchers can gain a deeper understanding of the techniques. Although these methods reveal the inner details of transparent specimens without labeling, they are simply intensity distributions, and the coupled phase information cannot be quantitatively retrieved in order to determine the local thickness or morphology of the specimen. Many complex biological processes pertaining to recent biomedical and diagnostic applications require quantitative phase information in order to extract unique biological information. A number of techniques exist for the quantitative extraction of phase from recorded intensity patterns, such as holography, interferometry, and coherent diffraction imaging. Recent applications for coherent diffraction imaging include wavefront reconstruction in short-wavelength imaging and optical element quality inspection. The Shack-Hartmann wavefront sensing and pyramid wavefront sensing techniques are widely used in astronomy and biology to quantify distorted wavefront. A detailed overview of quantitative phase imaging is presented in the following chapters, along with MATLAB simulations.

Table 2.1 Comparisons of qualitative phase imaging techniques

	Principle	Advantage	Disadvantage	Applications	Quantitative potential
Phase contrast microscopy	Interference and modulation	Microscope equipped; Polarization insensitive	Halos	Biological label-free microscopy	Yes [7, 8]
DIC microscopy	Interference	Microscope equipped; High-quality	Polarization sensitive; expensive	Biological label-free microscopy	Yes [14–16]
Spectrum modulation contrast imaging	Modulation	Simple system	Complicated spectrum filters	Biological label-free microscopy	Yes [24–27]
Hoffman modulation contrast microscopy	Modulation	Microscope equipped; Cheap; polarization insensitive	Illumination loss	Biological label-free microscopy	No
Schlieren photography	Modulation	Simple system		Turbulence and plasma observation	No

References

1. Zernike, F.: Phase contrast, a new method for the microscopic observation of transparent objects part I. Physica **9**, 686–698 (1942)
2. Zernike, F.: Phase contrast, a new method for the microscopic observation of transparent objects part II. Physica **9**, 974–986 (1942)
3. Zernike, F.: How I discovered phase contrast. Science **9**, 345–349 (1955)
4. Tsinopoulos, S.V., Polyzos, D.: Scattering of He–Ne laser light by an average-sized red blood cell. Appl. Opt. **38**, 5499–5510 (1999)
5. https://www.microscopyu.com/galleries/phase-contrast
6. https://www.olympus-lifescience.com.cn/en/microscope-resource/primer/techniques/phasegallery/
7. Wang, Z., Millet, L., Mir, M., Ding, H., Unarunotai, S., Rogers, J., Gillette, M.U., Popescu, G.: Spatial light interference microscopy (SLIM). Opt. Express **19**, 1016–1026 (2011)
8. Chen, X., Kandel, M.E., Popescu, G.: Spatial light interference microscopy: principle and applications to biomedicine. Adv. Opt. Photonics **13**, 353–425 (2021)
9. Nomarski, G.: Microinterfrometre differentiel a ondes polarisees. J. Phys. Radium **16**, S9 (1955)
10. Allen, R.D., David, G.B., Nomarski, G.: The Zeiss-Nomarski differential interference equipment for transmitted-light microscopy. Z. Wiss. Mikrosk. **69**, 193–221 (1969)
11. Pluta, M.: Nomarski's DIC microscopy: a review. Proc. SPIE **1846**, 10–25 (1992)
12. https://www.microscopyu.com/galleries/dic-phase-contrast
13. https://www.olympus-lifescience.com.cn/en/microscope-resource/primer/techniques/dic/dicgallery/
14. Creath, K.: Phase-shifting speckle interferometry. Appl. Opt. **24**, 3053–3058 (1985)
15. McIntyre, T.J., Maurer, C., Fassl, S., Khan, S., Bernet, S., Ritsch-Marte, M.: Quantitative SLM-based differential interference contrast imaging. Opt. Express **18**, 14063–14078 (2010)

16. Wei, Q., Li, Y., Vargas, J., Wang, J., Gong, Q., Kong, Y., Jiang, Z., Xue, L., Liu, C., Liu, F., Wang, S.: Principal component analysis-based quantitative differential interference contrast microscopy. Opt. Lett. **44**, 45–48 (2019)
17. Kim, T., Popescu, G.: Laplace field microscopy for label-free imaging of dynamic biological structures. Opt. Lett. **36**, 4704–4706 (2011)
18. Kim, T., Sridharan, S., Popescu, G.: Gradient field microscopy of unstained specimens. Opt. Express **20**, 6737–6745 (2012)
19. Kim, T., Sridharan, S., Kajdacsy-Balla, A., Tangella, K., Popescu, G.: Gradient field microscopy for label-free diagnosis of human biopsies. Appl. Opt. **52**, A92–A96 (2013)
20. Fanous, M.J., Li, Y.F., Kandel, M.E., Abdeen, A.A., Kilian, K.A., Popescu, G.: Effects of substrate patterning on cellular spheroid growth and dynamics measured by gradient light interference microscopy (GLIM). J. Biophotonics **12**, e201900178 (2019)
21. https://phioptics.com
22. Jesacher, A., Furhapter, S., Bernet, S., Ritsch-Marte, M.: Shadow effects in spiral phase contrast microscopy. Phys. Rev. Lett. **94**, 233902 (2005)
23. Furhapter, S., Jesacher, A., Bernet, S., Ritsch-Marte, M.: Spiral phase contrast imaging in microscopy. Opt. Express **13**, 689–694 (2005)
24. Nguyen, T.H., Kandel, M.E., Rubessa, M., Wheeler, M.B., Popescu, G.: Gradient light interference microscopy for 3D imaging of unlabeled specimens. Nat. Commun. **8**, 210 (2017)
25. Kandel, M.E., Hu, C., Kouzehgarani, G.N., Min, E., Sullivan, K.M., Kong, H., Li, J.M., Robson, D.N., Gillette, M.U., Best-Popescu, C., Popescu, G.: Epi-illumination gradient light interference microscopy for imaging opaque structures. Nat. Commun. **10**, 4691 (2019)
26. Wang, Y., Kandel, M.E., Fanous, M.J., Hu, C., Chen, H.Y., Lu, X., Popescu, G.: Harmonically decoupled gradient light interference microscopy (HD-GLIM). Opt. Lett. **45**, 1487–1490 (2020)
27. Bernet, S., Jesacher, A., Furhapter, S., Maurer, C., Ritsch-Marte, M.: Quantitative imaging of complex samples by spiral phase contrast microscopy. Opt. Express **14**, 3792–3805 (2006)
28. Hoffman, R., Gross, L.: Modulation contrast microscope. Appl. Opt. **14**, 1169–1176 (1975)
29. Hoffman, R.: The modulation contrast microscope: principles and performance. J. Microsc. **110**, 205–222 (1977)
30. https://www.olympus-lifescience.com.cn/en/microscope-resource/primer/techniques/hoffman/
31. https://www.olympus-lifescience.com.cn/zh/microscope-resource/primer/techniques/hoffmangallery/
32. Rienitz, J.: Schlieren experiment 300 years ago. Nature **254**, 293–295 (1975)
33. Traldi, E., Boselli, M., Simoncelli, E., Stancampiano, A., Gherardi, M., Colombo, V., Settles, G.S.: Schlieren imaging: a powerful tool for atmospheric plasma diagnostic. EPJ Tech. Instrum. **5**, 4 (2018)
34. Settles, G.S.: Schlieren and Shadowgraph Techniques. Springer, Berlin (2001)
35. Merzkirch, W.: Flow Visualization, 2nd edn. Academic Press, Waltham (1987)

Chapter 3
Interference-Based Quantitative Optical Phase Imaging

By converting phase variations of light caused by transparent specimens into changes in light amplitude visible to the human eye, optical phase contrast microscopy has helped scientists uncover the structure and function of cells and subcellular organelles by generating image contrast from transparent samples with insufficient intensity variations to generate a brightness contrast image. With phase contrast imaging, cellular studies have been elevated to a whole new level. However, images obtained by optical phase contrast are not able to provide any quantitative information about cell thicknesses or local refractive index modulation. In order to study cells structurally and functionally, it is necessary to understand spatial variations in the optical thickness of the samples. Recently developed Quantitative Phase Imaging (QPI) techniques are capable of extracting local phase shifts from the sample which is proportional to the optical thickness of the object, which allows for a more detailed analysis of the cells. Methods for quantitative optical phase imaging fall into two categories: interference-based (holography and interferometry) and non-interference-based (coherent diffraction imaging, transport of intensity phase sensing, Shack-Hartmann wavefront sensing and so on). Through interferometric and holographic techniques, it is possible to recover and quantify the phase of interference patterns caused by optical path differences between the reference and the object waves. Interferometry is a classical technique widely used in optical shop testing [1], and also extended to various applications such as gravitational wave detection [2]. Holography, an optical recording and encoding method invented by Dennis Gabor in 1947 [3], has revolutionized a number of fields [4–8]. As more advanced sensor arrays and high-volume data storage became available, it became much easier to acquire and process interference patterns, thereby quantifying differences in the beam optical path. Therefore, these digital holographic and interferometric techniques have been considered a promising concept in microscopy to determine the 3D structure, position, and orientation of specimens.

In this chapter, we review holography and interferometry that have proven successful for quantitative optical phase imaging. The chapter is organized as follows. Holography and interferometry are briefly described in Sect. 3.1. Section 3.2

© The Author(s), under exclusive license to Springer Nature Singapore Pte Ltd. 2022
C. Liu et al., *Computational Optical Phase Imaging*, Progress in Optical Science
and Photonics 21, https://doi.org/10.1007/978-981-19-1641-0_3

describes different types of holography and interferometry, and Sect. 3.3 describes several simulation examples using MATLAB®. In Sects. 3.4 and 3.5, we provide a summary of some recent improvements and extensions designed to make holography and interferometry more versatile optical tools.

3.1 Description of Holography and Interferometry

A Scalar light field can be represented by Eq. (3.1) and it is shown in Fig. 3.1a, in which A_O is the amplitude and φ_O is the phase. Classical imaging records only the intensity I as shown in Eq. (3.2) and Fig. 3.1b. However, holography and interferometry record both intensity I and phase φ_O by introducing a reference wavefront R as described in Eq. (3.3) and shown in Fig. 3.1a, where A_R is the amplitude and φ_R is the phase. Equation (3.4) and Fig. 3.1c represent the recorded interferogram. The interference patterns formed are sinusoidal modulation of the phase difference between specimen phase and reference phase, $\varphi_O - \varphi_R$. This process is discussed in the interference section in many optics textbooks (such as [9]). In other words, fringe pattern can be treated as superposition of object and reference intensity with modulated phase information. Therefore, it is possible for holography and interferometry to retrieve both amplitude and phase distributions of the object wavefront through

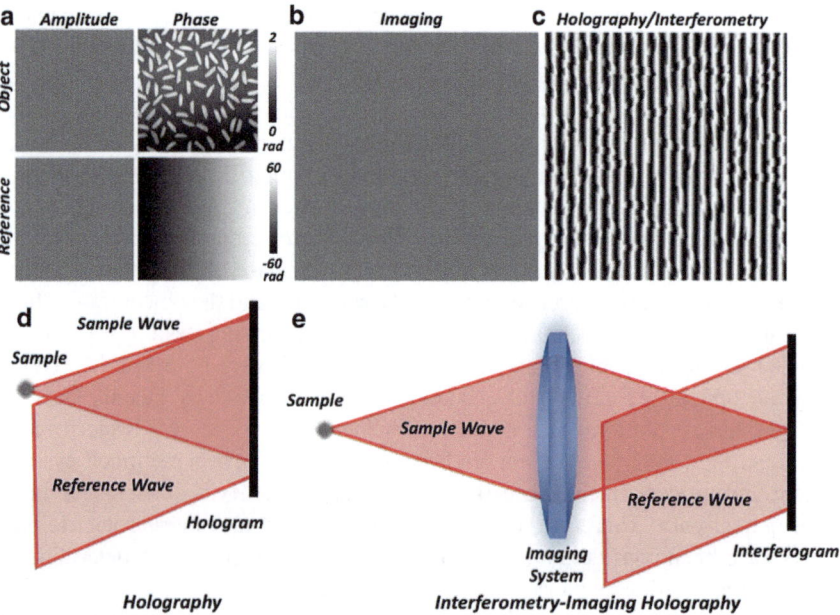

Fig. 3.1 Description of Holography and Interferometry. **a** Object and reference wavefronts; **b** imaging intensity; **c** hologram/interferogram; **d** holography scheme; **e** interferometry scheme

a reconstruction process. While both holography and interferometry are based on interference, holography often deals with the wavefront (both amplitude and phase) reconstruction in defocus condition (specimen and image recorder are not in conjugated planes) as shown in Fig. 3.1d, whereas interferometry often deals with the phase (amplitude is often not considered) retrieval in in-focus condition (specimen and image recorder are in conjugated planes) as shown in Fig. 3.1e. Interferometry, then, is a special case of holography (imaging holography). Interferometry is therefore mostly used in optical testing and metrology, while holography is widely used in imaging and display.

$$\boldsymbol{O} = A_O e^{-i\varphi_O} \tag{3.1}$$

$$I = |\boldsymbol{O}|^2 = \left| A_O e^{-i\varphi_O} \right|^2 = |A_O|^2 \tag{3.2}$$

$$\boldsymbol{R} = A_R e^{-i\varphi_R} \tag{3.3}$$

$$I = |\boldsymbol{O} + \boldsymbol{R}|^2 = |A_O|^2 + |A_R|^2 + 2|A_O A_R| \cos(\varphi_O - \varphi_R) \tag{3.4}$$

To extract the phase from holography and interferometry, two key steps are required: recording and reconstruction. $\boldsymbol{O}(x, y)$ and $\boldsymbol{R}(x, y)$ are object and reference wavefronts, interference of both are shown in Eq. (3.5), in which * represents the conjugation.

$$
\begin{aligned}
I(x, y) &= |\boldsymbol{O}(x, y) + \boldsymbol{R}(x, y)|^2 = |\boldsymbol{O}(x, y)|^2 + |\boldsymbol{R}(x, y)|^2 \\
&\quad + \boldsymbol{R}^*(x, y)\boldsymbol{O}(x, y) + \boldsymbol{R}(x, y)\boldsymbol{O}^*(x, y)
\end{aligned} \tag{3.5}
$$

The response of CCD/CMOS camera to the interference pattern is recorded as a hologram/interferogram as revealed in Eq. (3.6), where a and b are constants indicating background and modulation. Equations (3.5) and (3.6) explain the recording process of holography and interferometry.

$$H(x, y) = a + bI(x, y) \tag{3.6}$$

To extract both the amplitude and phase distributions of wavefront, a reconstruction process is implemented where the same reference wave $\boldsymbol{R}(x, y)$ is used to illuminate the recorded hologram. The reconstruction operation can be described through Eqs. (3.7)–(3.11).

$$H(x, y)\boldsymbol{R}(x, y) = \boldsymbol{U}_1(x, y) + \boldsymbol{U}_2(x, y) + \boldsymbol{U}_3(x, y) + \boldsymbol{U}_4(x, y) \tag{3.7}$$

$$\boldsymbol{U}_1(x, y) = \boldsymbol{R}(x, y)\left(a + b|\boldsymbol{R}(x, y)|^2\right) \tag{3.8}$$

$$U_2(x, y) = b\boldsymbol{R}(x, y)|\boldsymbol{O}(x, y)|^2 \tag{3.9}$$

$$U_3(x, y) = b\boldsymbol{R}(x, y)\boldsymbol{R}^*(x, y)\boldsymbol{O}(x, y) = b|\boldsymbol{R}(x, y)|^2\boldsymbol{O}(x, y) \tag{3.10}$$

$$U_4(x, y) = b\boldsymbol{R}(x, y)\boldsymbol{R}(x, y)\boldsymbol{O}^*(x, y) = b\boldsymbol{R}^2(x, y)\boldsymbol{O}^*(x, y) \tag{3.11}$$

Equations (3.8) and (3.9) are the 0th order information. Equations (3.10) and (3.11) represent +1st and −1st order information, representing real and virtual image. It should be noted that both amplitude and phase distributions of the object wavefront can be directly extracted from the $U_3(x, y)$ term. Considering the fact that the target and the image recorder are often not conjugated, additional wavefront propagation is required to retrieve the wavefront at the object plane. Interferometry often directly retrieves the wavefront at the image plane and it focuses more on phase reconstruction rather than amplitude reconstruction. Phase retrieval in interferometry can be demonstrated by

$$\begin{aligned} I(x, y) = a &+ b|\boldsymbol{O}(x, y)|^2 + b|\boldsymbol{R}(x, y)|^2 \\ &+ 2b|\boldsymbol{R}(x, y)||\boldsymbol{O}(x, y)|\cos(\varphi_O(x, y) - \varphi_R(x, y)) \end{aligned} \tag{3.12}$$

In interferometry, the phase difference $\varphi_O(x, y) - \varphi_R(x, y)$ can be extracted from the captured interferogram and since $\varphi_R(x, y)$ is known or previously calibrated, and it can be used to compute the object phase $\varphi_O(x, y)$. However, it should be noted that the wavefront reconstruction in holography as in Eqs. (3.7)–(3.11) and the phase retrieval in interferometry as in Eq. (3.12) are equivalent.

The amplitude and phase distributions of the object wavefront are acquired using holography and interferometry by first recording holograms/interferograms and extracting the information from them during the reconstruction process. Holograms and interferograms are mainly recorded with digital cameras, such as CCDs and CMOSs. Since there are so many phase retrieval algorithms available for holography and interferometry, reconstruction has many more options than recording, as exemplified in this chapter.

3.2   Classification of Holography and Interferometry

Holography and interferometry can be classified according to different perspectives, such as on-axis/off-axis, Fourier/Fresnel, shearing/non-shearing, common path/non-common path, reflection/transmission, binary/digital/analog, optical/computational and so on [10]. In this section, on-axis/off-axis, Fourier/Fresnel, and shearing/non-shearing are illustrated and discussed in detail.

3.2.1 On-Axis and Off-Axis

Figure 3.2 shows the on-axis scheme. The most classical on-axis one is Gabor holography, which is the interference between scattered (object) and directly transmitted (reference) wavefronts as shown in Fig. 3.2a. Slightly different from Gabor holography, another scheme introduces additional reference wavefront to interfere with the sample wavefront as shown in Fig. 3.2b. This allows on-axis holography and interferometry to handle samples with diffuse reflection condition, not only transparent samples with small variations in its transmittance.

Figure 3.2c reveals the reconstruction scheme of on-axis holography and interferometry. According to the reconstruction process demonstrated in Eqs. (3.7)–(3.11), $U_1(x, y)$, $U_2(x, y)$, $U_3(x, y)$, and $U_4(x, y)$ are mixed and cause an overlap of real and virtual images. Though a method has been proposed in 1951 to reduce the twin image using hologram subtraction [11], it was complicated and provided poor performance. For this reason, on-axis (Gabor) holography could not find much applications in the early years of its development.

New generation digital technologies brought a significant change in the field of holography and interferometry. Imaging devices such as CCD and CMOS cameras have replaced photographic films. Digital holography and interferometry

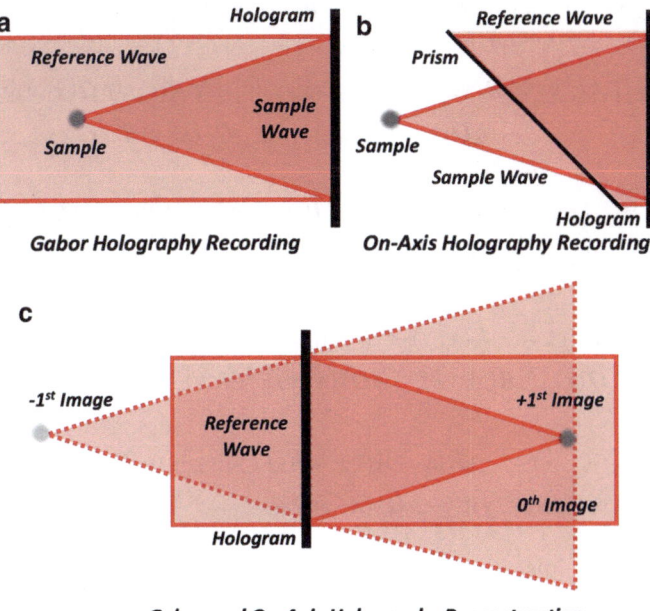

Fig. 3.2 Principles of on-axis holography and interferometry. **a** Recording scheme of Gabor holography; **b** recording scheme of on-axis holography and interferometry; **c** reconstruction scheme of on-axis holography and interferometry

have opened up new possibilities for processing holograms and interferograms by representing them as complex numbers and applying various mathematical transformations to remove or suppress unwanted images or aberrations, even in real time, which was previously not possible. By extracting phase and intensity data, it is now possible to obtain quantitative information about a sample. For example, using a phase-shifting algorithm in on-axis holography and interferometry, it is possible to suppress the 0th order and twin image (-1st order) terms [12] during the reconstruction. This is very useful for on-axis holography and interferometry techniques that are used in precise optical testing and all the pixels in the image recorder can be used for target imaging and reconstruction. Equations (3.13)–(3.21) describe how to obtain only $+1\text{st}$ order image using the most classical four-step phase-shifting holography by shifting the phase to $\alpha = 0, \frac{\pi}{2}, \pi$ and $\frac{3\pi}{2}$ respectively.

$$R_0(x, y) = R(x, y)e^{i0} \tag{3.13}$$

$$R_{\pi/2}(x, y) = R(x, y)e^{i\pi/2} \tag{3.14}$$

$$R_\pi(x, y) = R(x, y)e^{i\pi} \tag{3.15}$$

$$R_{3\pi/2}(x, y) = R(x, y)e^{i3\pi/2} \tag{3.16}$$

$$
\begin{aligned}
R_0 H_0''(x, y) = &\, R(x, y)\left(a + b|R(x, y)|^2\right) + bR(x, y)|O(x, y)|^2 \\
&+ bO(x, y)|R(x, y)|^2 + bO^*(x, y)R^2(x, y)
\end{aligned}
\tag{3.17}
$$

$$
\begin{aligned}
R_{\pi/2} H_{\pi/2}''(x, y) = &\, iR(x, y)\left(a + b|R(x, y)|^2\right) + ibR(x, y)|O(x, y)|^2 \\
&+ bO(x, y)|R(x, y)|^2 - bO^*(x, y)R^2(x, y)
\end{aligned}
\tag{3.18}
$$

$$
\begin{aligned}
R_\pi H_\pi''(x, y) = &\, -R(x, y)\left(a + b|R(x, y)|^2\right) - bR(x, y)|O(x, y)|^2 \\
&+ bO(x, y)|R(x, y)|^2 + bO^*(x, y)R^2(x, y)
\end{aligned}
\tag{3.19}
$$

$$
\begin{aligned}
R_{3\pi/2} H_{3\pi/2}''(x, y) = &\, -iR(x, y)\left(a + b|R(x, y)|^2\right) - ibR(x, y)|O(x, y)|^2 \\
&+ bO(x, y)|R(x, y)|^2 - bO^*(x, y)R^2(x, y)
\end{aligned}
\tag{3.20}
$$

The $+1\text{st}$ order image information is then extracted as Eq. (3.21) to reconstruct the target wavefront.

$$
\begin{aligned}
bO(x, y)|R(x, y)|^2 = &\, \frac{1}{4}R_0 H_0''(x, y) + \frac{1}{4}R_{\pi/2}H_{\pi/2}''(x, y) \\
&+ \frac{1}{4}R_\pi H_\pi''(x, y) + \frac{1}{4}R_{3\pi/2}H_{3\pi/2}''(x, y)
\end{aligned}
\tag{3.21}
$$

Similar tactic is also useful for phase retrieval in interferometry. Phase-shifted on-axis interferograms with phase-shifting angles as 0, $\frac{\pi}{2}$, π, and $\frac{3\pi}{2}$ are described by Eqs. (3.22)–(3.24), and (3.25), respectively.

$$H_0(x, y) = a + b|\boldsymbol{O}(x, y)|^2 + b|\boldsymbol{O}(x, y)| \cos[\varphi(x, y) + 0] \tag{3.22}$$

$$H_{\pi/2}(x, y) = a + b|\boldsymbol{O}(x, y)|^2 + b|\boldsymbol{O}(x, y)| \cos\left[\varphi(x, y) + \frac{\pi}{2}\right] \tag{3.23}$$

$$H_{\pi}(x, y) = a + b|\boldsymbol{O}(x, y)|^2 + b|\boldsymbol{O}(x, y)| \cos[\varphi(x, y) + \pi] \tag{3.24}$$

$$H_{3\pi/2}(x, y) = a + b|\boldsymbol{O}(x, y)|^2 + b|\boldsymbol{O}(x, y)| \cos\left[\varphi(x, y) + \frac{3\pi}{2}\right] \tag{3.25}$$

Therefore, the object phase φ can be obtained according to Eq. (3.26).

$$\varphi = \tan^{-1} \frac{H_{3\pi/2}(x, y) - H_{\pi/2}(x, y)}{H_0(x, y) - H_{\pi}(x, y)} \tag{3.26}$$

Leith and Upatnieks adopted the idea of carrier frequency technique used in signal processing and developed off-axis holography [4–6] which is illustrated in Fig. 3.3a. Ever since, holography was widely used in many fields. Equation (3.27) describes the off-axis holography and interferometry, in which $\boldsymbol{O}(x, y)$ and $\boldsymbol{R}(x, y)e^{ik_r \cdot r}$ are object and reference wavefronts as shown in Fig. 3.3a.

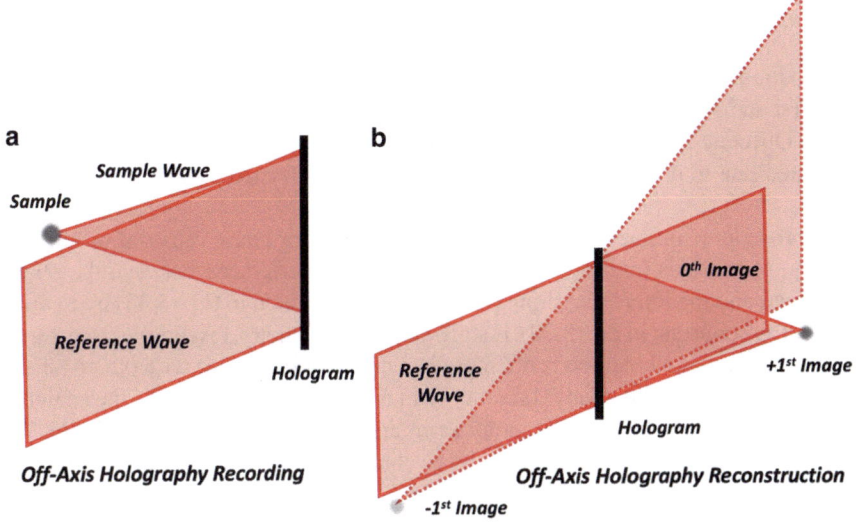

Fig. 3.3 Principles of off-axis holography and interferometry. **a** Recording scheme of off-axis holography and interferometry; **b** reconstruction scheme of off-axis holography and interferometry

$$I'(x, y) = \left| O(x, y) + R(x, y)e^{ik_r \cdot r} \right|^2 \tag{3.27}$$

The generated off-axis hologram and interferogram can be described by

$$H'(x, y) = a + bI'(x, y) \tag{3.28}$$

From (3.27) and (28), we can write the recorded hologram and interferogram as

$$\begin{aligned} H'^{(x,y)} &= a + b|R(x, y)|^2 + b|O(x, y)|^2 \\ &+ bO(x, y)R^*(x, y)e^{-ik_r \cdot r} + bO^*(x, y)R(x, y)e^{ik_r \cdot r} \end{aligned} \tag{3.29}$$

While in holography reconstruction, the reference wavefront is illuminated on the hologram as illustrated in Eqs. (3.30)–(3.34).

$$H'(x, y)R(x, y)e^{ik_r \cdot r} = V_1(x, y) + V_2(x, y) + V_3(x, y) + V_4(x, y) \tag{3.30}$$

$$V_1(x, y) = R(x, y)e^{ik_r \cdot r}\left(a + b|R(x, y)|^2\right) \tag{3.31}$$

$$V_2(x, y) = bR(x, y)e^{ik_r \cdot r}|O(x, y)|^2 \tag{3.32}$$

$$V_3(x, y) = bR(x, y)e^{ik_r \cdot r}R^*(x, y)e^{-ik_r \cdot r}O(x, y) = b|R(x, y)|^2O(x, y) \tag{3.33}$$

$$\begin{aligned} V_4(x, y) &= bR(x, y)e^{ik_r \cdot r}R(x, y)e^{ik_r \cdot r}O^*(x, y) \\ &= bR^2(x, y)O^*(x, y)e^{ik_r \cdot r} \end{aligned} \tag{3.34}$$

Equations (3.31) and (3.32) represent 0th order image, and Eqs. (3.33) and (3.34) are +1st and −1st orders, representing real and virtual images in Fig. 3.3b, respectively. Different from on-axis scheme, 0th, +1st, and −1st order information can be separated due to different propagation direction as indicated by $e^{ik_r \cdot r}$, $e^{-i0k_r \cdot r}$ and $e^{i2k_r \cdot r}$.

A procedure similar to the above is also suitable for phase retrieval in interferometry. An off-axis interferogram obtained by the interference between the object wavefront and the reference tilting plane wave is illustrated in Eq. (3.35) by treating $R(x, y)$ as a constant in Eq. (3.29). For off-axis interferometry, fast Fourier transform-based phase retrieval method is often used as shown in Eq. (3.36) with 0th, +1st and −1st order terms, respectively. 1st order term is extracted and moved to the center of the spectrum, while 0th and −1st order terms are removed as described by Eq. (3.37). The object amplitude A and phase φ can be obtained by an inverse Fourier transform as described in Eqs. (3.38) and (3.39).

$$H(x, y) = a + b|O(x, y)|^2 + bO(x, y)e^{-ik_r \cdot r} + bO^*(x, y)e^{ik_r \cdot r} \tag{3.35}$$

$$\mathcal{F}[H(x, y)] = \mathcal{F}\left[a + b|\mathbf{O}(x, y)|^2\right] + \mathcal{F}\left[b\mathbf{O}(x, y)e^{-ik_r \cdot r}\right]$$
$$+ \mathcal{F}\left[b\mathbf{O}^*(x, y)e^{ik_r \cdot r}\right] \tag{3.36}$$

$$\mathcal{F}[H_{\text{filtered}}(x, y)] = \mathcal{F}[b\mathbf{O}(x, y)] \tag{3.37}$$

$$A = \left|\mathcal{F}^{-1}\{[\mathcal{F}[H_{\text{filtered}}(x, y)]]\}\right| \tag{3.38}$$

$$\varphi = \tan^{-1} \frac{\text{Im } F^{-1}\{[\mathcal{F}[H_{\text{filtered}}(x, y)]]\}}{\text{Re } F^{-1}\{[\mathcal{F}[H_{\text{filtered}}(x, y)]]\}} \tag{3.39}$$

In both on-axis and off-axis modes, interferometry and holography can retrieve the wavefront of a target. A single-shot hologram/interferogram can provide the target complex amplitude in off-axis holography/interferometry, while phase-shifting holograms/interferograms can provide the target complex amplitude in on-axis holography/interferometry. Due to this, off-axis holography and interferometry often have a higher temporal resolution than on-axis holography and interferometry; however, off-axis holography and interferometry require a larger bandwidth than on-axis holography and interferometry, as shown in Fig. 3.4.

Equation (3.40) describes a hologram/interferogram. While $k_r = 0$ indicates on-axis mode, $k_r \neq 0$ indicates off-axis holography and interferometry.

$$H(x, y) = a + b|\mathbf{R}(x, y)|^2 + b|\mathbf{O}(x, y)|^2$$

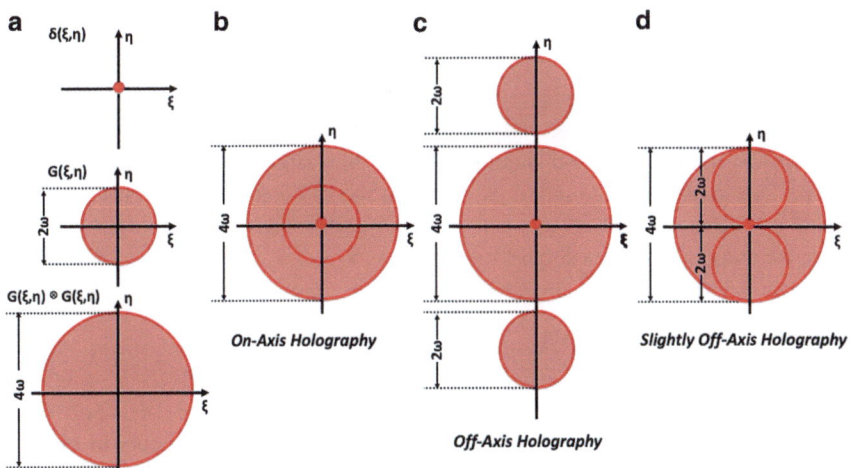

Fig. 3.4 Bandwidths of on-axis and off-axis holography and interferometry. a Bandwidth of reference and object; b bandwidth of on-axis holography and interferometry; c bandwidth of off-axis holography and interferometry; d bandwidth of slightly off-axis holography and interferometry

$$+ bO(x, y)R^*(x, y)e^{-ik_r \cdot r} + bO^*(x, y)R(x, y)e^{ik_r \cdot r} \tag{3.40}$$

Spectral domain distribution of each term in Eq. (3.40) is listed in Eqs. (3.41)–(3.44), in which $\delta(\xi, \eta)$ is the impulse function, $G(\xi, \eta)$ is the distribution of the object in the spectral domain, ω is the bandwidth of $G(\xi, \eta)$, ξ_r and η_r are the orthogonal components of $\boldsymbol{k}_r$, and $\otimes$ represents convolution.

$$\mathcal{F}\big[a + b|\boldsymbol{R}(x, y)|^2\big] \propto \delta(\xi, \eta) \tag{3.41}$$

$$\mathcal{F}\big[b|\boldsymbol{O}(x, y)|^2\big] \propto G(\xi, \eta) \otimes G(\xi, \eta) \tag{3.42}$$

$$\mathcal{F}\big[b\boldsymbol{O}(x, y)\boldsymbol{R}^*(x, y)e^{-ik_r \cdot r}\big] \propto G(\xi - \xi_r, \eta - \eta_r) \tag{3.43}$$

$$\mathcal{F}\big[b\boldsymbol{O}^*(x, y)\boldsymbol{R}(x, y)e^{ik_r \cdot r}\big] \propto G(\xi + \xi_r, \eta + \eta_r) \tag{3.44}$$

Figure 3.4a shows the distributions of reference and object wavefronts in the spectral domain. Figure 3.4b shows the bandwidth of the on-axis holography and interferometry, while Fig. 3.4c shows the bandwidth of the off-axis holography and interferometry. The bandwidth of off-axis holography and interferometry is twice that of on-axis holography and interferometry. It should be noted that the 1st order information can only be separated from the 0th information in off-axis holography and interferometry when $k_r > 3\omega$. Moreover, k_r is determined by the angle between the reference and object waves γ according to Eq. (3.45), in which λ is the wavelength.

$$k_r = \frac{\sin \gamma}{\lambda} \tag{3.45}$$

Therefore, the minimal angle $\gamma_{\min}$ should be $\sin^{-1} 3\lambda\omega$. Moreover, in order to accurately record the fringes, Nyquist sampling condition should be satisfied. In other words, the fringe period should be larger than or equal to two times the pixel size p. This gives the maximal angle between the reference and object waves $\gamma_{\max}$ as $\sin^{-1} \frac{\lambda}{2p}$. Therefore, in off-axis holography/interferometry, the angle between the reference and object waves should satisfy

$$\sin^{-1} 3\lambda\omega < \gamma < \sin^{-1} \frac{\lambda}{2p} \tag{3.46}$$

Additionally, slightly off-axis holography and interferometry [13, 14] are provided as shown in Fig. 3.4d in order to reduce the bandwidth requirements in off-axis holography and interferometry.

In both holography and interferometry, the phase and amplitude distributions can be reconstructed. Off-axis holography and interferometry have a higher temporal

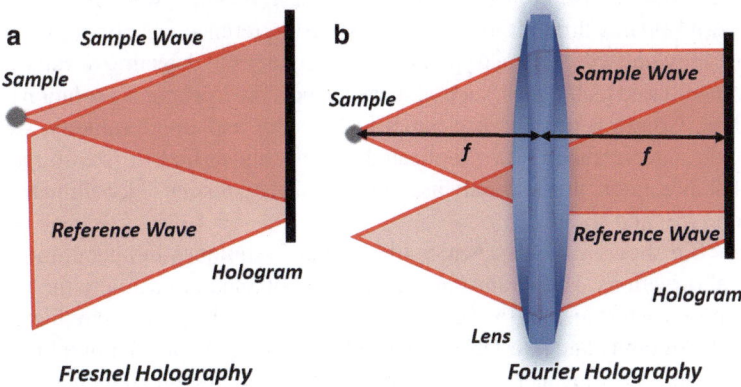

Fig. 3.5 **a** Scheme of Fresnel holography; **b** scheme of Fourier holography

resolution, but require a larger bandwidth, whereas on-axis holography and interferometry require only a smaller bandwidth, but can only retrieve phase from multiple phase-shifting holograms. Both on-axis and off-axis holography and interferometry are numerically simulated in the following sections.

3.2.2 Fresnel and Fourier

Holography can also be classified into Fresnel and Fourier holography, based on the position of the hologram plane relative to the sample plane. In Fresnel holography, the hologram plane lies within the Fresnel zone from the sample plane as shown in Fig. 3.5a. Fresnel propagation is used to reconstruct the sample and the pixel size can be adjusted to achieve higher resolution with smaller pixels or a wider field of view with larger pixels. As an alternative, holograms can record spectra of a sample in the back focal plane of a lens, while placing the sample in the front focal plane. This mode is known as Fourier holography, as shown in Fig. 3.5b. With the Fourier transform, it is possible to directly reconstruct the sample with less computational load than Fresnel holography. Both Fresnel holography and Fourier holography are numerically simulated in the following sections.

3.2.3 Shearing and Non-shearing

There is a major challenge in most of the interferometric and holographic techniques in that since the phase difference between the sample and reference wavefronts is displayed in the interferogram/hologram, it requires a priori known information on the specially designed reference wave to reconstruct the sample wavefront accurately. However, it remains difficult to obtain the precise reference wavefront information in holography and interferometry, inevitably reducing the detecting accuracy. Self-interference-based techniques, such as shearing, become a promising solution to such a problem. Contrary to reference-based holography and interferometry, shearing permits the interference of two replicated wavefronts shifted in lateral, radial, or azimuthal directions; this is widely used in optical shop testing [1], collimated beam detection [15], and so on.

Figure 3.6 shows various schemes of shearing techniques in lateral, radial, and azimuthal directions. Holograms/interferograms obtained are actually the result of phase difference between the wavefront to be detected and its sheared replica. While classical holography and interferometry need an additional optical arm for reference wavefronts, the shearing technique simplifies the optical system, making it a popular technology for optical phase sensing applications.

The interference between the sample wavefront and its sheared replica can be expressed as

$$
\begin{aligned}
I = \left| S(r) + S(r') \right|^2 &= |S(r)|^2 + \left| S(r') \right|^2 \\
&+ S(r)S^*(r') + S^*(r)S(r')
\end{aligned}
\tag{3.47}
$$

Lateral shearing can be written as

$$
\begin{aligned}
I = |S(r) + S(r + \Delta r)|^2 &= |S(r)|^2 + |S(r + \Delta r)|^2 \\
&+ S(r)S^*(r + \Delta r) + S^*(r)S(r + \Delta r)
\end{aligned}
\tag{3.48}
$$

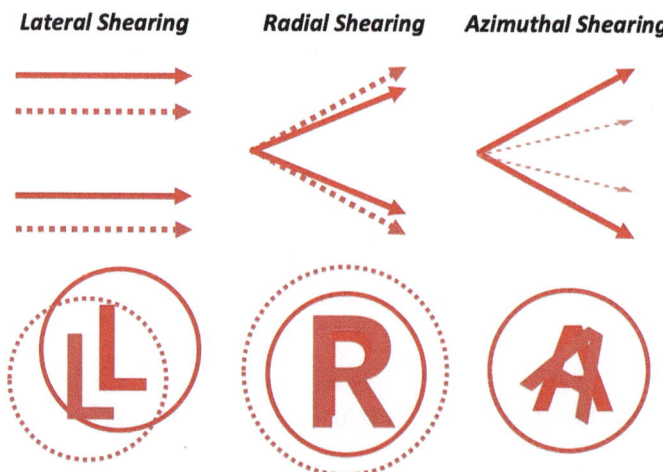

Lateral Shearing   **Radial Shearing**   **Azimuthal Shearing**

Fig. 3.6 Scheme of lateral, radial, and azimuthal shearing

Radial shearing can be expressed as

$$I = |S(r) + S(\alpha r)|^2 = |S(r)|^2 + |S(\alpha r)|^2$$
$$+ S(r)S^*(\alpha r) + S^*(r)S(\alpha r) \tag{3.49}$$

Finally, azimuthal shearing can be written as

$$I = |S(\theta) + S(\theta + \Delta\theta)|^2 = |S(\theta)|^2 + |S(\theta + \Delta\theta)|^2$$
$$+ S(\theta)S^*(\theta + \Delta\theta) + S^*(\theta)S(\theta + \Delta\theta) \tag{3.50}$$

Considering $S(r) = S(r)e^{i\varphi(r)}$, the phase difference $\Delta\varphi$ between sample wavefront and its replica, represented in the fringes of lateral, radial, and azimuthal shearing respectively can be formulated as

$$\Delta\varphi = \varphi(r) - \varphi(r') \tag{3.51}$$

$$\Delta\varphi = \varphi(r + \Delta r) - \varphi(r) \tag{3.52}$$

$$\Delta\varphi = \varphi(\alpha r) - \varphi(r) \tag{3.53}$$

$$\Delta\varphi = \varphi(\theta + \Delta\theta) - \varphi(\theta) \tag{3.54}$$

Lateral shearing is a widely used technique since it is easy to perform using glass plates, rectangular prisms, or rotating mirrors. Another reason is that lateral shearing can also calculate the phase derivative as follows:

$$\Delta\varphi = \varphi(r + \Delta r) - \varphi(r) = \frac{\partial\varphi}{\partial r} \tag{3.55}$$

The sample phase can be reconstructed by an integration of shearing phases obtained from shearing fringes in two (orthogonal preferred) directions as

$$\Delta\varphi_x = \varphi(x + \Delta x, y) - \varphi(x, y) = \frac{\partial\varphi}{\partial x} \tag{3.56}$$

$$\Delta\varphi_y = \varphi(x, y + \Delta y) - \varphi(x, y) = \frac{\partial\varphi}{\partial y} \tag{3.57}$$

Shearing holography and interferometry are self-interference techniques that use the same path configuration, so they are not sensitive to external vibrations. In addition, since no additional reference is required, shearing holography and interferometry often have simple designs although they require additional shearing devices. Since the path difference between two replicas of the wavefront is very small, even partially coherent light can produce high contrast fringes. Shearing interferometry overcomes the disadvantages of classical holography and interferometry, which are vibration-sensitive and require a reference system. Even though shearing holography and interferometry offer benefits, they also have disadvantages. The main difference is that the phase under detection is to be calculated from the retrieved phase. For example, when using lateral shearing interferometry, the retrieved phase is actually the phase derivative along the shearing direction. Initially, phase slopes in two directions should be extracted, followed by phase integration in order to reconstruct the sample phase.

3.3   Numerical Simulations on Holography and Interferometry

Numerical simulations on Holography and Interferometry are presented in this section along with MATLAB codes. This include Gabor digital holography, on-axis and off-axis digital holography, interferometry (imaging digital holography), Fresnel digital holography, Fourier digital holography, shearing interferometry, and phase unwrapping.

3.3.1   Numerical Simulation on Gabor Digital Holography

Gabor digital holography is numerically simulated using the following Matlab code. The simulation is according to the scheme shown in Fig. 3.2a, c, as well as Eqs. (3.5)–(3.11). In this numerical simulation, the wavelength is 632.8 nm, the pixel number is 512×512, the pixel size is 10 μm, the distance between the sample and hologram planes is 300 mm, and the numerical wavefront propagation is based on the classical angular spectrum method. Figure 3.7a shows the amplitude and phase distributions of the object, and Fig. 3.7b is the Gabor digital hologram of the interference between the object scattered wavefront and the directly transmitted wavefront (plane wave). Figure 3.7c demonstrates both the reconstructed amplitude and phase distributions of the object. Compared to those in Fig. 3.7a, the reconstructed results suffer from poor quality as all of the 0th, +1st, and −1st order images are not be separated according to the Gabor digital holography reconstruction principle demonstrated in Fig. 3.2c.

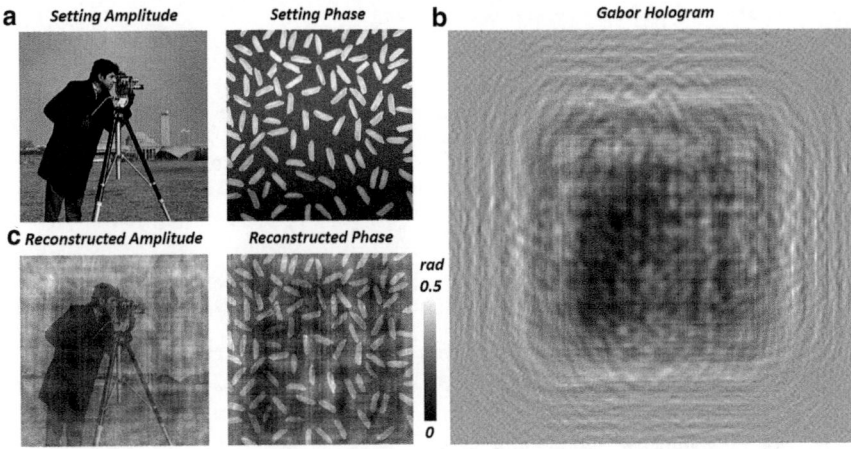

Fig. 3.7 Numerical simulation on Gabor digital holography. **a** Object amplitude and phase distributions; **b** Gabor hologram; **c** reconstructed object amplitude and phase distributions

Matlab Code for Gabor Digital Holography

```
1    % Computational Optical Phase Imaging
2    % Gabor Digital Holography
3
4    %% CCC
5    clear all;
6    close all;
7    clc;
8
9    %% Input Constants
10   lambda = 632.8*10^(-9); % Wavelength
11   PixelNum = 2^9; % Pixel Number
12   PixelSize = 10*10^(-6); % Pixel Size
13   d = 300*10^(-3); % Distance between sample and hologram
14   RI = 1; % Refractive Index
15
16   %% Other Constants
17   k = 2*pi/lambda; % Wave number
18   Freqency = 1/PixelSize; % Frequency
19   Fxvector = linspace(-Freqency/2,Freqency/2,PixelNum);
20   Fyvector = linspace(-Freqency/2,Freqency/2,PixelNum);
21   [FxMat, FyMat] = meshgrid(Fxvector,Fyvector); % Spectral
     coordinate
22   Hz = exp((1i*2*pi*d*RI/lambda)*(1-(lambda.*FxMat/RI).^2-
     (lambda.*FyMat/RI).^2).^0.5);
     % Transfer function-Propagation
23   Hmz = exp((1i*2*pi*(-d)*RI/lambda)*(1-(lambda.*FxMat/RI).^2-
     (lambda.*FyMat/RI).^2).^0.5);
     % Transfer function-Back Propagation
24
25   %% Sample Wavefront
26   Amplitude = double(imread('cameraman.tif'));
```

```
27   Phase = double(imread('rice.png'));
28   Amplitude = (Amplitude/max(max(Amplitude))+2)/3;  % For high
     transmittance
29   Phase = Phase/max(max(Phase))/2;
30   AmplitudeSample = ones(PixelNum,PixelNum);
31   PhaseSample = zeros(PixelNum,PixelNum);
32   AmplitudeSample(PixelNum/2-127:PixelNum/2+128,PixelNum/2-
     127:PixelNum/2+128) = Amplitude;  % Sample amplitude
33   PhaseSample(PixelNum/2-127:PixelNum/2+128,PixelNum/2-
     127:PixelNum/2+128) = Phase;
     % Sample phase
34   WavefrontSample = AmplitudeSample.*exp(1i*PhaseSample);  %
     Sample wavefront at sample plane
35   WavefrontSampleHologram =
     fftshift(ifft2(ifftshift(fftshift(fft2(ifftshift(WavefrontSamp
     le))).*Hz)));
     % Sample wavefront at hologram plane
36
37   %% Propagation Reference Wave
38   AmplitudeReference = ones(PixelNum,PixelNum);  % Reference
     amplitude
39   PhaseReference = zeros(PixelNum,PixelNum);  % Reference phase
40   WavefrontReference =
     AmplitudeReference.*exp(1i*PhaseReference);
     % Reference wavefront at sample plane
41   WavefrontReferenceHologram =
     fftshift(ifft2(ifftshift(fftshift(fft2(ifftshift(WavefrontRefe
     rence))).*Hz)));
     % Reference wavefront at hologram plane
42
43   %% Holography Recording
44   Hologram =
     (WavefrontSampleHologram+WavefrontReferenceHologram).*
     conj(WavefrontSampleHologram+WavefrontReferenceHologram);  %
     Hologram
45
46   %% Holography Reconstruction
47   BackPropagation =
     fftshift(ifft2(ifftshift(fftshift(fft2(ifftshift(
     Hologram.*WavefrontReferenceHologram))).*Hmz)));  %
     Backpropagation
48   AmplitudeSolve = sqrt(BackPropagation.*conj(BackPropagation));
     % Reconstructed sample amplitude
49   PhaseSolve =
     atan2(imag(BackPropagation),real(BackPropagation));  %
     Reconstructed sample phase
50   CutAmplitudeSolve = AmplitudeSolve(PixelNum/2-
     127:PixelNum/2+128,PixelNum/2-127:PixelNum/2+128);  %
     Reconstructed sample amplitude in field of interest
51   CutPhaseSolve = PhaseSolve(PixelNum/2-
     127:PixelNum/2+128,PixelNum/2-127:PixelNum/2+128);  %
     Reconstructed sample phase in field of interest
```

```
52
53  %% Plot
54  figure
55  subplot(2,3,1); imagesc(Amplitude); axis image; axis off;
    colormap(gray); title('Setting Sample Amplitude');
56  subplot(2,3,2); imagesc(Phase); axis image; axis off;
    colormap(gray); title('Setting Sample Phase');
57  subplot(2,3,3); imagesc(Hologram); axis image; axis off;
    colormap(gray); title('Hologram');
58  subplot(2,3,4); imagesc(CutAmplitudeSolve); axis image; axis
    off; colormap(gray); title('Reconstructed Sample Amplitude');
59  subplot(2,3,5); imagesc(CutPhaseSolve); axis image; axis off;
    colormap(gray); title('Reconstructed Sample Phase');
```

3.3.2 Numerical Simulation on On-Axis Digital Holography

Similar to Gabor digital holography, on-axis digital holography is also numerically simulated using the following Matlab code, according to the scheme shown in Fig. 3.2b, c as well as Eqs. (3.13)–(3.21). In this numerical simulation, the wavelength is 632.8 nm, the pixel number is 512×512, the pixel size is $10\,\mu$m, the distance between the sample and hologram planes is 300 mm, and the numerical wavefront propagation is based on the classical angular spectrum method. Figure 3.8a shows the amplitude and phase distributions of the object, and Fig. 3.8b lists the on-axis holograms of the interference between the object and the reference plane wavefront obtained with phase-shifting angles 0, $\frac{\pi}{2}$, π and $\frac{3\pi}{2}$ for the reference plane wavefront. According to Eqs. (3.17)–(3.21), the $+1$st order term can be extracted and both

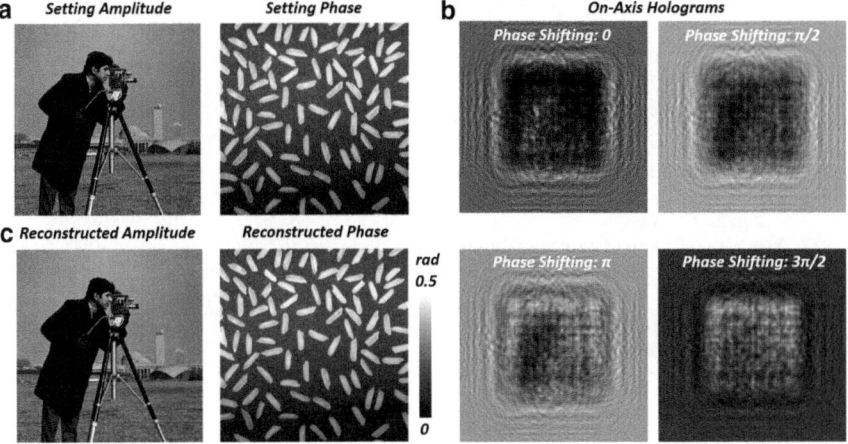

Fig. 3.8 Numerical simulation on on-axis digital holography. **a** Object amplitude and phase distributions; **b** on-axis phase-shifted holograms; **c** reconstructed object amplitude and phase distributions

the reconstructed amplitude and phase distributions of the object can be obtained as listed in Fig. 3.8c. Compared to the original amplitude and phase in Fig. 3.8a and to the reconstruction of the Gabor digital hologram in Fig. 3.7c, these reconstructions exhibit quite good quality, completely avoiding the influence of the 0th and −1st order terms.

Matlab Code for On-Axis Digital Holography

```
1    % Computational Optical Phase Imaging
2    % On-Axis Digital Holography
3
4    %% CCC
5    clear all;
6    close all;
7    clc;
8
9    %% Input Constants
10   lambda = 632.8*10^(-9);  % Wavelength
11   PixelNum = 2^9;  % Pixel Number
12   PixelSize = 10*10^(-6);  % Pixel Size
13   d = 300*10^(-3);  % Distance between sample and hologram
14   RI = 1;  % Refractive Index
15
16   %% Other Constants
17   k = 2*pi/lambda;  % Wave number
18   Freqency = 1/PixelSize;  % Frequency
19   Fxvector = linspace(-Freqency/2,Freqency/2,PixelNum);
20   Fyvector = linspace(-Freqency/2,Freqency/2,PixelNum);
21   [FxMat, FyMat] = meshgrid(Fxvector,Fyvector);  % Spectral
     coordinate
22   Hz = exp((1i*2*pi*d*RI/lambda)*(1-(lambda.*FxMat/RI).^2-
     (lambda.*FyMat/RI).^2).^0.5);
     % Transfer function-Propagation
23   Hmz = exp((1i*2*pi*(-d)*RI/lambda)*(1-(lambda.*FxMat/RI).^2-
     (lambda.*FyMat/RI).^2).^0.5);
     % Transfer function-Back Propagation
24
25   %% Sample Wavefront
26   Amplitude = double(imread('cameraman.tif'));
27   Phase = double(imread('rice.png'));
28   Amplitude = (Amplitude/max(max(Amplitude))+2)/3;  % For high
     transmittance
29   Phase = Phase/max(max(Phase))/2;
30   AmplitudeSample = ones(PixelNum,PixelNum);
```

```
31  PhaseSample = zeros(PixelNum,PixelNum);
32  AmplitudeSample(PixelNum/2-127:PixelNum/2+128,PixelNum/2-
    127:PixelNum/2+128) = Amplitude;
    % Sample amplitude
33  PhaseSample(PixelNum/2-127:PixelNum/2+128,PixelNum/2-
    127:PixelNum/2+128) = Phase;
    % Sample phase
34  WavefrontSample = AmplitudeSample.*exp(1i*PhaseSample);  %
    Sample wavefront at sample plane
35  WavefrontSampleHologram =
    fftshift(ifft2(ifftshift(fftshift(fft2(ifftshift(WavefrontSamp
    le))).*Hz)));
    % Sample wavefront at hologram plane

36
37  %% Propagation Reference Wave
38  AmplitudeReference = ones(PixelNum,PixelNum);  % Reference
    amplitude
39  PhaseReference1 = zeros(PixelNum,PixelNum);  % Reference
    phase-0
40  WavefrontReferenceHologram1 =
    AmplitudeReference.*exp(1i*PhaseReference1);
    % Reference wavefront-Plane wave with phase shifting angle of
    0
41  PhaseReference2 = pi/2*ones(PixelNum,PixelNum);  % Reference
    phase-pi/2
42  WavefrontReferenceHologram2 =
    AmplitudeReference.*exp(1i*PhaseReference2);
    % Reference wavefront-Plane wave with phase shifting angle of
    pi/2
43  PhaseReference3 = pi*ones(PixelNum,PixelNum);  % Reference
    phase-pi
44  WavefrontReferenceHologram3 =
    AmplitudeReference.*exp(1i*PhaseReference3);
    % Reference wavefront-Plane wave with phase shifting angle of
    pi
45  PhaseReference4 = 3*pi/2*ones(PixelNum,PixelNum);  % Reference
    phase-3*pi/2
46  WavefrontReferenceHologram4 =
    AmplitudeReference.*exp(1i*PhaseReference4);
    % Reference wavefront-Plane wave with phase shifting angle of
    3*pi/2

47
48  %% Holography Recording
49  Hologram1 =
    (WavefrontSampleHologram+WavefrontReferenceHologram1).*
    conj(WavefrontSampleHologram+WavefrontReferenceHologram1);  %
    Hologram-0
50  Hologram2 =
    (WavefrontSampleHologram+WavefrontReferenceHologram2).*
    conj(WavefrontSampleHologram+WavefrontReferenceHologram2);  %
    Hologram-pi/2
51  Hologram3 =
    (WavefrontSampleHologram+WavefrontReferenceHologram3).*
    conj(WavefrontSampleHologram+WavefrontReferenceHologram3);  %
    Hologram-pi
```

```
52   Hologram4 =
     (WavefrontSampleHologram+WavefrontReferenceHologram4).*
     conj(WavefrontSampleHologram+WavefrontReferenceHologram4);  %
     Hologram-3*pi/2
53
54   %% Holography Reconstruction
55   BackPropagation1 =
     fftshift(ifft2(ifftshift(fftshift(fft2(ifftshift(Hologram1.*
     WavefrontReferenceHologram1))).*Hmz)));  % Backpropagation-0
56   BackPropagation2 =
     fftshift(ifft2(ifftshift(fftshift(fft2(ifftshift(Hologram2.*
     WavefrontReferenceHologram2))).*Hmz)));  % Backpropagation-
     pi/2
57   BackPropagation3 =
     fftshift(ifft2(ifftshift(fftshift(fft2(ifftshift(Hologram3.*
     WavefrontReferenceHologram3))).*Hmz)));  % Backpropagation-pi
58   BackPropagation4 =
     fftshift(ifft2(ifftshift(fftshift(fft2(ifftshift(Hologram4.*
     WavefrontReferenceHologram4))).*Hmz)));  % Backpropagation-
     3*pi/2
59   BackPropagation =
     (BackPropagation1+BackPropagation2+BackPropagation3+BackPropag
     ation4)/4;
     % Backpropagation
60   AmplitudeSolve =
     sqrt(BackPropagation.*conj(BackPropagation));  % Reconstructed
     sample amplitude
61   PhaseSolve =
     atan2(imag(BackPropagation),real(BackPropagation));  %
     Reconstructed sample phase
62   CutAmplitudeSolve = AmplitudeSolve(PixelNum/2-
     127:PixelNum/2+128,PixelNum/2-127:PixelNum/2+128);
     % Reconstructed sample amplitude in field of interest
63   CutPhaseSolve = PhaseSolve(PixelNum/2-
     127:PixelNum/2+128,PixelNum/2-127:PixelNum/2+128);
     % Reconstructed sample phase in field of interest
64
65   %% Plot
66   figure
67   subplot(2,4,1); imagesc(Amplitude); axis image; axis off;
     colormap(gray); title('Setting Sample Amplitude');
68   subplot(2,4,2); imagesc(Phase); axis image; axis off;
     colormap(gray); title('Setting Sample Phase');
69   subplot(2,4,3); imagesc(Hologram1); axis image; axis off;
     colormap(gray); title('Hologram: 0');
70   subplot(2,4,4); imagesc(Hologram2); axis image; axis off;
     colormap(gray); title('Hologram: pi/2');
71   subplot(2,4,5); imagesc(CutAmplitudeSolve); axis image; axis
     off; colormap(gray);
     title('Reconstructed Sample Amplitude');
72   subplot(2,4,6); imagesc(CutPhaseSolve); axis image; axis off;
     colormap(gray);
     title('Reconstructed Sample Phase');
73   subplot(2,4,7); imagesc(Hologram3); axis image; axis off;
     colormap(gray); title('Hologram: pi');
74   subplot(2,4,8); imagesc(Hologram4); axis image; axis off;
     colormap(gray); title('Hologram: 3*pi/2');
```

3.3.3 Numerical Simulation on Off-Axis Digital Holography

The following numerical simulation and Matlab code are for off-axis digital holography. They are according to the scheme shown in Fig. 3.3b as well as Eqs. (3.29)–(3.34). In this numerical simulation, the wavelength is 632.8 nm, the pixel number is 2048 × 2048, the pixel size is 10 μm, the distance between the sample and hologram planes is 300 mm, the angle between sample and reference wave is 0.8°, and the numerical wavefront propagation is based on the classical angular spectrum method. The pixel number is much higher than that of on-axis and Gabor digital holography which is done with an intention to include and separate 0th, + 1st, and −1st order images. Figure 3.9a shows the amplitude and phase distributions of the object, and Fig. 3.9b is the hologram of the interference between the object and the reference wave front (tilted plane wave). Figure 3.9c shows reconstructed images corresponding to 0th, +1st, and −1st order terms. Object amplitude and phase distributions can be obtained from extracted +1st order term.

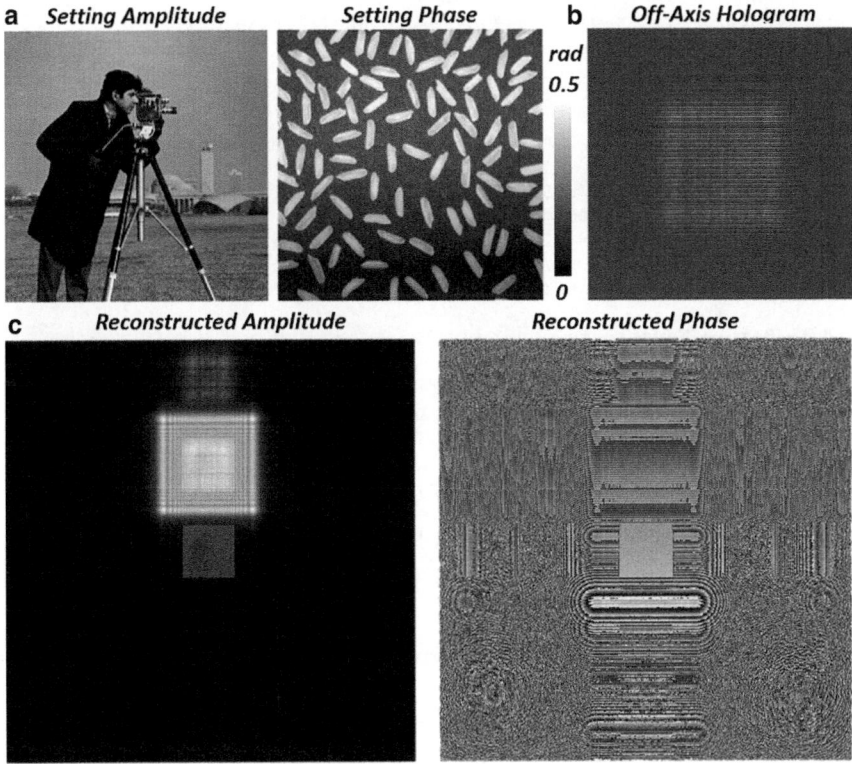

Fig. 3.9 Numerical simulation on off-axis digital holography. **a** Object amplitude and phase distributions; **b** off-axis hologram; **c** reconstructed object amplitude and phase distributions

Matlab Code for Off-Axis Digital Holography

```
1   % Computational Optical Phase Imaging
2   % Off-Axis Digital Holography: Backpropagation
3
4   %% CCC
5   clear all;
6   close all;
7   clc;
8
9   %% Input Constants
10  lambda = 632.8*10^(-9); % Wavelength
11  PixelNum = 2^9; % Pixel Number
12  PixelSize = 10*10^(-6); % Pixel Size
13  d = 300*10^(-3); % Distance between sample and hologram
14  RI = 1; % Refractive Index
15
16  %% Other Constants
17  k = 2*pi/lambda; % Wave number
18  Freqency = 1/PixelSize; % Frequency
19  Fxvector = linspace(-Freqency/2,Freqency/2,PixelNum);
20  Fyvector = linspace(-Freqency/2,Freqency/2,PixelNum);
21  [FxMat, FyMat] = meshgrid(Fxvector,Fyvector); % Spectral
    coordinate
22  Sxvector = linspace(-
    PixelNum*PixelSize/2,PixelNum*PixelSize/2,PixelNum);
23  Syvector = linspace(-
    PixelNum*PixelSize/2,PixelNum*PixelSize/2,PixelNum);
24  [SxMat, SyMat] = meshgrid(Sxvector,Syvector); % Spatial coordinat
25  Hz = exp((1i*2*pi*d*RI/lambda)*(1-(lambda.*FxMat/RI).^2-
    (lambda.*FyMat/RI).^2).^0.5);
    % Transfer function-Propagation
26
27  %% Sample Wavefront
28  Amplitude = double(imread('cameraman.tif'));
29  Phase = double(imread('rice.png'));
30  Amplitude = (Amplitude/max(max(Amplitude))+2)/3;  % For high
    transmittance
31  Phase = Phase/max(max(Phase))*0.5;
32  AmplitudeSample = zeros(PixelNum,PixelNum);
33  PhaseSample = zeros(PixelNum,PixelNum);
34  AmplitudeSample(PixelNum/2-127:PixelNum/2+128,PixelNum/2-
    127:PixelNum/2+128) = Amplitude; % Sample amplitude
35  PhaseSample(PixelNum/2-127:PixelNum/2+128,PixelNum/2-
    127:PixelNum/2+128) = Phase;
    % Sample phase
36  WavefrontSample = AmplitudeSample.*exp(1i*PhaseSample); % Sample
    wavefront at sample plane
37  WavefrontSampleHologram =
    fftshift(ifft2(ifftshift(fftshift(fft2(ifftshift(WavefrontSample)
    ).*Hz)));
    % Sample wavefront at hologram plane
38  WavefrontSampleHologram =
    WavefrontSampleHologram/max(max(WavefrontSampleHologram.*conj(Wave
    frontSampleHologram))); % Normalization
```

```
39
40 %% Propagation Reference Wave
41 AmplitudeReference = ones(PixelNum,PixelNum);  % Reference
   amplitude
42 Theta = 0.8;  % Tilting angle
43 PhaseReference = k*tand(Theta)*SyMat;  % Tilting phase
44 WavefrontReferenceHologram =
   AmplitudeReference.*exp(1i*PhaseReference);
   % Reference wavefront Plane Wave
45
46 %% Holography Recording
47 Hologram = (WavefrontSampleHologram+WavefrontReferenceHologram).*
   conj(WavefrontSampleHologram+WavefrontReferenceHologram);  %
   Hologram
48
49 %% Coordinate Extension
50 FxvectorExtension = linspace(-Freqency/2,Freqency/2,4*PixelNum);
51 FyvectorExtension = linspace(-Freqency/2,Freqency/2,4*PixelNum);
52 [FxMatExtension, FyMatExtension] =
   meshgrid(FxvectorExtension,FyvectorExtension);
   % Spectral coordinate extension
53 SxvectorExtension = linspace(-
   4*PixelNum*PixelSize/2,4*PixelNum*PixelSize/2,4*PixelNum);
54 SyvectorExtension = linspace(-
   4*PixelNum*PixelSize/2,4*PixelNum*PixelSize/2,4*PixelNum);
55 [SxMatExtension, SyMatExtension] =
   meshgrid(SxvectorExtension,SyvectorExtension);
   % Spatial coordinate extension
56 HzExtension = exp((1i*2*pi*d*RI/lambda)*(1-
   (lambda.*FxMatExtension/RI).^2-
   (lambda.*FyMatExtension/RI).^2).^0.5);  % Transfer function-
   Propagation
57 HmzExtension = exp((1i*2*pi*(-d)*RI/lambda)*(1-
   (lambda.*FxMatExtension/RI).^2-
   (lambda.*FyMatExtension/RI).^2).^0.5);  % Transfer function-Back
   Propagation
58 HologramExtension = zeros(4*PixelNum,4*PixelNum);
59 HologramExtension(4*PixelNum/2-255:4*PixelNum/2+256,4*PixelNum/2-
   255:4*PixelNum/2+256) = Hologram;  % Extended Hologram
60
61 %% Expanded Reference Wavefront
62 AmplitudeReferenceExtension = ones(4*PixelNum,4*PixelNum);  %
   Reference amplitude
63 PhaseReferenceExtension = k*tand(Theta)*SyMatExtension;  % Tilting
   phase
64 WavefrontReferenceHologramExtension =
   AmplitudeReferenceExtension.*exp(1i*PhaseReferenceExtension);  %
   Extended reference wavefront
65
66 %% Holography Reconstruction
67 BackPropagation =
   fftshift(ifft2(ifftshift(fftshift(fft2(ifftshift(HologramExtension
   .*
   WavefrontReferenceHologramExtension)))).*HmzExtension)));  %
   BackPropagation
```

```
68 AmplitudeSolve = sqrt(BackPropagation.*conj(BackPropagation));  %
   Reconstructed sample amplitude
69 PhaseSolve = atan2(imag(BackPropagation),real(BackPropagation));
   Reconstructed sample phase
70 CutAmplitudeSolve = AmplitudeSolve(4*PixelNum/2-
   127:4*PixelNum/2+128,4*PixelNum/2-127:4*PixelNum/2+128);  %
   Reconstructed sample amplitude in field of interest
71 CutPhaseSolve = PhaseSolve(4*PixelNum/2-
   127:4*PixelNum/2+128,4*PixelNum/2-127:4*PixelNum/2+128);  %
   Reconstructed sample phase in field of interest
72
73 %% Plot
74 figure
75 subplot(3,3,1); imagesc(Amplitude); axis image; axis off;
   colormap(gray); title('Setting Sample Amplitude');
76 subplot(3,3,2); imagesc(Phase); axis image; axis off;
   colormap(gray); title('Setting Sample Phase');
77 subplot(3,3,3); imagesc(Hologram); axis image; axis off;
   colormap(gray); title('Hologram');
78 subplot(3,3,4); imagesc(AmplitudeSolve); axis image; axis off;
   colormap(gray); title('Reconstructed Sample Amplitude');
79 subplot(3,3,5); imagesc(PhaseSolve); axis image; axis off;
   colormap(gray); title('Reconstructed Sample Phase');
80 subplot(3,3,7); imagesc(CutAmplitudeSolve); axis image; axis off;
   colormap(gray); title('Reconstructed Sample Amplitude');
81 subplot(3,3,8); imagesc(CutPhaseSolve); axis image; axis off;
   colormap(gray); title('Reconstructed Sample Phase');
```

An expanded field of view with more pixel numbers is required to include and separate all of the 0th, $+1$st, and -1st order images especially in off-axis mode, however, such procedure inevitably increases the computational burden and also slows down the computation. It is possible to solve such problems by preserving the $+1$st order spectrum before holographic reconstruction while removing the 0th and -1st order spectra; in other words, only the information represented by Eq. (3.33) is retained while removing those represented in Eqs. (3.31), (3.32), and (3.34). A field of view expansion is not required in this case. Using the same parameters, off-axis digital holography is numerically simulated via spectral filtering using the following Matlab code. Figure 3.10a shows the amplitude and phase distributions of the object used, and Fig. 3.10b reveals the hologram of the interference between the object and the reference wavefront (tilted plane wave). In this method, the hologram spectrum shown in Fig. 3.10c is first filtered to remove the 0th and -1st order terms. Illuminating the filtered hologram with a reference wavefront shows both the object amplitude and phase distributions as shown in Fig. 3.10d.

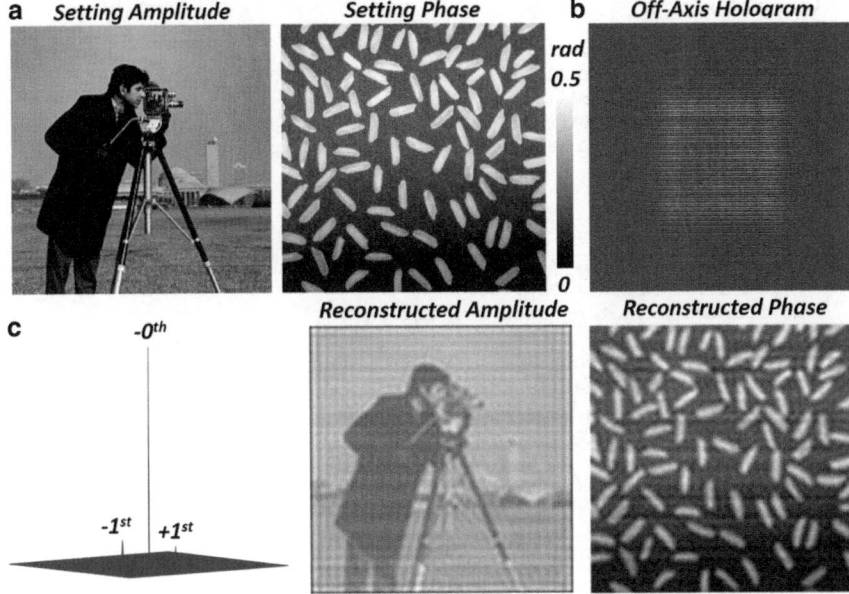

Fig. 3.10 Numerical simulation on off-axis digital holography using spectral filtering. **a** Object amplitude and phase distributions; **b** off-axis hologram; **c** hologram spectrum; **d** reconstructed object amplitude and phase distributions

Matlab Code for Off-Axis Digital Holography Using Spectral Filtering

```
1   % Computational Optical Phase Imaging
2   % Off-Axis Digital Holography: Filtering
3
4   %% CCC
5   clear all;
6   close all;
7   clc;
8
9   %% Input Constants
10  lambda = 632.8*10^(-9); % Wavelength
11  PixelNum = 2^9; % Pixel Number
12  PixelSize = 10*10^(-6); % Pixel Size
13  d = 300*10^(-3); % Distance between sample and hologram
14  RI = 1; % Refractive Index
15
16  %% Other Constants
17  k = 2*pi/lambda; % Wave number
18  Freqency = 1/PixelSize; % Frequency
19  Fxvector = linspace(-Freqency/2,Freqency/2,PixelNum);
20  Fyvector = linspace(-Freqency/2,Freqency/2,PixelNum);
```

```matlab
21  [FxMat, FyMat] = meshgrid(Fxvector,Fyvector);  % Spectral
    coordinate
22  Sxvector = linspace(-
    PixelNum*PixelSize/2,PixelNum*PixelSize/2,PixelNum);
23  Syvector = linspace(-
    PixelNum*PixelSize/2,PixelNum*PixelSize/2,PixelNum);
24  [SxMat, SyMat] = meshgrid(Sxvector,Syvector);  % Spatial
    coordinate
25  Hz = exp((1i*2*pi*d*RI/lambda)*(1-(lambda.*FxMat/RI).^2-
    (lambda.*FyMat/RI).^2).^0.5);
    % Transfer function-Propagation
26  Hmz = exp((1i*2*pi*(-d)*RI/lambda)*(1-(lambda.*FxMat/RI).^2-
    (lambda.*FyMat/RI).^2).^0.5);
    % Transfer function-Back Propagation
27
28  %% Sample Wavefront
29  Amplitude = double(imread('cameraman.tif'));
30  Phase = double(imread('rice.png'));
31  Amplitude = (Amplitude/max(max(Amplitude))+2)/3;  % For high
    transmittance
32  Phase = Phase/max(max(Phase))*0.5;
33  AmplitudeSample = zeros(PixelNum,PixelNum);
34  PhaseSample = zeros(PixelNum,PixelNum);
35  AmplitudeSample(PixelNum/2-127:PixelNum/2+128,PixelNum/2-
    127:PixelNum/2+128) = Amplitude;  % Sample amplitude
36  PhaseSample(PixelNum/2-127:PixelNum/2+128,PixelNum/2-
    127:PixelNum/2+128) = Phase;
    % Sample phase
37  WavefrontSample = AmplitudeSample.*exp(1i*PhaseSample);  %
    Sample wavefront at sample plane
38  WavefrontSampleHologram =
    fftshift(ifft2(ifftshift(fftshift(fft2(ifftshift(WavefrontSampl
    e))).*Hz)));
    % Sample wavefront at hologram plane
39  WavefrontSampleHologram =
    WavefrontSampleHologram/max(max(WavefrontSampleHologram.*conj(W
    avefrontSampleHologram)));  % Normalization
40
41  %% Propagation Reference Wave
42  AmplitudeReference = ones(PixelNum,PixelNum);  % Reference
    amplitude
43  Theta = 0.8;  % Tilting angle
44  PhaseReference = k*tand(Theta)*SyMat;  % Tilting phase
45  WavefrontReferenceHologram =
    AmplitudeReference.*exp(1i*PhaseReference);
    % Reference wavefront Plane Wave
46
47  %% Holography Recording
48  Hologram =
    (WavefrontSampleHologram+WavefrontReferenceHologram).*conj(Wave
    frontSampleHologram+WavefrontReferenceHologram);  % Hologram
49
50  %% Spectral Filtering
51  Spectrum = fftshift(fft2(ifftshift(Hologram)));
52  NewSpectrum = zeros(PixelNum,PixelNum);
```

```
53  NewSpectrum(144-60:144+60,257-60:257+60) = Spectrum(144-
    60:144+60,257-60:257+60);
    % Extract 1st spectrum
54  NewHologram = fftshift(ifft2(ifftshift(NewSpectrum)));
55
56  %% Holography Reconstruction
57  BackPropagation =
    fftshift(ifft2(ifftshift(fftshift(fft2(ifftshift(NewHologram.*W
    avefrontReferenceHologram))).*Hmz)));
    % BackPropagation
58  AmplitudeSolve =
    sqrt(BackPropagation.*conj(BackPropagation));  % Reconstructed
    sample amplitude
59  PhaseSolve =
    atan2(imag(BackPropagation),real(BackPropagation));  %
    Reconstructed sample phase
60  CutAmplitudeSolve = AmplitudeSolve(PixelNum/2-
    127:PixelNum/2+128,PixelNum/2-127:PixelNum/2+128);  %
    Reconstructed sample amplitude in field of interest
61  CutPhaseSolve = PhaseSolve(PixelNum/2-
    127:PixelNum/2+128,PixelNum/2-127:PixelNum/2+128);
    % Reconstructed sample phase in field of interest
62
63  %% Plot
64  figure
65  subplot(3,3,1); imagesc(Amplitude); axis image; axis off;
    colormap(gray); title('Setting Sample Amplitude');
66  subplot(3,3,2); imagesc(Phase); axis image; axis off;
    colormap(gray); title('Setting Sample Phase');
67  subplot(3,3,3); imagesc(Hologram); axis image; axis off;
    colormap(gray); title('Hologram');
68  subplot(3,3,4); mesh(abs(Spectrum)); axis off; colormap(gray);
    title('Hologram Spectrum');
69  subplot(3,3,5); imagesc(AmplitudeSolve); axis image; axis off;
    colormap(gray); title('Reconstructed Sample Amplitude');
70  subplot(3,3,6); imagesc(PhaseSolve); axis image; axis off;
    colormap(gray); title('Reconstructed Sample Phase');
71  subplot(3,3,7); imagesc(CutAmplitudeSolve); axis image; axis
    off; colormap(gray); title('Reconstructed Sample Amplitude');
72  subplot(3,3,8); imagesc(CutPhaseSolve); axis image; axis off;
    colormap(gray); title('Reconstructed Sample Phase');
```

As compared to on-axis digital holography, the reconstruction quality of off-axis digital holography cannot be as good because of the wide bandwidth requirements, whereas on-axis digital holography retains more high-frequency information (details) than off-axis digital holography.

3.3.4   Numerical Simulation on Interferometry

In interferometry (imaging holography), the object and recording planes are conjugated unlike other general holography techniques discussed above. Therefore, phase retrieval for both on-axis and off-axis digital interferometry is simplified without using back propagation since the reconstructed wavefront at the imaging plane directly reflects that at the object plane.

On-axis digital interferometry is numerically simulated using the following Matlab code. In this numerical simulation, the pixel number is 256×256.

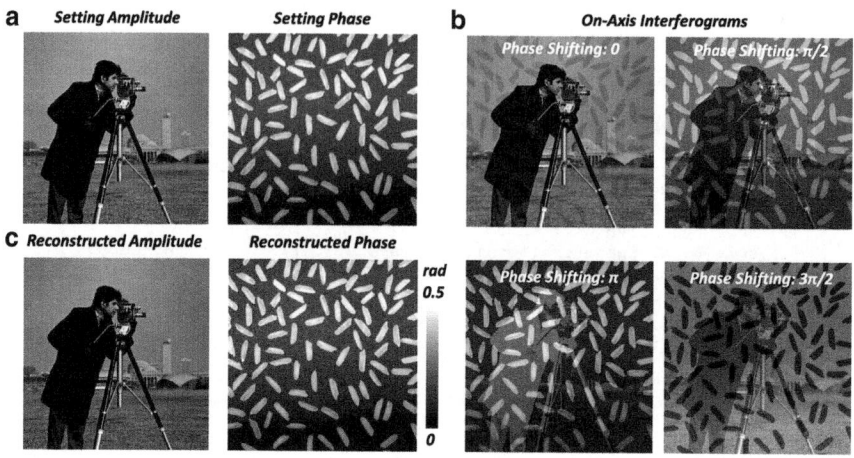

Fig. 3.11 Numerical simulation on on-axis digital interferometry. **a** Object amplitude and phase distributions; **b** on-axis phase-shifting interferograms; **c** reconstructed object amplitude and phase distributions

Figure 3.11a shows the original amplitude and phase distributions of the object, and Fig. 3.11b lists the on-axis phase-shifting interferograms of the interference between the object wavefront and phase shifted reference plane wavefront at angles 0, $\frac{\pi}{2}$, π, and $\frac{3\pi}{2}$ as described by Eqs. (3.22)–(3.25), respectively. The object phase φ obtained by Eq. (3.26) is shown in Fig. 3.11c.

Matlab Code for On-Axis Interferometry

```
1    % Computational Optical Phase Imaging
2    % On-Axis Interferometry (Imaging Digital Holography)
3
4    %% CCC
5    clear all;
6    close all;
7    clc;
8
9    %% Input Constants
10   PixelNum = 2^8;  % Pixel Number
11
12   %% Sample Wavefront
13   AmplitudeSample = double(imread('cameraman.tif'));
14   PhaseSample = double(imread('rice.png'));
15   AmplitudeSample =
     (AmplitudeSample/max(max(AmplitudeSample))+2)/3;  % For high
     transmittance
```

```
16   PhaseSample = PhaseSample/max(max(PhaseSample))/2;
17   WavefrontSampleHologram =
     AmplitudeSample.*exp(1i*PhaseSample);
     % Sample wavefront at hologram plane
18
19   %% Propagation Reference Wave
20   AmplitudeReference = ones(PixelNum,PixelNum); % Reference
     amplitude
21   PhaseReference1 = zeros(PixelNum,PixelNum);  % Reference phase-
     0
22   WavefrontReferenceHologram1 =
     AmplitudeReference.*exp(1i*PhaseReference1);
     % Reference wavefront-Plane wave with phase shifting angle of
     0
23   PhaseReference2 = pi/2*ones(PixelNum,PixelNum);  % Reference
     phase-pi/2
24   WavefrontReferenceHologram2 =
     AmplitudeReference.*exp(1i*PhaseReference2);
     % Reference wavefront-Plane wave with phase shifting angle of
     pi/2
25   PhaseReference3 = pi*ones(PixelNum,PixelNum);  % Reference
     phase-pi
26   WavefrontReferenceHologram3 =
     AmplitudeReference.*exp(1i*PhaseReference3);
     % Reference wavefront-Plane wave with phase shifting angle of
     pi
27   PhaseReference4 = 3*pi/2*ones(PixelNum,PixelNum);  % Reference
     phase-3*pi/2
28   WavefrontReferenceHologram4 =
     AmplitudeReference.*exp(1i*PhaseReference4);
     % Reference wavefront-Plane wave with phase shifting angle of
     3*pi/2
29
30   %% Holography Recording
31   Hologram1 =
     (WavefrontSampleHologram+WavefrontReferenceHologram1).*
     conj(WavefrontSampleHologram+WavefrontReferenceHologram1);  %
     Hologram-0
32   Hologram2 =
     (WavefrontSampleHologram+WavefrontReferenceHologram2).*
     conj(WavefrontSampleHologram+WavefrontReferenceHologram2);  %
     Hologram-pi/2
33   Hologram3 =
     (WavefrontSampleHologram+WavefrontReferenceHologram3).*
     conj(WavefrontSampleHologram+WavefrontReferenceHologram3);  %
     Hologram-pi
34   Hologram4 =
     (WavefrontSampleHologram+WavefrontReferenceHologram4).*
     conj(WavefrontSampleHologram+WavefrontReferenceHologram4);  %
     Hologram-3*pi/2
35
36   %% Holography Reconstruction
37   AmplitudeSolve = (Hologram1+Hologram2+Hologram3+Hologram4)/4;
38   PhaseSolve = atan2((Hologram1-Hologram3),(Hologram4-
     Hologram2));
```

```
39
40    %% Plot
41    figure
42    subplot(2,4,1); imagesc(AmplitudeSample); axis image; axis
      off; colormap(gray); title('Setting Sample Amplitude');
43    subplot(2,4,2); imagesc(PhaseSample); axis image; axis off;
      colormap(gray); title('Setting Sample Phase');
44    subplot(2,4,3); imagesc(Hologram1); axis image; axis off;
      colormap(gray); title('Hologram: 0');
45    subplot(2,4,4); imagesc(Hologram2); axis image; axis off;
      colormap(gray); title('Hologram: pi/2');
46    subplot(2,4,5); imagesc(AmplitudeSolve); axis image; axis off;
      colormap(gray); title('Reconstructed Sample Amplitude');
47    subplot(2,4,6); imagesc(PhaseSolve); axis image; axis off;
      colormap(gray); title('Reconstructed Sample Phase');
48    subplot(2,4,7); imagesc(Hologram3); axis image; axis off;
      colormap(gray); title('Hologram: pi');
49    subplot(2,4,8); imagesc(Hologram4); axis image; axis off;
      colormap(gray); title('Hologram: 3*pi/2');
```

Off-axis digital interferometry is numerically simulated using the following Matlab code. In this numerical simulation, the pixel number is 256×256, and the angle between sample and reference wave is set to $1°$. Figure 3.12a is the original amplitude and phase distributions of the object, and Fig. 3.12b is the off-axis interferogram of the interference between the object wavefront and the reference wavefront

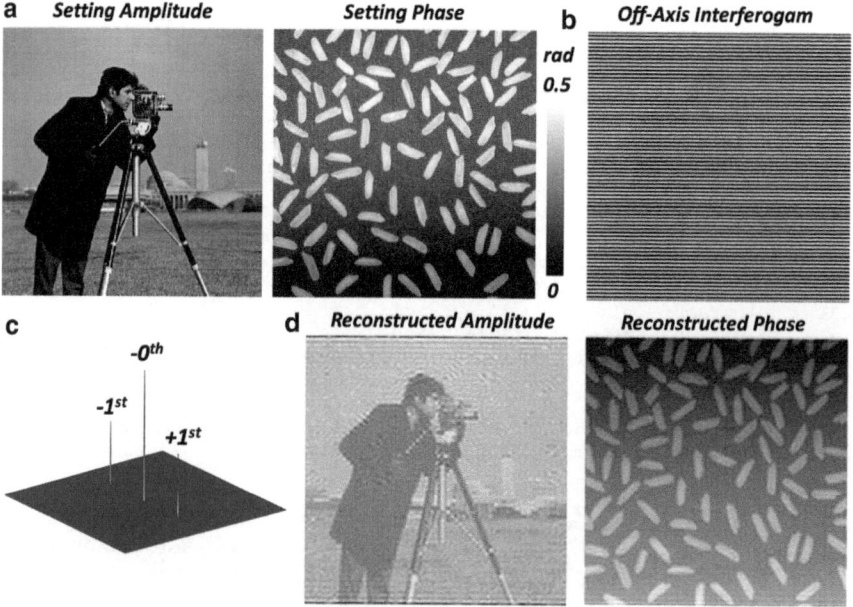

Fig. 3.12 Numerical simulation on off-axis digital interferometry. **a** Original object amplitude and phase distributions; **b** off-axis interferogram; **c** interferogram spectrum; **d** reconstructed object amplitude and phase distributions

(tilted plane wave) as described in Eq. (3.35). For off-axis interferometry, fast Fourier transform-based phase retrieval method is often used. Equation (3.35) is the simplified form of Eq. (3.29) obtained by treating $R(x, y)$ as a constant. Figure 3.12c shows the spectrum of the off-axis interferogram as described in Eq. (3.36). Three peaks represent the 0th, $+1$st, and -1st order terms, respectively. 1st order term is extracted and moved to the center of the spectrum, while 0th and -1st order terms are removed as described by Eq. (3.37). The object amplitude A and phase φ obtained are shown in Fig. 3.12d, which are according to Eqs. (3.38) and (3.39) following an inverse Fourier transform.

Matlab Code for Off-Axis Interferometry

```
1    % Computational Optical Phase Imaging
2    % Off-Axis Interferometry (Imaging Digital Holography)
3
4    %% CCC
5    clear all;
6    close all;
7    clc;
8
9    %% Input Constants
10   lambda = 632.8*10^(-9);  % Wavelength
11   PixelNum = 2^8;  % Pixel Number
12   PixelSize = 10*10^(-6);  % Pixel Size
13   k = 2*pi/lambda;  % Wave number
14   Sxvector = linspace(-
     PixelNum*PixelSize/2,PixelNum*PixelSize/2,PixelNum);
15   Syvector = linspace(-
     PixelNum*PixelSize/2,PixelNum*PixelSize/2,PixelNum);
16   [SxMat, SyMat] = meshgrid(Sxvector,Syvector);  % Spatial
     coordinate
17
18   %% Sample Wavefront
19   AmplitudeSample = double(imread('cameraman.tif'));
20   PhaseSample = double(imread('rice.png'));
21   AmplitudeSample =
     (AmplitudeSample/max(max(AmplitudeSample))+9)/10;  % For high
     transmittance
22   PhaseSample = PhaseSample/max(max(PhaseSample))*0.5;
23   WavefrontSampleHologram =
     AmplitudeSample.*exp(1i*PhaseSample);
     % Sample wavefront at sample plane
24
25   %% Propagation Reference Wave
26   AmplitudeReference = ones(PixelNum,PixelNum);  % Reference
     amplitude
27   Theta = 1;  % Tilting angle
28   PhaseReference = k*tand(Theta)*SyMat;  % Tilting phase
29   WavefrontReferenceHologram =
     AmplitudeReference.*exp(1i*PhaseReference);
     % Reference wavefront Plane Wave
30
```

```
31   %% Holography Recording
32   Hologram =
     (WavefrontSampleHologram+WavefrontReferenceHologram).*
     conj(WavefrontSampleHologram+WavefrontReferenceHologram);  %
     Hologram
33
34   %% Spectral Filtering
35   Spectrum = fftshift(fft2(ifftshift(Hologram)));
36   Spectrum1st = zeros(PixelNum,PixelNum);
37   Spectrum1st(PixelNum/2+1-50:PixelNum/2+1+50,PixelNum/2+1-
     50:PixelNum/2+1+50) = Spectrum(58-50:58+50,129-50:129+50);  %
     Extract 1st spectrum
38   Hologram1st = fftshift(ifft2(ifftshift(Spectrum1st)));
39
40   %% Holography Reconstruction
41   AmplitudeSolve = abs(Hologram1st);
42   PhaseSolve = atan2(imag(Hologram1st),real(Hologram1st));
43
44   %% Plot
45   figure
46   subplot(2,3,1); imagesc(AmplitudeSample); axis image; axis
     off; colormap(gray); title('Setting Sample Amplitude');
47   subplot(2,3,2); imagesc(PhaseSample); axis image; axis off;
     colormap(gray); title('Setting Sample Phase');
48   subplot(2,3,3); imagesc(Hologram); axis image; axis off;
     colormap(gray); title('Hologram');
49   subplot(2,3,4); mesh(abs(Spectrum)); axis off; colormap(gray);
     title('Hologram Spectrum');
50   subplot(2,3,5); imagesc(AmplitudeSolve); axis image; axis off;
     colormap(gray); title('Reconstructed Sample Amplitude');
51   subplot(2,3,6); imagesc(PhaseSolve); axis image; axis off;
     colormap(gray); title('Reconstructed Sample Phase');
```

3.3.5   Numerical Simulation on Fresnel Digital Holography

Fresnel digital holography can adjust the magnification of reconstruction via introducing an extra spherical wavefront. Here, Fresnel digital holography is numerically simulated using the following Matlab code, according to the scheme shown in Fig. 3.5a. The wavelength for the simulation is 632.8 nm, the pixel number is 512×512, the pixel size is 10 μm, and the angle between sample and reference wave is $0.8°$. Figure 3.13a is the amplitude and phase distributions of the object, and Fig. 3.13b is the off-axis hologram of the interference between the object wavefront and the reference wavefront (tilted plane wave). Figure 3.13c lists the reconstructed object amplitude and phase, which have the same size of the setting object; while Fig. 3.13d lists the reconstructed object amplitude and phase with a magnification of 1.5. It is worth noting that extra phase is introduced in Fig. 3.13d since extra spherical wavefront is used for reconstruction.

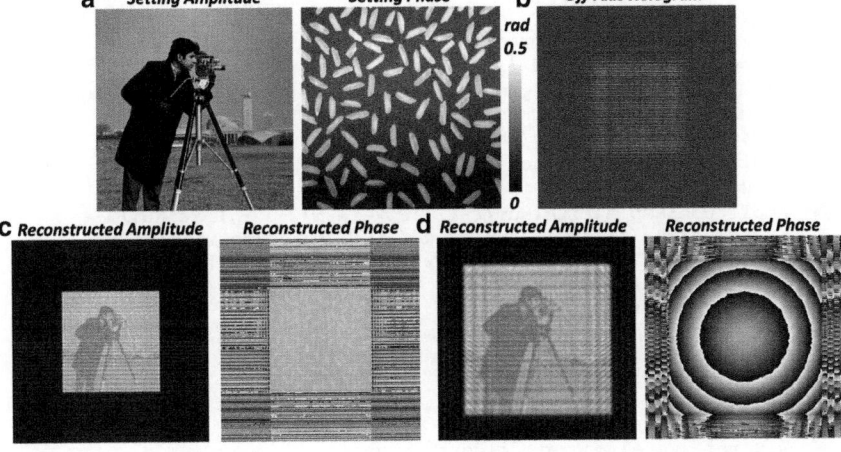

Fig. 3.13 Numerical simulation on Fresnel digital holography. **a** Object amplitude and phase distributions; **b** off-axis hologram; **c** reconstructed object amplitude and phase distributions with unit magnification; **d** reconstructed object amplitude and phase distributions with 1.5 times magnification

Matlab Code for Fresnel Digital Holography

```
1   % Computational Optical Phase Imaging
2   % Fresnel Digital Holography
3
4   %% CCC
5   clear all;
6   close all;
7   clc;
8
9   %% Input Constants
10  lambda = 632.8*10^(-9); % Wavelength
11  PixelNum = 2^9; % Pixel Number
12  PixelSize = 10*10^(-6); % Pixel Size
13  d = 300*10^(-3); % Distance between sample and hologram
14  RI = 1; % Refractive Index
15
16  %% Other Constants
17  k = 2*pi/lambda; % Wave number
18  Freqency = 1/PixelSize; % Frequency
19  Fxvector = linspace(-Freqency/2,Freqency/2,PixelNum);
20  Fyvector = linspace(-Freqency/2,Freqency/2,PixelNum);
21  [FxMat, FyMat] = meshgrid(Fxvector,Fyvector); % Spectral
    coordinate
22  Sxvector = linspace(-
    PixelNum*PixelSize/2,PixelNum*PixelSize/2,PixelNum);
23  Syvector = linspace(-
    PixelNum*PixelSize/2,PixelNum*PixelSize/2,PixelNum);
24  [SxMat, SyMat] = meshgrid(Sxvector,Syvector); % Spatial
    coordinate
25  Hz = exp((1i*2*pi*d*RI/lambda)*(1-(lambda.*FxMat/RI).^2-
    (lambda.*FyMat/RI).^2).^0.5);
    % Transfer function-Propagation
```

```
26
27  %% Sample Wavefront
28  Amplitude = double(imread('cameraman.tif'));
29  Phase = double(imread('rice.png'));
30  Amplitude = (Amplitude/max(max(Amplitude))+2)/3;  % For high
    transmittance
31  Phase = Phase/max(max(Phase))*0.5;
32  AmplitudeSample = zeros(PixelNum,PixelNum);
33  PhaseSample = zeros(PixelNum,PixelNum);
34  AmplitudeSample(PixelNum/2-127:PixelNum/2+128,PixelNum/2-
    127:PixelNum/2+128) = Amplitude;  % Sample amplitude
35  PhaseSample(PixelNum/2-127:PixelNum/2+128,PixelNum/2-
    127:PixelNum/2+128) = Phase;
    % Sample phase
36  WavefrontSample = AmplitudeSample.*exp(1i*PhaseSample);  %
    Sample wavefront at sample plane
37  WavefrontSampleHologram =
    fftshift(ifft2(ifftshift(fftshift(fft2(ifftshift(WavefrontSampl
    e))).*Hz)));
    % Sample wavefront at hologram plane
38  WavefrontSampleHologram =
    WavefrontSampleHologram/max(max(WavefrontSampleHologram.*conj(W
    avefrontSampleHologram)));  % Normalization
39
40  %% Propagation Reference Wave
41  AmplitudeReference = ones(PixelNum,PixelNum);  % Reference
    amplitude
42  Theta = 0.8;  % Tilting angle
43  PhaseReference = k*tand(Theta)*SyMat;  % Tilting phase
44  WavefrontReferenceHologram =
    AmplitudeReference.*exp(1i*PhaseReference);
    % Reference wavefront Plane Wave
45
46  %% Holography Recording
47  Hologram =
    (WavefrontSampleHologram+WavefrontReferenceHologram).*
    conj(WavefrontSampleHologram+WavefrontReferenceHologram);  %
    Hologram
48
49  %% Holography Reconstruction
50  Magnification = 1.5;
51  switch Magnification
52      case 1
53          PhaseFactor = fftshift(fft2(ifftshift(exp(1i*k/2/(-
    d)*(SxMat.^2+SyMat.^2)))));
54          Transmission = fftshift(fft2(ifftshift(Hologram.*
    WavefrontReferenceHologram)));
55          BackPropagation = exp(1i*k*(-d))/(1i*lambda*(-d)).*
                  fftshift(ifft2(ifftshift(PhaseFactor.*Transmission
    )));
56          AmplitudeSolve =
    sqrt(BackPropagation.*conj(BackPropagation));
57          PhaseSolve =
    atan2(imag(BackPropagation),real(BackPropagation));
```

```
58      otherwise
59          dspherical = Magnification*(-d)/(Magnification-1);
60          dobject = Magnification*(-d);
61          SphericalWave =
    exp(1i*k*dspherical).*exp(1i*k*(SxMat.^2+SyMat.^2)/2/dspherical
    );
62          PhaseFactor =
    fftshift(fft2(ifftshift(exp(1i*k/2/dobject*(SxMat.^2+SyMat.^2))
    )));
63          Transmission = fftshift(fft2(ifftshift(Hologram.*
                WavefrontReferenceHologram.*SphericalWave)));
64          BackPropagation = exp(1i*k*dobject)/(1i*lambda*dobject).*
                fftshift(ifft2(ifftshift(PhaseFactor.*Transmission
    )));
65          AmplitudeSolve =
    sqrt(BackPropagation.*conj(BackPropagation));
66          PhaseSolve =
    atan2(imag(BackPropagation),real(BackPropagation));
67  end
68
69  %% Plot
70  figure
71  subplot(2,3,1); imagesc(Amplitude); axis image; axis off;
    colormap(gray);
    title('Setting Sample Amplitude');
72  subplot(2,3,2); imagesc(Phase); axis image; axis off;
    colormap(gray); title('Setting Sample Phase');
73  subplot(2,3,3); imagesc(Hologram); axis image; axis off;
    colormap(gray); title('Hologram');
74  subplot(2,3,4); imagesc(AmplitudeSolve); axis image; axis off;
    colormap(gray);
    title('Reconstructed Sample Amplitude');
75  subplot(2,3,5); imagesc(PhaseSolve); axis image; axis off;
    colormap(gray);
    title('Reconstructed Sample Phase');
```

3.3.6 Numerical Simulation on Fourier Digital Holography

In contrast to other digital holography techniques those record object wavefronts directly, Fourier digital holography records the complex amplitude of object spectrum. Fourier digital holography is numerically simulated using the following Matlab code, according to the scheme shown in Fig. 3.5b. The object is located in the front focal plane, and the hologram is located in the back focal plane. In this numerical simulation, the wavelength is 632.8 nm, the pixel number is 1024×1024, the pixel size is 10 μm, and the angle between sample and reference wave is 0.8°. The pixel number is much more than that in on-axis and Gabor digital holography to accommodate 0th, +1st, and −1st order images. Figure 3.14a is the setting amplitude and phase distributions of the object, and Fig. 3.14b is the Fourier hologram recording of the interference between the object spectrum and the reference wavefront (tilted plane wave). Both amplitude and phase can be obtained directly by implementing

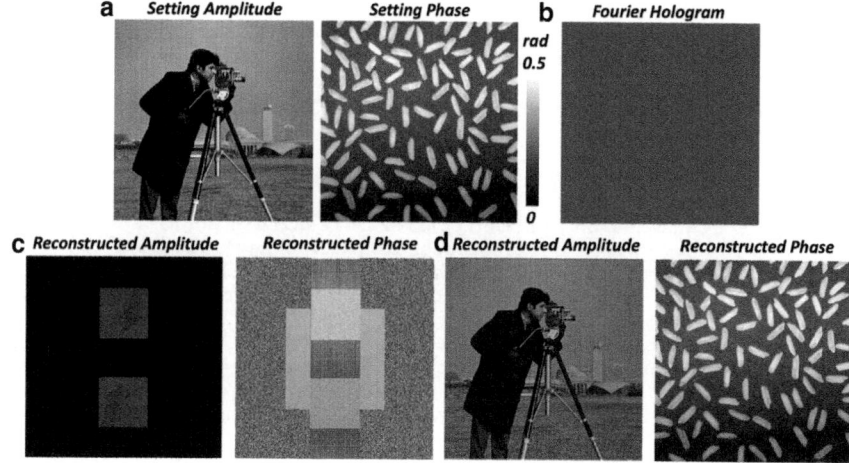

Fig. 3.14 Numerical simulation on Fourier digital holography. **a** Setting object amplitude and phase distributions; **b** Fourier hologram; **c** reconstructed object amplitude and phase distributions in the whole field of view; **d** reconstructed object amplitude and phase distributions in the field of interest

the Fourier transform of the hologram. Three images can be obtained in the imaging plane corresponding to 0th, +1st, and −1st order images, as shown in Fig. 3.14c. Figure 3.14d lists both the amplitude and phase corresponding to the +1st order term.

Matlab Code for Fourier Digital Holography

```
1    % Computational Optical Phase Imaging
2    % Fourier Digital Holography: With Lens
3
4    %% CCC
5    clear all;
6    close all;
7    clc;
8
9    %% Input Constants
10   lambda = 632.8*10^(-9); % Wavelength
11   PixelNum = 2^10; % Pixel Number
12   PixelSize = 10*10^(-6); % Pixel Size
13   k = 2*pi/lambda; % Wave number
14   Sxvector = linspace(-
     PixelNum*PixelSize/2,PixelNum*PixelSize/2,PixelNum);
15   Syvector = linspace(-
     PixelNum*PixelSize/2,PixelNum*PixelSize/2,PixelNum);
16   [SxMat, SyMat] = meshgrid(Sxvector,Syvector); % Spatial
     coordinate
17
18   %% Sample Wavefront
```

```
19  Amplitude = double(imread('cameraman.tif'));
20  Phase = double(imread('rice.png'));
21  Amplitude = (Amplitude/max(max(Amplitude))+2)/3;  % For high
    transmittance
22  Phase = Phase/max(max(Phase))*0.5;
23  AmplitudeSample = zeros(PixelNum,PixelNum);
24  PhaseSample = zeros(PixelNum,PixelNum);
25  AmplitudeSample(PixelNum/2-127:PixelNum/2+128,PixelNum/2-
    127:PixelNum/2+128) = Amplitude;  % Sample amplitude
26  PhaseSample(PixelNum/2-127:PixelNum/2+128,PixelNum/2-
    127:PixelNum/2+128) = Phase;
    % Sample phase
27  WavefrontSample = AmplitudeSample.*exp(1i*PhaseSample);  %
    Sample wavefront at sample plane
28  WavefrontSampleHologram =
    fftshift(fft2(ifftshift(WavefrontSample)));  % Fourier
    transform using lens
29  WavefrontSampleHologram =
    WavefrontSampleHologram/max(max(WavefrontSampleHologram.*
    conj(WavefrontSampleHologram)));  % Normalization
30
31  %% Propagation Reference Wave
32  AmplitudeReference = ones(PixelNum,PixelNum);  % Reference
    amplitude
33  Theta = 0.8;  % Tilting angle
34  PhaseReference = k*tand(Theta)*SyMat;  % Tilting phase
35  WavefrontReferenceHologram =
    AmplitudeReference.*exp(1i*PhaseReference);
    % Reference wavefront Plane Wave
36
37  %% Holography Recording
38  Hologram =
    (WavefrontSampleHologram+WavefrontReferenceHologram).*
    conj(WavefrontSampleHologram+WavefrontReferenceHologram);  %
    Hologram
39
40  %% Holography Reconstruction
41  BackPropagation = fftshift(fft2(ifftshift(Hologram)));
42  AmplitudeSolve =
    sqrt(BackPropagation.*conj(BackPropagation));  % Reconstructed
    sample amplitude
43  PhaseSolve = -
    atan2(imag(BackPropagation),real(BackPropagation));  %
    Reconstructed sample phase
44  CutAmplitudeSolve = AmplitudeSolve(611:866,385:640);
    % Reconstructed sample amplitude in field of interest
45  CutPhaseSolve = PhaseSolve(611:866,385:640);  % Reconstructed
    sample phase in field of interest
46
47  %% Plot
48  figure
49  subplot(3,3,1); imagesc(Amplitude); axis image; axis off;
    colormap(gray);
    title('Setting Sample Amplitude');
```

```
49   subplot(3,3,1); imagesc(Amplitude); axis image; axis off;
     colormap(gray);
     title('Setting Sample Amplitude');
50   subplot(3,3,2); imagesc(Phase); axis image; axis off;
     colormap(gray); title('Setting Sample Phase');
51   subplot(3,3,3); imagesc(Hologram); axis image; axis off;
     colormap(gray); title('Hologram');
52   subplot(3,3,4); imagesc(AmplitudeSolve); axis image; axis off;
     colormap(gray);
     title('Reconstructed Sample Amplitude');
53   subplot(3,3,5); imagesc(PhaseSolve); axis image; axis off;
     colormap(gray);
     title('Reconstructed Sample Phase');
54   subplot(3,3,7); imagesc(CutAmplitudeSolve); axis image; axis
     off; colormap(gray);
     title('Reconstructed Sample Amplitude');
55   subplot(3,3,8); imagesc(CutPhaseSolve); axis image; axis off;
     colormap(gray);
     title('Reconstructed Sample Phase');
```

Moreover, Fourier digital holography can also be realized in a lensless case as revealed in Fig. 3.15a, in which a spherical wave from a point source in the same plane of the object functions as the reference wave. Lensless Fourier digital holography is also numerically simulated using the following Matlab code. In this numerical simulation, the wavelength is 632.8 nm, the pixel number is 2048×2048, the pixel size is 10 μm, the distance between the sample and hologram planes is 300 mm, and the numerical wavefront propagation is based on the classical angular spectrum method. With the same setting amplitude and phase distributions of the object as Fig. 3.14a, Fourier hologram recording of the interference between the object wavefront and the spherical reference wavefront is obtained in Fig. 3.15b. Figure 3.15c exhibits reconstructed amplitude, in which 0th, $+$1st, and $-$1st order images can be directly obtained by implementing Fourier transform of the hologram.

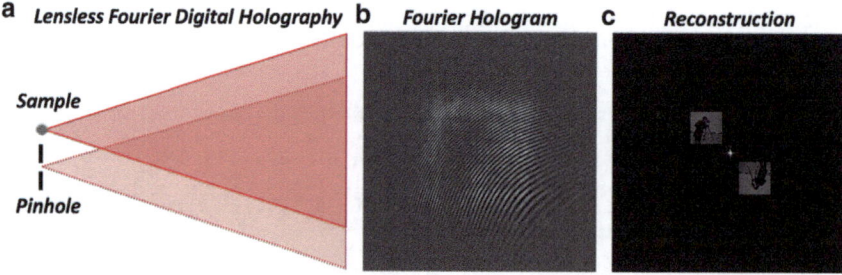

Fig. 3.15 Numerical simulation on lensless Fourier digital holography. **a** Lensless Fourier digital holography scheme; **b** lensless Fourier hologram; **c** reconstructed object amplitude distribution

Matlab Code for Lensless Fourier Digital Holography

```
1    % Computational Optical Phase Imaging
2    % Fourier Digital Holography: Lensless
3
4    %% CCC
5    clear all;
6    close all;
7    clc;
8
9    %% Input Constants
10   lambda = 632.8*10^(-9); % Wavelength
11   PixelNum = 2^9; % Pixel Number
12   PixelSize = 10*10^(-6); % Pixel Size
13   d = 300*10^(-3); % Distance between sample and hologram
14   RI = 1; % Refractive Index
15
16   %% Other Constants
17   k = 2*pi/lambda; % Wave number
18   Freqency = 1/PixelSize; % Frequency
19   Fxvector = linspace(-Freqency/2,Freqency/2,PixelNum);
20   Fyvector = linspace(-Freqency/2,Freqency/2,PixelNum);
21   [FxMat, FyMat] = meshgrid(Fxvector,Fyvector); % Spectral
     coordinate
22   Sxvector = linspace(-
     PixelNum*PixelSize/2,PixelNum*PixelSize/2,PixelNum);
23   Syvector = linspace(-
     PixelNum*PixelSize/2,PixelNum*PixelSize/2,PixelNum);
24   [SxMat, SyMat] = meshgrid(Sxvector,Syvector); % Spatial
     coordinate
25   Hz = exp((1i*2*pi*d*RI/lambda)*(1-(lambda.*FxMat/RI).^2-
     (lambda.*FyMat/RI).^2).^0.5);
     % Transfer function-Propagation
26
27   %% Sample Wavefront
28   Amplitude = double(imread('cameraman.tif'));
29   Phase = double(imread('rice.png'));
30   Amplitude = (Amplitude/max(max(Amplitude))); % For high
     transmittance
31   Phase = Phase/max(max(Phase))*0.5;
32   AmplitudeSample = zeros(PixelNum,PixelNum);
33   PhaseSample = zeros(PixelNum,PixelNum);
34   AmplitudeSample(PixelNum/2-127:PixelNum/2+128,PixelNum/2-
     127:PixelNum/2+128) = Amplitude; % Sample amplitude
35   PhaseSample(PixelNum/2-127:PixelNum/2+128,PixelNum/2-
     127:PixelNum/2+128) = Phase;
     % Sample phase
36   WavefrontSample = AmplitudeSample.*exp(1i*PhaseSample); %
     Sample wavefront at sample plane
37   WavefrontSampleHologram =
     fftshift(ifft2(ifftshift(fftshift(fft2(ifftshift(WavefrontSamp
     le))).*Hz)));
     % Sample wavefront at hologram plane
38   WavefrontSampleHologram =
     WavefrontSampleHologram/max(max(WavefrontSampleHologram.*
     conj(WavefrontSampleHologram))); % Normalization
```

```
39
40   %% Propagation Reference Wave
41   X = 200*PixelSize;
42   Y = 200*PixelSize;
43   WavefrontReferenceHologram = exp(1i*k*d).*exp(1i*k*((SxMat-
     X).^2+(SyMat-Y).^2)/2/d);
     % Spherical wave

44
45   %% Holography Recording
46   Hologram =
     (WavefrontSampleHologram+WavefrontReferenceHologram).*
     conj(WavefrontSampleHologram+WavefrontReferenceHologram);  %
     Hologram

47
48   %% Holography Reconstruction
49   BackPropagation = fftshift(fft2(ifftshift(Hologram)));
50   AmplitudeSolve =
     sqrt(BackPropagation.*conj(BackPropagation)); % Reconstructed
     sample amplitude
51   CutAmplitudeSolve = AmplitudeSolve(169:238,169:238);
     % Reconstructed sample amplitude in field of interest

52
53   %% Plot
54   figure
55   subplot(2,3,1); imagesc(Amplitude); axis image; axis off;
     colormap(gray);
     title('Setting Sample Amplitude');
56   subplot(2,3,2); imagesc(Phase); axis image; axis off;
     colormap(gray);title('Setting Sample Phase');
57   subplot(2,3,3); imagesc(Hologram); axis image; axis off;
     colormap(gray); title('Hologram');
58   subplot(2,3,4); imagesc(AmplitudeSolve); axis image; axis off;
     colormap(gray);
     title('Reconstructed Sample Amplitude');
59   subplot(2,3,5); imagesc(CutAmplitudeSolve); axis image; axis
     off; colormap(gray);
     title('Reconstructed Sample Amplitude');
```

3.3.7   Numerical Simulation on Lateral Shearing Interferometry

Figure 3.16 shows the simulation results on lateral shearing interferometry. Figure 3.16a shows the setting phase of the wavefront with unique amplitude. Figure 3.16b shows a lateral shearing interferogram with a horizontal shear of one-pixel size, and Fig. 3.16c, a vertical shear of the same size. Using phase-shifting-based phase retrieval method, the phase slopes in both horizontal and vertical directions are reconstructed as illustrated in Fig. 3.16d, e. Finally, with the known shearing, the sample phase can be reconstructed through 2-D integration as shown in Fig. 3.16f.

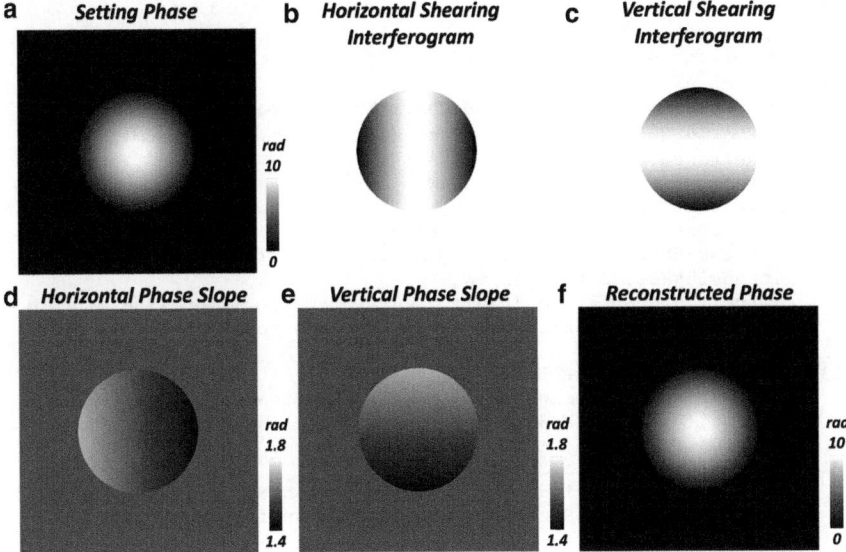

Fig. 3.16 Numerical simulation on lateral shearing interferometry. **a** Setting sample phase distributions; **b** horizontal lateral shearing interferogram; **c** vertical lateral shearing interferogram; **d** horizontal phase slope; **e** vertical phase slope; **f** reconstructed sample phase

Matlab Code for Lateral Shearing Interferometry

```
1    % Computational Optical Phase Imaging
2    % Lateral Shearing Interferometry
3
4    %% CCC
5    clear all;
6    close all;
7    clc;
8
9    %% Input Constants
10   lambda = 632.8*10^(-9); % Wavelength
11   k = 2*pi/lambda; % Wave number
12   PixelNum = 2^9; % Pixel number
13   PixelSize = 10*10^(-6); % Pixel size
14   RI = 1; % Refractive Index
15
16   %% Wavefront Under Detection
17   Sxvector = linspace(-
     (PixelNum+1)*PixelSize/2,(PixelNum+1)*PixelSize/2,PixelNum+1);
18   Syvector = linspace(-
     (PixelNum+1)*PixelSize/2,(PixelNum+1)*PixelSize/2,PixelNum+1);
19   [SxMat, SyMat] = meshgrid(Sxvector,Syvector); % Microlens
     coordinate
20   Phase = 6*10^6*(SxMat.^2+SyMat.^2); % Phase
21   Phase = Phase(1,1)-Phase;
22   Phase = Phase-Phase(PixelNum/2,PixelNum/4);
```

```matlab
23    Phase(Phase<0) = 0;  % Phase surrounding should be 0
24    Amplitude = ones(PixelNum+1,PixelNum+1);  % Amplitude
25    Wavefront = Amplitude.*exp(1i*Phase);  % Wavefront
26
27    %% Shearing Wavefronts and Their Interferograms
28    WavefrontNoShearing = Wavefront(1:PixelNum,1:PixelNum);
29    WavefrontHorizontalShearing =
      Wavefront(1:PixelNum,2:PixelNum+1);
30    WavefrontVerticalShearing =
      Wavefront(2:PixelNum+1,1:PixelNum);
31    % Phase-Shifting Horizontal Shearing Interferograms
32    IntensityHorizontalShearing1 =
      (WavefrontNoShearing.*exp(1i*0)+WavefrontHorizontalShearing).*
      conj(WavefrontNoShearing.*exp(1i*0)+WavefrontHorizontalShearin
      g);
33    IntensityHorizontalShearing2 =
      (WavefrontNoShearing.*exp(1i*pi/2)+WavefrontHorizontalShearing
      ).*conj(WavefrontNoShearing.*exp(1i*pi/2)+WavefrontHorizontalS
      hearing);
34    IntensityHorizontalShearing3 =
      (WavefrontNoShearing.*exp(1i*pi)+WavefrontHorizontalShearing).
      *conj(WavefrontNoShearing.*exp(1i*pi)+WavefrontHorizontalShear
      ing);
35    IntensityHorizontalShearing4 =
      (WavefrontNoShearing.*exp(1i*3*pi/2)+WavefrontHorizontalSheari
      ng).*conj(WavefrontNoShearing.*exp(1i*3*pi/2)+WavefrontHorizon
      talShearing);
36    % Phase-Shifting Vertical Shearing Interferograms
37    IntensityVerticalShearing1 =
      (WavefrontNoShearing.*exp(1i*0)+WavefrontVerticalShearing).*co
      nj(WavefrontNoShearing.*exp(1i*0)+WavefrontVerticalShearing);
38    IntensityVerticalShearing2 =
      (WavefrontNoShearing.*exp(1i*pi/2)+WavefrontVerticalShearing).
      *conj(WavefrontNoShearing.*exp(1i*pi/2)+WavefrontVerticalShear
      ing);
39    IntensityVerticalShearing3 =
      (WavefrontNoShearing.*exp(1i*pi)+WavefrontVerticalShearing).*c
      onj(WavefrontNoShearing.*exp(1i*pi)+WavefrontVerticalShearing)
      ;
40    IntensityVerticalShearing4 =
      (WavefrontNoShearing.*exp(1i*3*pi/2)+WavefrontVerticalShearing
      ).*conj(WavefrontNoShearing.*exp(1i*3*pi/2)+WavefrontVerticalS
      hearing);
41
42    %% Phase-Shifting Phase Retrieval
43    PhaseSlopeHorizontalShearing =
      atan2((IntensityHorizontalShearing1-
      IntensityHorizontalShearing3),(IntensityHorizontalShearing4-
      IntensityHorizontalShearing2));
44    PhaseSlopeVerticalShearing =
      atan2((IntensityVerticalShearing1-
      IntensityVerticalShearing3),(IntensityVerticalShearing4-
      IntensityVerticalShearing2));
45
46    %% Wavefront Under Detection Slope Retrieval
```

```
47    RetrievalPhase = zeros(PixelNum,PixelNum);  % Retrieved phase
48    for ii = 1:1:PixelNum
49        if ii == 1
50            RetrievalPhase(ii,1) = PhaseSlopeVerticalShearing(ii,1);
51        else
52            RetrievalPhase(ii,1) = RetrievalPhase(ii-
      1,1)+PhaseSlopeVerticalShearing(ii,1);
      % Integration in x axis
53        end
54    end
55    for ii = 1:1:PixelNum
56        for jj = 2:1:PixelNum
57            RetrievalPhase(ii,jj) = RetrievalPhase(ii,jj-
      1)+PhaseSlopeHorizontalShearing(ii,jj);
      % Integration in y axis
58        end
59    end
60
61    %% Background Tilting Removal
62    % Phase of The First Volumn Should Be 0
63    Volumn = zeros(PixelNum,3);
64    Volumn(:,1) = (Sxvector(1,1:PixelNum))';
65    Volumn(:,2) = Syvector(1,1);
66    Volumn(:,3) = RetrievalPhase(:,1);
67    % Phase of The First Row Should Be 0
68    Row = zeros(PixelNum,3);
69    Row(:,1) = Sxvector(1,1);
70    Row(:,2) = (Syvector(1,1:PixelNum))';
71    Row(:,3) = (RetrievalPhase(1,:))';
72    % Tilting Regression and Removing
73    Surrounding = vertcat(Volumn,Row);
74    XRegression = Surrounding(:,1);
75    YRegression = Surrounding(:,2);
76    PhaseRegression = Surrounding(:,3);
77    Coordinate =
      [ones(size(XRegression)),XRegression,YRegression];
78    Coefficient = regress(PhaseRegression,Coordinate);
79    Tilting =
      Coefficient(1,1)+Coefficient(2,1)*SxMat+Coefficient(3,1)*SyMat
      ;
80    RetrievalPhase = RetrievalPhase-
      Tilting(1:PixelNum,1:PixelNum);
81    RetrievalPhase = RetrievalPhase-RetrievalPhase(1,1);
82
83    %% Plot
84    figure
85    subplot(2,3,1); imagesc(Phase); axis image; axis off;
      colormap(gray); title('Phase Under Detection');
86    subplot(2,3,2); imagesc(IntensityHorizontalShearing1); axis
      image; axis off; colormap(gray); title('Horizontal Shearing
      Interferogram');
87    subplot(2,3,3); imagesc(IntensityVerticalShearing1); axis
      image; axis off; colormap(gray); title('Vertical Shearing
      Interferogram');
88    subplot(2,3,4); imagesc(PhaseSlopeHorizontalShearing); axis
      image; axis off; colormap(gray); title('Horizontal Phase
      Slope');
```

```
89   subplot(2,3,5); imagesc(PhaseSlopeVerticalShearing); axis
     image; axis off; colormap(gray); title('Vertical Phase
     Slope');
90   subplot(2,3,6); imagesc(RetrievalPhase); axis image; axis off;
     colormap(gray); title('Reconstructed Phase');
```

3.3.8 Numerical Simulation on Phase Unwrapping

In the above numerical simulations, the phase range is set within $[-\pi, \pi)$. For samples that exceed this range, phase extraction based on arc tangent function still wraps the retrieved phases within $[-\pi, \pi)$. Therefore, phase unwrapping is crucial to obtaining an unambiguous optical thickness profile by transforming discontinuous phase distributions to continuous ones. The book "Two-Dimensional Phase Unwrapping: Theory, Algorithms, and Software" summarizes several phase unwrapping methods, mainly classified as path-dependent and path-independent [16]. The following Matlab code is used to demonstrate phase unwrapping in two examples that are shown in Fig. 3.17, each corresponding to path dependent and independent tactics. Figure 3.17a shows the example continuous phase exceeding $[-\pi, \pi)$ and

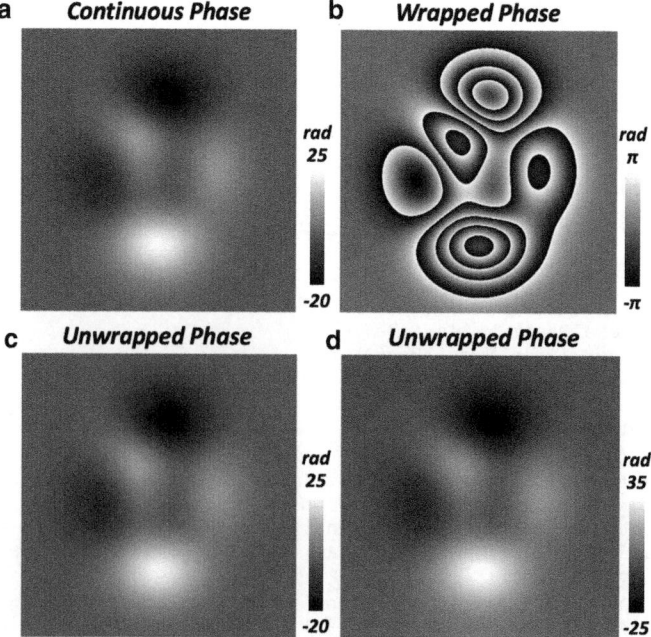

Fig. 3.17 Numerical simulation on phase unwrapping. **a** Continuous phase and **b** wrapped phase used for simulation; Unwrapped phase obtained by **c** path dependent scheme **d** path independent scheme

Fig. 3.17b, its corresponding wrapped phase within $[-\pi, \pi)$. Figure 3.17c shows the unwrapped phase extracted using the path dependent scheme and Fig. 3.17d, that obtained by path independent scheme.

Matlab Code for Path Dependent Phase Unwrapping

```
1   % Computational Optical Phase Imaging
2   % Phase Unwrapping: Path Dependent
3
4   %% CCC
5   clear all;
6   close all;
7   clc;
8
9   %% Continuous Phase and Wrapped Phase
10  Size = 1024;
11  ContinuousPhase = 3*peaks(Size); % Geneare continuous phase
12  WrappedPhase =
    atan2(imag(exp(1i*ContinuousPhase)),real(exp(1i*ContinuousPhas
    e))); % Geneare wrapped phase
13
14  %% Phase Unwrapping-Step 1: Unwrap the First Row
15  [Xsize, Ysize] = size(WrappedPhase);
16  UnwrappedPhase = WrappedPhase;
17  for nn = 2:1:Ysize
18  derivative = WrappedPhase(1,nn)-WrappedPhase(1,nn-1); % Phase
    derivative
19      if abs(derivative) <= pi
20          UnwrappedPhase(1,nn) = UnwrappedPhase(1,nn-
    1)+derivative;
21      elseif derivative < 0
22          UnwrappedPhase(1,nn) = UnwrappedPhase(1,nn-
    1)+derivative+2*pi; % 2*pi compensation
23      else
24          UnwrappedPhase(1,nn) = UnwrappedPhase(1,nn-
    1)+derivative-2*pi; % 2*pi compensation
25      end
26  end
27
28  %% Phase Unwrapping-Step 2: Unwrap All the Columns
29  for nn = 1:1:Ysize
30      for mm = 2:1:Xsize
31          derivative = WrappedPhase(mm,nn)-WrappedPhase(mm-
    1,nn); % Phase derivative
32          if abs(derivative)<=pi
33              UnwrappedPhase(mm,nn) = UnwrappedPhase(mm-
    1,nn)+derivative;
34          elseif derivative < 0
35              UnwrappedPhase(mm,nn) = UnwrappedPhase(mm-
    1,nn)+derivative+2*pi; % 2*pi compensation
36          else
37              UnwrappedPhase(mm,nn) = UnwrappedPhase(mm-
    1,nn)+derivative-2*pi; % 2*pi compensation
38          end
39      end
40  end
```

```
41
42   %% Plot
43   figure
44   subplot(1,3,1); imagesc(ContinuousPhase); axis image; axis
     off; colormap(gray); title('Setting Continuous Phase');
45   subplot(1,3,2); imagesc(WrappedPhase); axis image; axis off;
     colormap(gray); title('Wrapped Phase');
46   subplot(1,3,3); imagesc(UnwrappedPhase); axis image; axis off;
     colormap(gray); title('Unwrapped Phase');
```

Matlab Code for Path Independent Phase Unwrapping

```
1    % Computational Optical Phase Imaging
2    % Phase Unwrapping: Path Independent
3
4    %% CCC
5    clear all;
6    close all;
7    clc;
8
9    %% Continuous Phase and Wrapped Phase
10   Size = 1024;
11   ContinuousPhase = 4*peaks(Size); % Geneare continuous phase
12   WrappedPhase =
     atan2(imag(exp(1i*ContinuousPhase)),real(exp(1i*ContinuousPhas
     e)));
     % Geneare wrapped phase
13
14   %% 1-D Derivatives
15   [Xsize, Ysize] = size(WrappedPhase);
16   [XMat,YMat] = meshgrid(1:1:Xsize,1:1:Ysize);
17   Derivative1DX = zeros(Xsize,Ysize);
18   Derivative1DY = zeros(Xsize,Ysize);
19   Derivative1DX(2:Xsize,:) = WrappedPhase(2:Xsize,:)-
     WrappedPhase(1:Xsize-1,:);
     % 1D Derivative in x axis
20   Derivative1DY(:,2:Ysize) = WrappedPhase(:,2:Ysize)-
     WrappedPhase(:,1:Ysize-1);
     % 1D Derivative in y axis
21   Derivative1DX = Derivative1DX-pi*round(Derivative1DX/pi); %
     Compensate in [-pi,pi]
22   Derivative1DY = Derivative1DY-pi*round(Derivative1DY/pi); %
     Compensate in [-pi,pi]
23
24   %% 2-D Derivatives
25   Derivative2DX = zeros(Xsize,Ysize);
26   Derivative2DY = zeros(Xsize,Ysize);
27   Derivative2DX(2:Xsize,:) = Derivative1DX(2:Xsize,:)-
     Derivative1DX(1:Xsize-1,:);
     % 2D Derivative in x axis
28   Derivative2DY(:,2:Ysize) = Derivative1DY(:,2:Ysize)-
     Derivative1DY(:,1:Ysize-1);
     % 2D Derivative in x axis
29   Derivative2D = Derivative2DX+Derivative2DY;  % 2D Derivative
30
```

```
31  %% Solve Poisson Equation Using DCT
32  UnwrappedPhase = idct2(dct2(Derivative2D)./(2*cos(pi*(XMat-
    1)/Xsize)+2*cos(pi*(YMat-1)/Ysize)-4+eps));
33  UnwrappedPhase = UnwrappedPhase-UnwrappedPhase(1,1);
34
35  %% Plot
36  figure
37  subplot(1,3,1); imagesc(ContinuousPhase); axis image; axis
    off; colormap(gray); title('Setting Continuous Phase');
38  subplot(1,3,2); imagesc(WrappedPhase); axis image; axis off;
    colormap(gray); title('Wrapped Phase');
39  subplot(1,3,3); imagesc(UnwrappedPhase); axis image; axis off;
    colormap(gray); title('Unwrapped Phase');
```

3.4 Improvements in Holography and Interferometry

The recent advancements in holography and interferometric techniques have led to greater accuracy in wavefront reconstruction, faster wavefront reconstruction, higher reconstruction resolution, and a simpler optical system. This section briefly reviews the development of phase retrieval methods and optical systems in holography and interferometry.

3.4.1 Improvements in Phase Retrieval Methods

The function of phase retrieval is to extract the sample complex amplitude from interference pattern. A large number of phase retrieval algorithms (including phase extraction and unwrapping) have been developed over the past few years. In this section, we briefly discuss a few selected algorithms with good performance and high popularity.

Gabor holography has remarkably simplified optical systems. Nevertheless, sample reconstruction by back propagation is unable to produce high-quality sample images as all of the 0th, +1st, and −1st order terms are not separated, which restrict Gabor holography to particle localization while limiting its possible applications in imaging. Ozcan group applied Gabor holography to construct lensless holographic microscopes for different kinds of samples [17, 18]. Their early works used direct back propagation, error reduction, and hybrid input–output methods, but these reconstruction results still suffer from limited quality [19, 20]. In order to reconstruct high-quality sample images in Gabor holography, multi-height iterative phase recovery approach [21–23] has been used, where during iterations, wavefront under detection is updated among multi-height holograms. As a result, the reconstruction quality as well as the spatial resolution of the reconstructed image is improved.

Phase-shifting algorithms are often used in on-axis holography and interferometry. However, most of these phase-shifting methods require prior knowledge of phase-shifting angles. If the phase-shifting angles are not precise, the retrieved phase may suffer from errors. In order to solve these problems, blind phase-shifting methods without prior knowledge of phase-shifting angles are preferred, especially in situations when phase-shifting device can hardly guarantee the phase-shifting precision. Among different blind phase-shifting methods, principal component analysis-based phase-shifting algorithm [24–28] is a preferred choice for phase extraction because this method has no restrictions on background, modulation, or phase shifts. Since it is also based on processing two quadrature signals, it is faster with lower computational requirements. Moreover, it also works well even while dealing with noisy fringes.

Besides FFT-based phase extraction algorithm in off-axis digital holography, many other algorithms have been reported. Hilbert transform (HT)-based algorithm has been designed to preserve more high-frequency information, thus to pursue higher phase extraction accuracy [29–34]. Unfortunately, HT-based method is easily influenced by noises since it keeps more high-frequency information, but at the same time, more noises are inevitably introduced. HT actually does a 90-degree phase-shifting of off-axis holograms. Therefore, HT-based algorithm can be treated as digital phase-shifting method. Moreover, derivative is also often used to extract the phase from off-axis holograms in order to accelerate the processing speed [35, 36] by introducing some approximations that slightly reduce the phase extraction accuracy.

A basic problem in holography and interferometry is that elevations and depressions cannot be distinguished from each other. Ritsch-Marte group designed spiral interferometry [37] which introduces spiral phase element in the beam path, thus generating spiral interferograms that are different from closed contour lines of traditional interferograms. It is now possible distinguish between elevations and depressions using spiral interferogram analysis [38].

There are many phase extraction methods available in the literature which are designed to accomplish different goals, such as improving the phase retrieval accuracy, accelerating the processing speed, increasing the signal to noise ratio, or even reducing the phase-shifting hologram captures. Phase extraction method is chosen based on the particular advantage that is targeted in a specific application. For example, classical four-step phase-shifting method is often used in systems with precise phase-shifting capability to simplify the retrieval operations as well as to accelerate the processing speed. However, in systems which are not equipped with precise phase-shifting devices, blind phase-shifting algorithms are preferred though they often require more phase-shifting hologram captures and processing time. Moreover, HT-based method can be used to pursue higher retrieval accuracy whenever captured off-axis holograms have higher signal to noise ratios; whereas in situations of reduced signal to noise ratios, FFT-based method is preferred to suppress the noise.

The reconstructed phase obtained after phase extraction is not continuous but wrapped in the range of $[--\pi, \pi)$. Phase unwrapping is employed to unwrap the discontinuous phase into continuous one by recognizing the phase discontinuities, and then compensating them by introducing extra $\pm 2\pi$ stages. Many phase unwrapping methods have been proposed and are mainly classified as either path dependent or independent which are aimed at both faster processing speed and robust functionality. While most of these methods are based on phase unwrapping from extracted phase corresponding to a single wavelength, a dual-wavelength-based method has been proposed [39, 40] to significantly accelerate the phase unwrapping speed by recording and processing hologram for two wavelengths (λ_1 and λ_2). Actually, this method increases the wavelength from λ_1 (or λ_2) to $\frac{\lambda_1\lambda_2}{|\lambda_1-\lambda_2|}$ which can be large while λ_1 and λ_2 are closer, to cover the optical thickness of the sample under detection to avoid a wrapped phase. Furthermore, multiple-wavelength-based phase unwrapping methods have also been designed to improve the unwrapping performance while dealing with thick samples and to reduce the phase noises [41, 42].

Besides these reported methods, deep learning-based phase retrieval methods have been designed and applied recently [43–54]. This is also introduced and discussed in Chapter 7. Deep learning can significantly accelerate the phase retrieval speed, however, it requires an extremely large quantity of training data to guarantee the phase retrieval accuracy and robustness in various conditions.

Improvements in Optical Systems

Improvements in holographic and interferometric systems have played a significant role in keeping such technique up to date. Gabor holography or lensfree holography has been designed and widely used, especially in microscopy. While conventional holographic systems rely on many optical elements, lensfree holographic system is much simpler since it only requires a source and an image recorder as shown in Fig. 3.18a. Moreover, samples can be set above the image recorder without introducing any lens, thus it completely avoids the aberrations. Ozcan group [17, 18] and Allier group [55–57] have designed various point-of-care holographic microscopes especially focusing on cell imaging to pursue extremely large field of view. Moreover, cell tracking is also possible with dual-wavelength lensfree holographic systems [58–63].

For on-axis holography, phase-shifting can be implemented using piezo and polarizers, often based on the application requirements: piezo can adjust the phase-shifting angles with a higher accuracy, and polarizers can adjust the phase-shifting angles at a lower cost. However, both of them suffer from lower speed. In order to obtain high phase-shifting speed, Popescu et al. designed Fourier phase microscopy [64, 65] as shown in Fig. 3.18b, which relies on programmable phase modulator for single pixel phase-shifting to obtain faster speed. Moreover, using a four-channel polarization-imaging camera facilitates simultaneous phase-shifting by taking over the role of rotating polarizers [66, 67].

With a common path configuration, point diffraction interferometer can significantly simplify the digital holographic system. Most of the point diffraction interferometers are on-axis systems, limiting their high-quality dynamic phase imaging capability. Popescu et al. developed diffraction-based common-path interferometer [68, 69] as shown in Fig. 3.18c. Using a grating, the wavefront under detection is separated into different diffraction orders, where the 0th and 1st diffraction orders are respectively used as reference wavefront while passing through a spatial filter and a sample wavefront. These two wavefront generate the off-axis hologram.

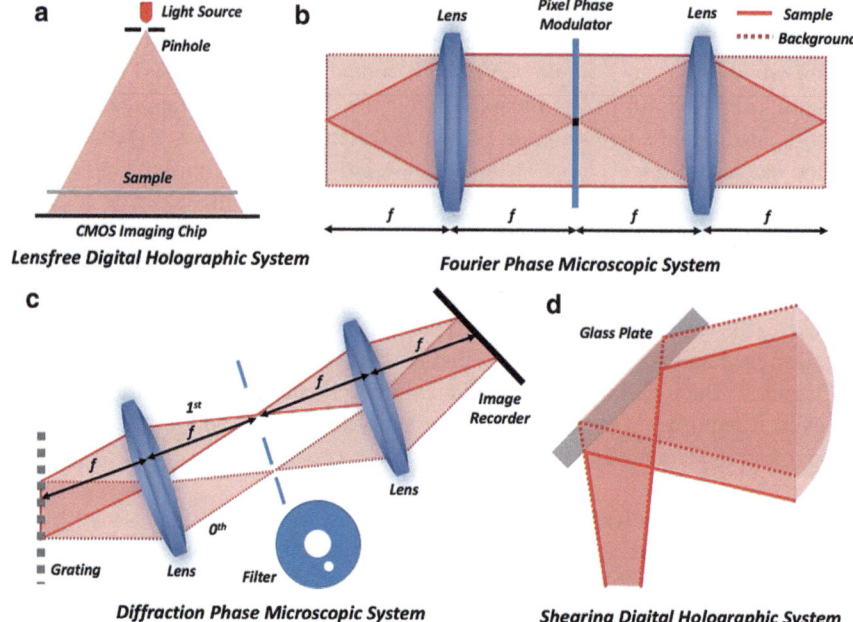

Fig. 3.18 Improvements on holographic systems. **a** Lensfree digital holographic system; **b** Fourier phase microscopic system; **c** diffraction phase microscopic system; **d** shearing digital holographic systems

Such diffraction-based common-path interferometer has the off-axis configuration and can be extended for multi-wavelength interferometry [60–72]. The design has been widely used in many applications, especially in quantitative dynamic phase microscopy [73–75].

Inspired by shearing interferometry, many shearing digital holographic systems have been reported. Often using an optical glass plate, the wavefront under detection is divided into two but with a shearing as shown in Fig. 3.18d. Using an image recorder, shearing hologram can be recorded. Often samples (such as cells) in the field of view are in sparse space and hence the hologram actually describes the interference between sample wavefront and background wavefront. Therefore, sample wavefront can be directly retrieved without using further phase integration. In such systems, one sample corresponds to two retrieved distributions with opposite phase distributions. Based on such systems, Anand and Javidi groups reported many applications on cell imaging [76–79]. In these applications, cells often had low concentrations to remain in a compressed location.

3.5 Extensions on Holography and Interferometry

We have briefly discussed conventional holography and interferometry techniques as well as their improvements in both phase retrieval methods and optical systems. However, there are many related techniques developed from conventional holography and interferometry techniques. For example, classical qualitative phase contrast and differential interference contrast microscopy techniques have been updated to quantitative ones by spatial light interference microscopy and quantitative differential interference contrast microscopy. Quadriwave lateral shearing interferometry is an updated version of shearing interferometry in two orthogonal directions. Most of the conventional digital holography relies on coherent light to pursue high accuracy in phase imaging or partially (low) coherent light aiming at tomography or reducing speckle noise. There are still techniques focusing on incoherent digital holography. Optical scanning holography (OSH) combining with heterodyne interferometry and field-of-view scanning can deal with incoherent conditions for sample recording, imaging, coding, and so on. Besides, Fresnel incoherent correlation holography (FINCH) is designed with a spatial light modulator that separates the wavefront from fluorophore and an image recorder to record their self-interference hologram. Furthermore, coded aperture correlation holography (COACH) with the aid of coded phase-only mask is also reported as a general model of FINCH since the quadratic phase mask used in FINCH is replaced by the coded phase-only mask used in COACH. Additionally, computer-generated holography and fringe projection profilometry are often used in optical testing. These techniques can be treated as an extension of holography and interferometry, and are briefly mentioned here in order to make this section complete.

3.5.1 Spatial Light Interference Microscopy (SLIM)

The spatial light interference microscopy (SLIM) developed by Popescu group [80] combines phase contrast microscopy with phase-shifting interferometry (or Fourier phase microscopy [64, 65]). In contrast with classical phase contrast microscopy discussed in Chap. 2, SLIM not only enhances the imaging contrast for transparent specimens but also transforms it from qualitative to quantitative; and SLIM does not heavily rely on coherent light sources, which greatly reduces speckle noise. The advantages of SLIM have led to its widespread use in many fields, including biological observations and surface metrology.

Figure 3.19a reveals the scheme of SLIM, which consists of two key systems as phase contrast microscopy and phase-shifting interferometry. A condenser ring aperture is inserted to generate the ring illumination. After passing through the sample, directly transmitted light still propagates along the original path, while the specimen deflected light which carries the specimen information (both amplitude and phase) deviates from the directly transmitted light, however, both of them are collected by

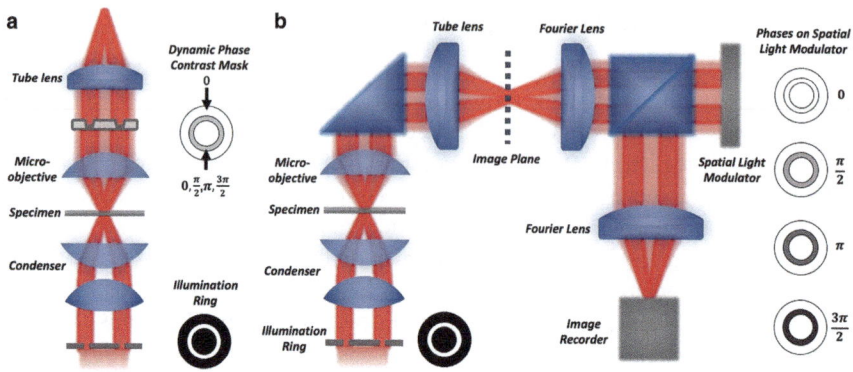

Fig. 3.19 SLIM. **a** SLIM scheme; **b** SLIM system in practical applications

the micro-objective (different from dark-field microscopy). When a phase contrast modulator is set at the plane conjugated to the condenser ring aperture plane, it is the classical phase contrast microscopy, and the generated intensity I in the imaging plane is actually the interference between the directly transmitted light Ψ_0 and the specimen deflected light $\Psi_s(x, y)$ as illustrated in Eq. (3.58), in which, $\varphi(x, y)$ is the specimen phase, and Δ is the phase-shifting angle induced by the phase contrast modulator. It is worth noting that Eq. (3.58) can be simplified into Eqs. (2.7) and (2.10) in Chap. 2 when introducing $e^{i\varphi(x,y)} \approx 1 + i\varphi(x, y)$ and Δ is $\frac{\pi}{2}$ or $\frac{3\pi}{2}$.

$$I = |\Psi_0|^2 + |\Psi_s(x, y)|^2 + 2|\Psi_0||\Psi_s(x, y)|\cos[\varphi(x, y) + \Delta] \qquad (3.58)$$

While in SLIM, the fixed phase contrast modulator is replaced by the adjustable phase contrast modulator, therefore, the phase shifting Δ can be precisely set as 0, $\frac{\pi}{2}$, π and $\frac{3\pi}{2}$, and Eq. (3.58) can be further developed into Eqs. (3.59)–(3.62) to correspond to different phase shifts Δ.

$$I_1 = |\Psi_0|^2 + |\Psi_s(x, y)|^2 + 2|\Psi_0||\Psi_s(x, y)|\cos[\varphi(x, y) + 0]$$

$$= |\Psi_0|^2 + |\Psi_s(x, y)|^2 + 2|\Psi_0||\Psi_s(x, y)|\cos[\varphi(x, y)] \qquad (3.59)$$

$$I_2 = |\Psi_0|^2 + |\Psi_s(x, y)|^2 + 2|\Psi_0||\Psi_s(x, y)|\cos\left[\varphi(x, y) + \frac{\pi}{2}\right]$$

$$= |\Psi_0|^2 + |\Psi_s(x, y)|^2 - 2|\Psi_0||\Psi_s(x, y)|\sin[\varphi(x, y)] \qquad (3.60)$$

$$I_3 = |\Psi_0|^2 + |\Psi_s(x, y)|^2 + 2|\Psi_0||\Psi_s(x, y)|\cos[\varphi(x, y) + \pi]$$

$$= |\Psi_0|^2 + |\Psi_s(x, y)|^2 - 2|\Psi_0||\Psi_s(x, y)|\cos[\varphi(x, y)] \qquad (3.61)$$

$$I_4 = |\Psi_0|^2 + |\Psi_s(x, y)|^2 + 2|\Psi_0||\Psi_s(x, y)| \cos\left[\varphi(x, y) + \frac{3\pi}{2}\right]$$

$$= |\Psi_0|^2 + |\Psi_s(x, y)|^2 + 2|\Psi_0||\Psi_s(x, y)| \sin[\varphi(x, y)] \qquad (3.62)$$

Therefore, using the classical four-step phase-shifting phase retrieval algorithm, the specimen phase can be reconstructed according to Eq. (3.63).

$$\varphi = \tan^{-1}\left(\frac{I_4 - I_2}{I_1 - I_3}\right) \qquad (3.63)$$

Equations (3.58) to (3.63) briefly illustrate that the SLIM is the combination of phase contrast microscopy and phase-shifting interferometry [80]. However, in practical systems, it is rather difficult to directly set an adjustable phase contrast modulator in the commercial microscope. Therefore, in practical applications, the phase shifting function often realized by a spatial light modulator is set outside the C-mount of the commercial microscope as shown in Fig. 3.19b.

After the development of SLIM, Popescu group has improved the SLIM in different perspectives. The phase-shifting module can be obtained using either reflective [80] or transmissive [81] systems. Considering the accuracy of SLIM, it can also be adopted in precision measurements [82]. Moreover, as the phase-shifting can be rapidly modulated using liquid crystal, SLIM has been used in dynamic phase imaging [83, 84]. Additionally, in order to overcome the halos generated in SLIM, algorithms have also been designed to compensate them [85, 86] in order to pursue high-quality imaging as well as high-accurate phase reconstruction.

Several applications have been reported based on SLIM as well as its improvements and most of them can be divided into two categories: static measurements for tissue slices, aiming at pathological analysis and dynamic measurements mostly for live cells, aiming at biological analysis. Considering the high accuracy of SLIM, it is more suitable for tissue observations [87]. Moreover, diseases, especially different type of cancers [88] can be diagnosed even without staining, which includes breast cancer [89–91], prostate cancer [92, 93], and pancreatic ductal adenocarcinoma [94]. Moreover, considering the fact that the phase shifting is extremely fast, SLIM also supports dynamic phase imaging, and it has been successfully adopted in live cell imaging. For example, SLIM was adopted to monitor the dynamic beating of cardiomyocyte cells [95], to study the dynamics of fibrin clot formation [96], to measure the fluctuations of red blood cells for banked blood quality inspection [97], to study the active intracellular transport in metastatic cells [98], and to assess the neural outgrowth [99]. Besides a series of biological applications, SLIM has also been used in label-free imaging of single microtubule dynamics [100].

As a combination of phase contrast microscopy and phase-shifting interferometry, SLIM not only provides high-contrast images, but also reconstructs high-accurate specimen phases. Moreover, SLIM has been successfully used in both static and dynamic measuring conditions. Therefore, SLIM is a promising tool among various quantitative phase imaging techniques.

3.5.2 Quantitative Differential Interference Contrast (DIC) Microscopy

DIC microscopy reveals the interference between the specimen wavefront and its sheared one, and the DIC image can remarkably improve the imaging contrast. Unfortunately, conventional DIC microscopy still cannot provide quantitative specimen phase distributions. In order to solve the problem, many improvements have been reported [27, 101–108], and most of these techniques are based on the combination between DIC microscopy and phase-shifting interferometry.

According to Fig. 3.20 and Sect. 2.2 in Chap. 2, the DIC image can be described by Eq. (3.64), in which A is the amplitude distribution, $\partial\varphi$ is defined as the phase gradient along the shearing direction, and Δ is the bias induced by the wollaston prism while causing a phase shift between the ordinary and extraordinary light. This is explained in polarization part in nearly all optics textbooks.

$$I_\Delta = A^2 + A^2 \cos(\partial\varphi - 2\Delta) \tag{3.64}$$

Fig. 3.20 Quantitative DIC microscopy

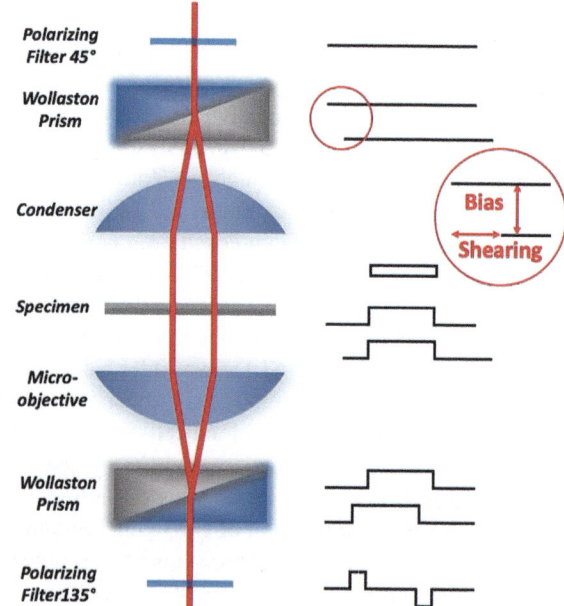

In Eq. (3.64), Δ can be modulated by adjusting the Wollaston prism, therefore, $\partial\varphi$ can be obtained according to phase-shifting interferometry almost the same way as φ in SLIM. The specimen phase φ can be finally reconstructed via 2-D integration from $\partial\varphi$ since the shearing ∂r is known in advance [104]. It should be noted that DIC image stacks in two directions (orthogonal directions are preferred) and should be obtained for specimen phase reconstruction via 2-D integration [107].

Unfortunately, it is very difficult to manually adjust the Wollaston prism to obtain the precise phase-shifting angles. In order to solve the problem, Ritsch-Marte group adopted a spatial light modulator to precisely modulate the phase shifting angles [108]. However, it makes the system more complicated with an expensive spatial light modulator. In order to simplify the system, Wang group proposed principal component analysis-based quantitative DIC microscopy [27], in which the phase gradient can be retrieved from a series of phase-shifting DIC images using principal component analysis [24–28] even the phase-shifting angles are not known in advance. Therefore, the quantitative DIC microscopy can be implemented even using commercial microscopes without any further modification. More details on the mathematical derivations related to the phase retrieval method used in principal component analysis-based quantitative DIC microscopy can be found in the reference [27].

3.5.3 Quadriwave Lateral Shearing Interferometry

Quadriwave lateral shearing interferometry can retrieve quantitative phase of wavefront using quite compact optical system with a chessboard grating in front of an imaging sensor [109]. The chessboard grating shown in Fig. 3.21 consists of multiple unit cells with phase distributions of 0 and π respectively and surrounded by opaque grids. In this design, the directly transmitted light as 0th order light can be reduced, mainly ±1st orders of diffraction light are left [109]. Therefore, such chessboard grating functions as 2-D Fresnel biprism, and patterns on the image recorder are actually the shearing interference between four diffraction orders of the grating. Different from lateral shearing interferometry, both horizontal and vertical lateral shearing patterns are both coded in a single-shot interferogram of the quadriwave lateral shearing interferometry as shown in Fig. 3.21. However, horizontal and vertical lateral shearing phases as phase derivatives in these orthogonal directions can still be extracted using classical fast Fourier transform-based phase retrieval method [109]. Finally, the phase can be reconstructed by integrating phase derivatives in two dimensions.

In principle, quadriwave lateral shearing interferometry is a typical two-dimensional lateral shearing interferometry method. For more details, the review on quadriwave lateral shearing interferometry by Baffou [109] is an important reference for deeply understanding quadriwave lateral shearing interferometry. Despite quadriwave lateral shearing interferometry's advantages, such as insensitivity to vibration and high phase retrieval accuracy, and compact design as many shearing interferometry methods, it suffers from higher cost due to the specially designed and fabricated

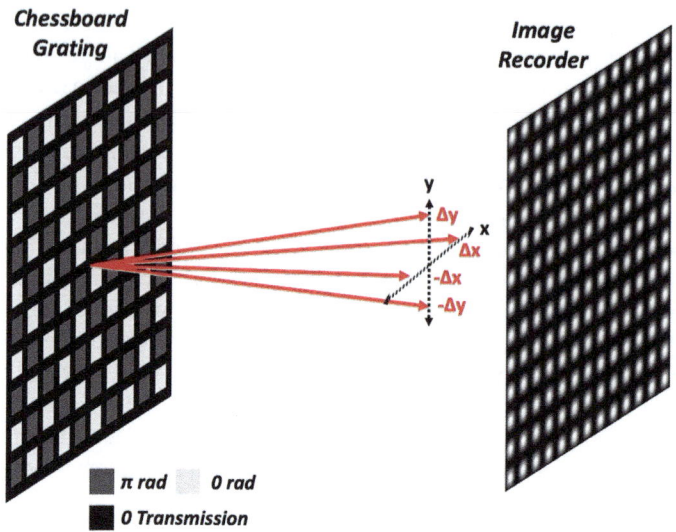

Fig. 3.21 Principle of quadriwave lateral shearing interferometry

chessboard grating. However, it still has many applications. Monneret group adopted such method in biological phase imaging [110–114]. The technique has also been used in laser beam detection [115, 116] and optical element testing [117]. Recently, Baffou group applied quadriwave lateral shearing interferometry in 2-D material [118], nanoparticle [119], and metasurface [120] characterization.

3.5.4 Optical Scanning Holography (OSH)

Different from the above mentioned coherent or partially coherent holography and interferometry techniques, an incoherent holographic method, optical scanning holography (OSH) can also obtain specimen complex amplitude via optical hetero-dyne detection, 2-D structured light scanning and single pixel detector recording in both coherent and incoherent conditions. OSH was first designed by Poon and Korpel [121], then the idea was formulated [122] and finally experimentally demonstrated both by Poon group [123]. Figure 3.22 briefly describes the OSH principle, and math-ematical derivations are briefly provided mainly referring to [124, 125]. Light beam from the laser with a frequency ω is divided by a beam splitting prism. One of beam passes through an acousto-optic modulator (AOM) to shift its frequency to $\omega + \Omega$. Two beams then pass through two pupils as $p_1(x, y)$ and $p_2(x, y)$ as well as two lens (Lens 1 and 2), respectively, and finally combined with another beam splitting prism to reach a scanning lens. $p_1(x, y)$ and the scanning lens are located at the front and back focal planes of Lens 1; and $p_2(x, y)$ and the scanning lens are located at front and back focal planes of Lens 2. With the scanning lens, the generated heterodyne

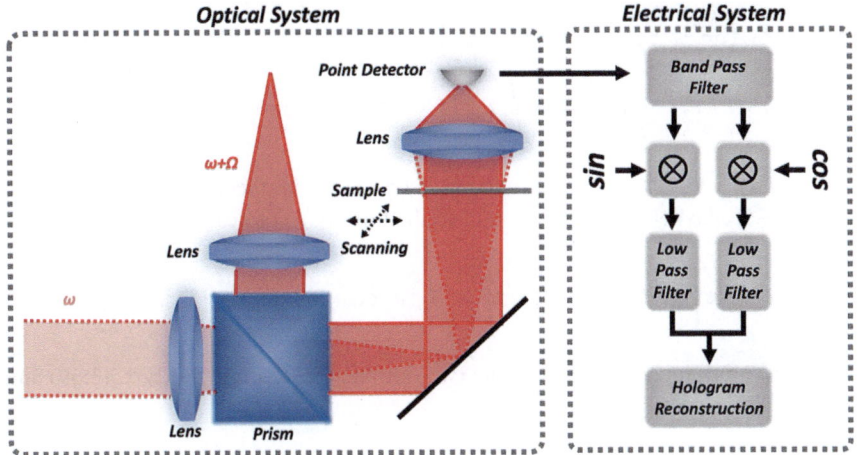

Fig. 3.22 Scheme of optical scanning holography

interference scans the sample, and the detecting light is collected using a single-pixel detector, which follows the processing circuits.

The key principle of OSH can be briefly described according to Eqs. (3.65)–(3.76) [124, 125], and more details can be found in the mathematical OSH model [124, 125]. The collected current can be represented by Eq. (3.65), which can be further simplified to Eqs. (3.66)–(3.67) representing direct current (i_{DC}) and heterodyne current (i_Ω). In these equations, $T(x, y, z)$ is the sample as a real function, $P_{1z}\left(\frac{k_0 x}{f}, \frac{k_0 y}{f}\right)$ and $P_{2z}\left(\frac{k_0 x}{f}, \frac{k_0 y}{f}\right)$ are pinhole propagating through a distance z corresponding to $p_1(x, y)$ and $p_2(x, y)$, respectively, and $\odot$ donates cross correlation.

$$
i \propto \int \left| \left[P_{1z}\left(\frac{k_0 x}{f}, \frac{k_0 y}{f}\right) e^{i\omega t} \right. \right.
$$
$$
\left. \left. + P_{2z}\left(\frac{k_0 x}{f}, \frac{k_0 y}{f}\right) e^{i(\omega+\Omega)t} \right] \times T(x + x', y + y', z) \right|^2 dxdydz \qquad (3.65)
$$

$$
i_{DC} \propto \int \left| P_{1z}\left(\frac{k_0 x}{f}, \frac{k_0 y}{f}\right) \right|^2 \cdot |T(x + x', y + y', z)|^2 dxdydz
$$
$$
+ \int \left| P_{2z}\left(\frac{k_0 x}{f}, \frac{k_0 y}{f}\right) \right|^2 \cdot |T(x + x', y + y', z)|^2 dxdydz
$$
$$
\propto \int |P_{1z}|^2 \odot |T|^2 dz + \int |P_{2z}|^2 \odot |T|^2 dz \qquad (3.66)
$$

$$
i_\Omega \propto \int P_{2z}\left(\frac{k_0 x}{f}, \frac{k_0 y}{f}\right) P_{1z}^*\left(\frac{k_0 x}{f}, \frac{k_0 y}{f}\right) e^{i\Omega t} \cdot |T(x + x', y + y', z)|^2 dxdydz
$$

$$+ \int P_{1z}\left(\frac{k_0 x}{f}, \frac{k_0 y}{f}\right) P_{2z}^*\left(\frac{k_0 x}{f}, \frac{k_0 y}{f}\right) e^{-i\Omega t} \cdot \left|T\left(x + x', y + y', z\right)\right|^2 \mathrm{d}x\mathrm{d}y\mathrm{d}z$$

$$\propto \int P_{2z} P_{1z}^* e^{i\Omega t} \odot |T|^2 \mathrm{d}z + \int P_{1z} P_{2z}^* e^{-i\Omega t} \odot |T|^2 \mathrm{d}z \qquad (3.67)$$

Define $P = P_{1z} P_{2z}^* = |P|e^{i\Phi}$; and based on the circuit theory and bandpass filter tuned to $-\Omega$, the heterodyne current described in Eq. (3.68) can be extracted.

$$i_\Omega \propto \mathrm{Re}\left\{\int P e^{i\Omega t} \odot |T|^2 \mathrm{d}z\right\} = \int |P| \cos(\Phi - \Omega t) \odot |T|^2 \mathrm{d}z \qquad (3.68)$$

According to lock-in-amplifier, the two outputs are described in Eqs. (3.69) and (3.70).

$$H_{\cos} \propto \int |P| \sin \Phi \odot |T|^2 \mathrm{d}z \qquad (3.69)$$

$$H_{\sin} \propto \int |P| \cos \Phi \odot |T|^2 \mathrm{d}z \qquad (3.70)$$

The hologram H in incoherent condition can be reconstructed according to Eq. (3.71), and the object can be numerically computed from such reconstructed hologram.

$$H = H_{\cos} + i H_{\sin} \propto \int |P| \odot |T|^2 \mathrm{d}z \qquad (3.71)$$

As a special example, when $p_1(x, y) = 1$ as a pupil and $p_2(x, y) = \delta$ as a pinhole, P can be described in Eq. (3.72).

$$P(x, y; z_0) = h(x, y; z_0) = \frac{k_0}{2\pi z_0} e^{\left[\frac{-ik_0}{2z_0}(x^2 + y^2)^2\right]} \qquad (3.72)$$

Therefore, Eq. (3.71) can be rewritten to Eq. (3.73), which is the reconstructed hologram corresponding to $|T|^2$ from heterodyne current via 2-D scanning.

$$H \propto h(x, y; z_0) \odot |T|^2 \qquad (3.73)$$

Therefore, the specimen as $|T|^2$ can be reconstructed according to Eq. (3.74).

$$h \otimes H = h \otimes h \odot |T|^2 = |T|^2 \qquad (3.74)$$

While in coherent condition, additional pinhole is set in front of the single-pixel detector, and $p_1(x, y) = \delta$, therefore, the heterodyne current can be explained by Eq. (3.75), and the hologram H in coherent condition can be reconstructed according to Eq. (3.76). More details on coherent condition can be found in reference [124].

$$i_\Omega \propto \int P_{2z} \odot T \, dz \tag{3.75}$$

$$H = H_{\cos} + i H_{\sin} \propto \int P_{2z} \odot T \, dz \tag{3.76}$$

Compared to conventional holography, OSH can work in both coherent and incoherent conditions (but it is often used in incoherent condition), and avoids the influence from background and twin image. Therefore, optical scanning holography has many applications mostly used in depth imaging in incoherent condition [126–132]; however, it has also been successfully used in optical phase imaging in coherent condition, as discussed in [133–135] and applied in [136].

3.5.5 Fresnel Incoherent Correlation Holography (FINCH)

Rosen and Brooker first proposed the idea of Fresnel incoherent correlation holography (FINCH) [137], in which the key idea is that the recorded hologram is the interference of incoherent monochromatic light beams originated from an object after it passes through a spatial light modulator (SLM) with two coded phase gratings acting as beam splitters as demonstrated in Fig. 3.23. The key principle of FINCH can be briefly described according to Eqs. (3.77)–(3.81) [137, 138]. According to the FINCH scheme, the captured hologram of a point source located at $(x_s, y_s, -z_s)$ can be described in Eq. (3.77), in which λ is the wavelength, d_1 is the distance between the lens, and the SLM, d_2 is the distance between the SLM and the image recorder, r_1 and r_2 are spherical waves in different radii coded on the SLM demonstrated by Eq. (3.78), and $Q(\cdot)$ is the function defined as $e^{\frac{i\pi}{\lambda}[(x-x_s)^2+(y-y_s)^2]}$ [137].

H_{PSF}

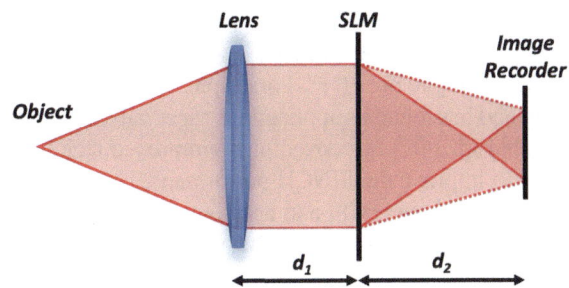

Fig. 3.23 FINCH scheme

$$= \left| Q\left(\frac{1}{z_s}\right) \cdot Q\left(-\frac{1}{f}\right) \otimes Q\left(\frac{1}{d_1}\right) \cdot \left[Q\left(-\frac{1}{r_1}\right) \cdot e^{i\theta} + Q\left(-\frac{1}{r_2}\right) \right] \otimes Q\left(\frac{1}{d_2}\right) \right|^2$$

$$\tag{3.77}$$

$$\varphi_{SLM} = Q\left(-\frac{1}{r_1}\right) \cdot e^{i\theta} + Q\left(-\frac{1}{r_2}\right) \tag{3.78}$$

Therefore, Eq. (3.77) is the recorded point-source hologram, and can be further simplified to Eq. (3.79), in which C_0 and C_1 are constants and C_1^* is conjugated to C_1; M is the magnification, z_r is the reconstruction distance.

$$H_{PSF} = C_0 + C_1 \cdot e^{\frac{i\pi}{\lambda z_r}[(x-Mx_s)^2+(y-My_s)^2]} \cdot e^{i\theta} + C_1^* \cdot e^{-\frac{i\pi}{\lambda z_r}[(x-Mx_s)^2+(y-My_s)^2]} \cdot e^{-i\theta}$$

$$\tag{3.79}$$

Therefore, the recorded hologram of actual specimen $o(x_s, y_s, z_s)$ composed of point sources can be demonstrated according to Eq. (3.80).

$$\begin{aligned} H = C_0 &\iiint o(x_s, y_s, z_s) dx_s dy_s dz_s \\ &+ C_1 \iiint o(x_s, y_s, z_s) e^{\frac{i\pi}{\lambda z_r}[(x-Mx_s)^2+(y-My_s)^2]} \cdot e^{i\theta} dx_s dy_s dz_s \\ &+ C_1^* \iiint o(x_s, y_s, z_s) e^{\frac{i\pi}{\lambda z_r}[(x-Mx_s)^2+(y-My_s)^2]} \cdot e^{-i\theta} dx_s dy_s dz_s \end{aligned} \tag{3.80}$$

In order to avoid the influence from background and twin image, classical phase-shifting method is adopted via capturing three phase-shifting holograms in three phase-shifting angles θ_1, θ_2, and θ_3, and the filtered hologram is demonstrated by Eq. (3.81).

$$\begin{aligned} H_F &= H_{\theta_1}\left(e^{-i\theta_3} - e^{-i\theta_2}\right) + H_{\theta_2}\left(e^{-i\theta_1} - e^{-i\theta_3}\right) + H_{\theta_3}\left(e^{-i\theta_2} - e^{-i\theta_1}\right) \\ &= C \iiint o(x_s, y_s, z_s) e^{\frac{i\pi}{\lambda z_r}[(x-Mx_s)^2+(y-My_s)^2]} dx_s dy_s dz_s \end{aligned} \tag{3.81}$$

Finally, a 3D image can be reconstructed from H_F using the Fresnel propagation as the convolution between H_F and $e^{\frac{i\pi}{\lambda z_r}(x^2+y^2)}$.

According to the FINCH principle, it can obtain 3D imaging by using incoherent light, and has found its applications especially in microscopy [139–144] and adaptive optics [145, 146]. Moreover, many improved techniques have also been designed in order to improve the FINCH performance, such as enhanced resolution [146–149], suppressed noise [150], and accelerated performance [151]. Additionally, progress and roadmaps to FINCH are also provided in the reviews [152–155].

3.5.6 Coded Aperture Correlation Holography (COACH)

Though FINCH has higher lateral resolution compared to many coherent and incoherent holography techniques, it still suffers from relatively low axial resolution. In
order to solve such problem, Rosen group designed coded aperture correlation holography (COACH) [156]: 3D object can be reconstructed from two incoherent holograms generated by object/pinhole wavefront and their modulated wavefront using
the same coded phase mask (CPM) with a random-like complex function $G(x, y)$ on
a SLM. Figure 3.24a demonstrates the COACH principle, and the key principle of
COACH can be briefly described according to Eqs. (3.82)–(3.87) according to [156].
The wave from a point source is collimated by a lens and transmitted by a polarizer. The polarization direction of the transmitted light is 45° with respect to SLM
modulation direction. The transmitted light is then separated by the SLM: one part
with the parallel polarization direction of the SLM is modulated by the CPM, while
another part with the perpendicular polarization direction of the SLM is not. Both
waves are then combined with another polarizer and recorded using an image sensor
with a distance of d_2 as demonstrated by Eq. (3.82), in which A is the amplitude of
the wave not modulated by the SLM, B is the amplitude of the wave modulated by
the SLM, and θ is the attached phase for further phase shift, similar to FINCH as
demonstrated by Eq. (3.83).

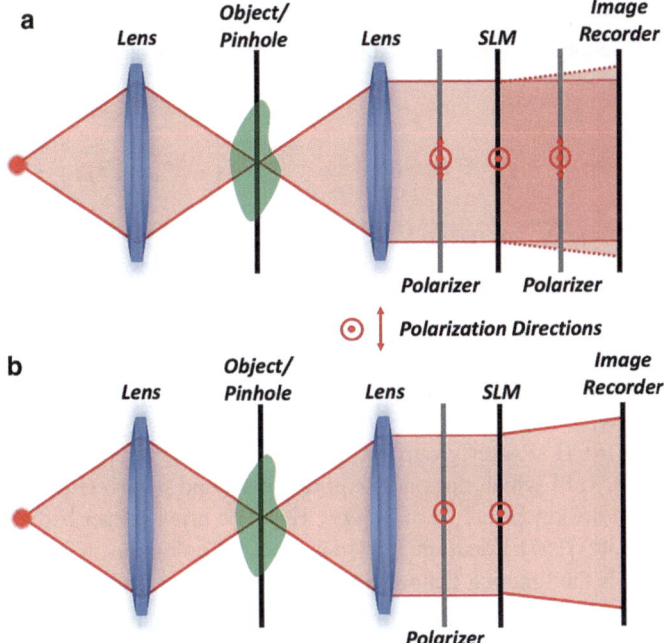

Fig. 3.24 COACH scheme. **a** COACH scheme; **b** I-COACH scheme

$$H_{PSF,\theta} = \left| A + B \cdot G(x, y) \cdot e^{i\theta} \right|^2 \tag{3.82}$$

$$H_{PSF} = H_{PSF,\theta_1}\left(e^{-i\theta_3} - e^{-i\theta_2}\right) + H_{PSF,\theta_2}\left(e^{-i\theta_1} - e^{-i\theta_3}\right)$$
$$+ H_{PSF,\theta_3}\left(e^{-i\theta_2} - e^{-i\theta_1}\right) \propto G(x, y) \tag{3.83}$$

Instead of the pinhole, when the sample actually as the combination of point sources is set at the same position as Eq. (3.84), the object hologram can be written as Eq. (3.85), in which $e^{\frac{i\pi(x_n x + y_n y)}{\lambda f}}$ indicates the tilting phase.

$$o(x, y) = \sum_n a_n \delta(x - x_n, y - y_n) \tag{3.84}$$

$$H_\theta = \sum_n \left| A_n e^{\frac{i\pi(x_n x + y_n y)}{\lambda f}} + B_n e^{\frac{i\pi(x_n x + y_n y)}{\lambda f}} G(x - x_n, y - y_n) e^{i\theta} \right|^2 \tag{3.85}$$

Similarly, using phase shifting, both background and twin image can be cancelled still as demonstrated in Eq. (3.86).

$$H = H_{\theta_1}\left(e^{-i\theta_3} - e^{-i\theta_2}\right) + H_{\theta_2}\left(e^{-i\theta_1} - e^{-i\theta_3}\right) + H_{\theta_3}\left(e^{-i\theta_2} - e^{-i\theta_1}\right)$$
$$= \sum_n A_n^* B_n G(x - x_n, y - y_n) \tag{3.86}$$

The image reconstruction can be implemented using correlation ($\odot$) according to Eq. (3.87).

$$H \odot H_{PSF} = \iint \left[\sum_n A_n^* B_n G(x - x_n, y - y_n) \right] \cdot G^*(x - x_n, y - y_n) du dv$$
$$\propto \sum_n A_n^* B_n \delta(x - x_n, y - y_n) \tag{3.87}$$

Object is reconstructed according to Eq. (3.87). In order to reconstruct a 3D object, the pinhole should be scanned in the axial axis to obtain H_{PSF} dataset for object reconstruction at different axial positions. COACH can be treated as a general model of FINCH.

Besides COACH, Rosen group further designed interferenceless COACH (I-COACH) [157, 158] which does not require the second polarizer demonstrated in Fig. 3.24b, and the key principle of I-COACH can be briefly described according to Eqs. (3.88)–(3.94) [157]. Equation (3.88) describes the hologram on the image sensor, in which $Q(\cdot)$ is the function defined as $e^{\frac{i\pi}{\lambda}[(x-x_s)^2+(y-y_s)^2]}$, and $L(\cdot)$ is the function defined as $e^{\frac{i2\pi}{\lambda z}xx_s+yy_s}$, and it can be further simplified as $H\left(r_0 - \frac{d_2}{d_1}r_s, d_1\right)$, r_0, and r_s are coordinates of object and image recorder, and d_1 as well as d_2 are distances between object and lens as well as between SLM and image recorder. Therefore, the hologram corresponding to pinhole as $H_{PSF,\theta}$ is Eq. (3.89) with phase-shifting angle θ, and the background-free one H_{PSF} can be obtained as Eq. (3.90) using phase-shifting tactics.

$$h = \left| Q\left(\frac{1}{d_1}\right) \cdot L\left(\frac{r_s}{d_1}\right) \cdot Q\left(-\frac{1}{f}\right) \cdot \varphi_{SLM} \otimes Q\left(\frac{1}{d_2}\right) \right|^2 = h\left(r_0 - \frac{d_2}{d_1}r_s, d_1\right) \tag{3.88}$$

$$H_{PSF,\theta} = h\left(r_0 - \frac{d_2}{d_1}r_s; d_1\right) \cdot e^{i\theta} \tag{3.89}$$

$$H_{PSF} = \sum_{\theta} h\left(r_0 - \frac{d_2}{d_1}r_s; d_1\right) \cdot e^{i\theta} \tag{3.90}$$

Considering the sample actually as the combination of point sources is set at the same position as Eq. (3.91), the captured hologram H_θ can be described as Eq. (3.92), and the background-free one H can be obtained as Eq. (3.93) using phase-shifting tactics.

$$o(x, y) = \sum_n a_n \delta(r - r_s) \tag{3.91}$$

$$H_\theta = \sum_n a_n h\left(r_0 - \frac{d_2}{d_1}r_s; d_1\right) \cdot e^{i\theta} \tag{3.92}$$

$$H = \sum_\theta \sum_n a_n h\left(r_0 - \frac{d_2}{d_1}r_s; d_1\right) \cdot e^{i\theta} \tag{3.93}$$

Therefore, the object can be finally obtained using 2-D correlation as Eq. (3.94), in which M is the transverse magnifiaction.

$$H \odot H_{PSF} = \iint \sum_n a_n H_{PSF}\left(r_0 - \frac{d_2}{d_1}r_s; z_s\right) \cdot H_{PSF}^* du dv$$

$$\propto \sum_n \delta\left(r_0 - \frac{d_2}{d_1}r_s\right) = o\left(\frac{r_{s,t}}{M}\right) \tag{3.94}$$

Very similar to classical coded imaging, the hologram of I-COACH can be represented by Eq. (3.95), and the object can be reconstructed according to Eq. (3.96) with the previously measured hologram corresponding to point spread function [158].

$$H = \int o \otimes H_{PSF} dz \tag{3.95}$$

$$\tilde{o} = H \odot H_{PSF} \tag{3.96}$$

According to the COACH principle, it can obtain 3D imaging [159–162] using incoherent light but with higher axial resolution compared to FINCH, though it sacrifices lateral resolution. Besides, in order to improve the COACH performances, updated techniques have also been designed to enhance the resolution [163], suppress the noise [164–166], accelerated performance [167–169], and to extend the field of view [170]. Additionally, progress and roadmaps to FINCH are also provided in reviews [155, 171].

3.5.7 Computer-Generated Holography (CGH)

Since computer-generated holography (CGH) can obtain complex optical field through numerical computation, it has been widely used in many applications such as display and optical testing. After numerically coding the digital data of 3D scene to fringes in CGH, the complex light field of 3D scene can be obtained through diffraction, thus to reconstruct a virtual 3D scene. Moreover, in asphere testing using null optics, a refractive null is often needed to generate a wavefront matching with the desired surface of the asphere under test. While comparing to refractive null fabrication, CGH fabrication is much easier. Therefore, CGH has been a widely used tool in especially asphere testing. Recently, metasurface-based CGH is also proposed and applied in many fields [172]. Different from digital holography for wavefront recording and sensing, CGH focuses on wavefront generation. Interested readers can refer to [173–175] to understand the CGH details and track their state-of-art techniques and applications.

Fringe Projection Profilometry
Different from holography and interferometry which obtains fringes via wavefront interference, fringe projection profilometry directly projects fringes on to the surface under detection and records the fringe pattern in another direction according to the scheme in Fig. 3.25. When the surface under detection is a planar surface, the captured

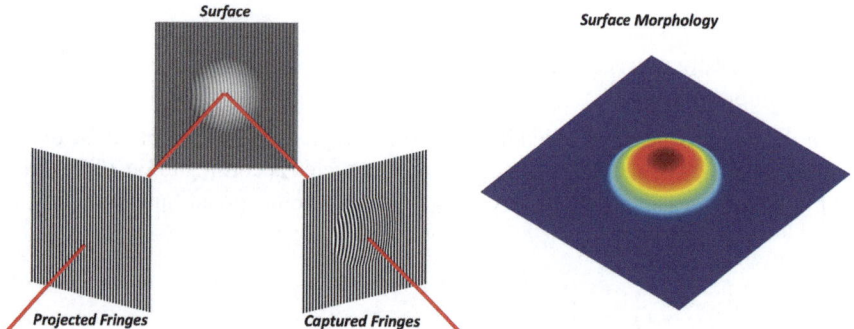

Fig. 3.25 Scheme of fringe projection profilometry

fringe image is very similar to the projected one; while when the surface under detection is a non-planar surface, the captured fringe image is distorted compared to the projected one. The surface morphology can be measured according to the distortion information of the captured fringe image using various structured light patterns and their corresponding algorithms. In particular, when the projected fringes are sinusoidal ones, the surface morphology can be measured using classical phase retrieval algorithms such as phase-shifting method and FFT-based method. In other words, the surface morphology is coded in the phase of the sinusoidal fringes. However, it should be noted that the extracted phase is not that measured by reflection interferometry. Therefore, strictly speaking, fringe projection profilometry does not belong to optical phase imaging, but phase retrieval for surface morphology measurements shares the same to those in interferometry when using sinusoidal fringe patterns. Therefore, in order to make this book complete, fringe projection profilometry is very briefly mentioned here, but for more details, readers can refer to many books and reviews [176–181] to understand the details of fringe projection profilometry and track their state-of-art techniques.

3.6 Summary

In this Section, we introduce the concept of holography and interferometry as well as their classical phase retrieval algorithms in both the spatial and frequency domains for both on-axis and off-axis modes, along with numerical simulations. For the reader to gain a deeper understanding of the state of the art, a summary on advances in holography and interferometry has been provided by including algorithms and systems. We also discuss extensive approaches to holography and interferometry. As digital holography and interferometry offer many advantages, including high accuracy, simple and cost-effective systems, and single-shot phase imaging capability, they have become widely used phase imaging techniques. Many techniques have already been developed and commercialized, such as digital holographic microscopy [182], quantitative interferometry diffraction imaging [183], SLIM [184], quadriwave lateral shearing interferometry [185], FINCH [186], and so forth.

Among various available quantitative phase imaging techniques, holography and interferometry are the mostly widely used, mainly because of the fact that they only rely on simple and cost-effective optical elements. In addition, phase retrieval algorithms evolved over the years are successful in phase extraction and unwrapping by eliminating problems of stagnation, noise, and positional inaccuracies. Many applications have been reported using digital holography and interferometry, such as cell imaging and tracking in biological fields, microbe sensing and particle monitoring in environmental fields, strain detection and optical element inspection in engineering fields and so on. Holography and interferometry still have some shortcomings. One such thing is that they require an extra reference beam and (partially) coherent light sources are not compatible with many commercial microscopes. Besides, phase unwrapping is still a time-consuming process, limiting its real-time imaging and display applications. Since it is not practically possible to address all aspects of holography and interferometry in our chapter, interested readers may refer to books that extensively discuss digital holography and its applications such as "Introduction to Modern Digital Holography: With Matlab" [187] and "Digital Holographic Microscopy: Principles, Techniques, and Applications" [188]. Various phase extraction and unwrapping methods are discussed in books "Interferogram Analysis for Optical Testing" [189] and "Two-dimensional Phase Unwrapping: Theory, Algorithms, and Software" [16]. "Optical Shop Testing" [1] is a valuable book on applications of digital holography (interferometry) in optical testing. "Quantitative phase imaging of cells and tissues" [190] is recommended for readers interested in biological imaging applications.

References

1. Malacara, D. (ed.): Optical Shop Testing, 3rd edn. Wiley-Interscience, New Jersey (2007)
2. https://www.nobelprize.org
3. Gabor, D.: A new microscopic principle. Nature **161**, 777–778 (1948)
4. Leith, E.N., Upatnieks, J.: Reconstructed wavefronts and communication theory. J. Opt. Soc. Am. **52**, 1123–1130 (1962)
5. Leith, E.N., Upatnieks, J.: Wavefront reconstruction with continuous-tone objects. J. Opt. Soc. Am. **53**, 1377–1381 (1963)
6. Leith, E.N., Upatnieks, J.: Wavefront reconstruction with diffused illumination and three-dimensional objects. J. Opt. Soc. Am. **54**, 1295–1301 (1964)
7. Brown, B.R., Lohmann, A.W.: Complex spatial filtering with binary masks. Appl. Opt. **5**, 967–969 (1966)
8. Goodman, J.W., Lawrence, R.W.: Digital image formation from electronically detected holograms. Appl. Phys. Lett. **11**, 77–79 (1967)
9. Goodman, J.W.: Introduction to Fourier Optics, 4th edn. W. H. Freeman, San Francisco (2017)
10. Blanche, P.-A.: Field Guide to Holography. SPIE press, Bellingham (2014)
11. Bragg, W.L., Rogers, G.L.: Elimination of the unwanted image in diffraction microscopy. Nature **167**, 190–191 (1951)
12. Yamaguchi, I., Zhang, T.: Phase-shifting digital holography. Opt. Lett. **22**, 1268–1270 (1997)
13. Shaked, N.T., Zhu, Y., Rinehart, M.T., Wax, A.: Two-step-only phase-shifting interferometry with optimized detector bandwidth for microscopy of live cells. Opt. Express **17**, 15585–15591 (2009)
14. Xue, L., Lai, J., Wang, S., Li, Z.: Single-shot slightly-off-axis interferometry based Hilbert phase microscopy of red blood cells. Biomed. Opt. Express **2**, 987–995 (2011)
15. Thorlabs Shearing Interferometers: https://www.thorlabs.com/newgrouppage9.cfm?object group_id=2970
16. Ghiglia, D.C., Pritt, M.D.: Two-Dimensional Phase Unwrapping: Theory, Algorithms, and Software. Wiley-Interscience, New Jersey (1998)
17. Greenbaum, A., Luo, W., Su, T.-W., Göröcs, Z., Xue, L., Isikman, S.O., Coskun, A.F., Mudanyali, O., Ozcan, A.: Imaging without lenses: achievements and remaining challenges of wide-field on-chip microscopy. Nat. Methods **9**, 889–895 (2012)
18. Ozcan, A., McLeod, E.: Lensless imaging and sensing. Annu. Rev. Biomed. Eng. **18**, 77–102 (2016)
19. Mudanyali, O., Tseng, D., Oh, C., Isikman, S.O., Sencan, I., Bishara, W., Oztoprak, C., Seo, S., Khademhosseini, B., Ozcan, A.: Compact, light-weight and cost-effective microscope based on lensless incoherent holography for telemedicine applications. Lab Chip **10**, 1417–1428 (2010)

20. Tseng, D., Mudanyali, O., Oztoprak, C., Isikman, S.O., Sencan, I., Yaglidere, O., Ozcan, A.: Lensfree microscopy on a cellphone. Lab Chip **10**, 1787–1792 (2010)
21. Greenbaum, A., Ozcan, A.: Maskless imaging of dense samples using pixel super-resolution based multi-height lensfree on-chip microscopy. Opt. Express **20**, 3129–3143 (2012)
22. Greenbaum, A., Sikora, U., Ozcan, A.: Field-portable wide-field microscopy of dense samples using multi-height pixel super-resolution based lensfree imaging. Lab Chip **12**, 1242–1245 (2012)
23. Greenbaum, A., Akbari, N., Feizi, A., Luo, W., Ozcan, A.: Field-portable pixel super-resolution colour microscope. PLos ONE **8**, e76475 (2013)
24. Vargas, J., Quiroga, J.A., Belenguer, T.: Phase-shifting interferometry based on principal component analysis. Opt. Lett. **36**, 1326–1328 (2011)
25. Vargas, J., Quiroga, J.A., Belenguer, T.: Analysis of the principal component algorithm in phase-shifting interferometry. Opt. Lett. **36**, 2215–2217 (2011)
26. Vargas, J., Sorzano, C.O.S.: Quadrature component analysis for interferometry. Opt. Lasers Eng. **51**, 637–641 (2013)
27. Wei, Q., Li, Y., Vargas, J., Wang, J., Gong, Q., Kong, Y., Jiang, Z., Xue, L., Liu, C., Liu, F., Wang, S.: Principal component analysis-based quantitative differential interference contrast microscopy. Opt. Lett. **44**, 45–48 (2019)
28. Vargas, J., Wang, S., Gómez-Pedrero, J.A., Estrada, J.C.: Robust weighted principal components analysis demodulation algorithm for phase-shifting interferometry. Opt. Express **29**, 16534–16546 (2021)
29. Ikeda, T., Popescu, G., Dasari, R.R., Feld, M.S.: Hilbert phase microscopy for investigating fast dynamics in transparent systems. Opt. Lett. **30**, 1165–1167 (2005)
30. Popescu, G., Ikeda, T., Best, C., Badizadegan, K., Dasari, R.R., Feld, M.S.: Erythrocyte structure and dynamics quantified by Hilbert phase microscopy. J. Biomed. Opt. **10**, 060503 (2005)
31. Lue, N., Bewersdorf, J., Lessard, M.D., Badizadegan, K., Dasari, R.R., Feld, M.S., Popescu, G.: Tissue refractometry using Hilbert phase microscopy. Opt. Lett. **32**, 3522–3524 (2007)
32. Wang, S., Xue, L., Lai, J., Li, Z.: An improved phase retrieval method based on Hilbert transform in interferometric microscopy. Optik **124**, 1897–1901 (2013)
33. Wang, S., Sun, N., Xue, L., Li, H., Lai, J., Song, Y., Li, Z.: Radial Hilbert transform phase retrieval algorithm for circular carrier interferogram. Opt. Commun. **304**, 148–152 (2013)
34. Wang, S., Yan, K., Xue, L.: Quantitative interferometric microscopy with two dimensional Hilbert transform based phase retrieval method. Opt. Commun. **383**, 537–544 (2017)
35. Debnath, S.K., Park, Y.K.: Real-time quantitative phase imaging with a spatial phase-shifting algorithm. Opt. Lett. **36**, 4677–4679 (2011)
36. Bhaduri, B., Popescu, G.: Derivative method for phase retrieval in off-axis quantitative phase imaging. Opt. Lett. **37**, 1868–1870 (2012)
37. Fürhapter, S., Jesacher, A., Bernet, S., Ritsch-Marte, M.: Spiral interferometry. Opt. Lett. **30**, 1953–1955 (2005)
38. Jesacher, A., Fürhapter, S., Bernet, S., Ritsch-Marte, M.: Spiral interferogram analysis. J. Opt. Soc. Am. A **23**, 1400–1409 (2006)
39. Gass, J., Dakoff, A., Kim, M.K.: Phase imaging without 2π ambiguity by multiwavelength digital holography. Opt. Lett. **28**, 1141–1143 (2003)
40. Parshall, D., Kim, M.K.: Digital holographic microscopy with dual-wavelength phase unwrapping. Appl. Opt. **45**, 451–459 (2006)
41. Warnasooriya, N., Kim, M.K.: Quantitative phase imaging using three-wavelength optical phase unwrapping. J. Mod. Opt. **56**, 67–74 (2009)
42. Li, Y., Xiao, W., Pan, F.: Multiple-wavelength-scanning-based phase unwrapping method for digital holographic microscopy. Appl. Opt. **53**, 979–987 (2014)
43. Rivenson, Y., Zhang, Y., Gunaydin, H., Teng, D., Ozcan, A.: Phase recovery and holographic image reconstruction using deep learning in neural networks. Light-Sci. Appl. **7**, 17141 (2018)
44. Wu, Y., Rivenson, Y., Zhang, Y., Wei, Z., Gunaydin, H., Lin, X., Ozcan, A.: Extended depth-of-field in holographic imaging using deep-learning-based autofocusing and phase recovery. Optica **5**, 704–710 (2018)

45. Rivenson, Y., Wu, Y., Ozcan, A.: Deep learning in holography and coherent imaging. Light-Sci. Appl. **8**, 85 (2019)
46. Ren, Z., Xu, Z., Lam, E.Y.: Learning-based nonparametric autofocusing for digital holography. Optica **5**, 337–344 (2018)
47. Ren, Z., Xu, Z., Lam, E.Y.: End-to-end deep learning framework for digital holographic reconstruction. Adv. Photonics **1**, 016004 (2019)
48. Zeng, T., So, H.K.H., Lam, E.Y.: RedCap: residual encoder-decoder capsule network for holographic image reconstruction. Opt. Express **28**, 4876–4887 (2020)
49. Wang, H., Lyu, M., Situ, G.: eHoloNet: a learning-based end-to-end approach for in-line digital holographic reconstruction. Opt. Express **26**, 22603–22614 (2018)
50. Jaferzadeh, K., Hwang, S.H., Moon, I., Javidi, B.: No-search focus prediction at the single cell level in digital holographic imaging with deep convolutional neural network. Biomed. Opt. Express **10**, 4276–4289 (2019)
51. Moon, I., Jaferzadeh, K., Kim, Y., Javidi, B.: Noise-free quantitative phase imaging in Gabor holography with conditional generative adversarial network. Opt. Express **28**, 26284–26301 (2020)
52. Wang, K., Dou, J., Qian, K., Di, J., Zhao, J.: Y-Net: a one-to-two deep learning framework for digital holographic reconstruction. Opt. Lett. **44**, 4765–4768 (2019)
53. Wang, K., Qian, K., Di, J., Zhao, J.: Y4-Net: a deep learning solution to one-shot dual-wavelength digital holographic reconstruction. Opt. Lett. **45**, 4220–4223 (2020)
54. Zhang, Z., Zheng, Y., Xu, T., Upadhya, A., Lim, Y.J., Mathews, A., Xie, L., Lee, W.M.: Holo-UNet: hologram-to-hologram neural network restoration for high fidelity low light quantitative phase imaging of live cells. Biomed. Opt. Express **11**, 5478–5487 (2020)
55. Kesavan, S.V., Momey, F., Cioni, O., David-Watine, B., Dubrulle, N., Shorte, S., Sulpice, E., Freida, D., Chalmond, B., Dinten, J.M., Gidrol, X., Allier, C.: High-throughput monitoring of major cell functions by means of lensfree video microscopy. Sci. Rep. **4**, 5942 (2014)
56. Kesavan, S.V., Navarro, F.P., Menneteau, M., Mittler, F., David-Watine, B., Dubrulle, N., Shorte, S.L., Chalmond, B., Dinten, J.-M., Allier, C.P.: Real-time label-free detection of dividing cells by means of lensfree video-microscopy. J. Biomed. Opt. **19**, 036004 (2014)
57. Momey, F., Coutard, J.-G., Bordy, T., Navarro, F., Menneteau, M., Dinten, J.-M., Allier, C.: Dynamics of cell and tissue growth acquired by means of extended field of view lensfree microscopy. Biomed. Opt. Express **7**, 512–524 (2016)
58. Su, T.-W., Xue, L., Ozcan, A.: High-throughput lensfree 3D tracking of human sperms reveals rare statistics of helical trajectories. Proc. Natl. Acad. Sci. U.S.A. **109**, 16018–16022 (2012)
59. Su, T.-W., Choi, I., Feng, J., Huang, K., McLeod, E., Ozcan, A.: Sperm trajectories form chiral ribbons. Sci. Rep. **3**, 1664 (2013)
60. Daloglu, M.U., Luo, W., Shabbir, F., Lin, F., Kim, K., Lee, I., Jiang, J.Q., Cai, W.J., Ramesh, V., Yu, M.-Y., Ozcan, A.: Label-free 3D computational imaging of spermatozoon locomotion, head spin and flagellum beating over a large volume. Light-Sci. Appl. **7**, 17121 (2018)
61. Su, T.W., Choi, I., Feng, J., Huang, K., Ozcan, A.: High-throughput analysis of horse sperms' 3D swimming patterns using computational on-chip imaging. Anim. Reprod. Sci. **169**, 45–55 (2016)
62. Daloglu, M.U., Ozcan, A.: Computational imaging of sperm locomotion. Biol. Reprod. **97**, 182–188 (2017)
63. Daloglu, M.U., Lin, F., Chong, B., Chien, D., Veli, M., Luo, W., Ozcan, A.: 3D imaging of sex-sorted bovine spermatozoon locomotion, head spin and flagellum beating. Sci. Rep. **8**, 15650 (2018)
64. Popescu, G., Deflores, L.P., Vaughan, J.C., Badizadegan, K., Iwai, H., Dasari, R.R., Feld, M.S.: Fourier phase microscopy for investigation of biological structures and dynamics. Opt. Lett. **29**, 2503–2505 (2004)
65. Bhaduri, B., Tangella, K., Popescu, G.: Fourier phase microscopy with white light. Biomed. Opt. Express **4**, 1434–1441 (2013)
66. Tahara, T., Ito, Y., Xia, P., Awatsuji, Y., Nishio, K., Ura, S., Kubota, T., Matoba, O.: Space-bandwidth extension in parallel phase-shifting digital holography using a four-channel polarization-imaging camera. Opt. Lett. **38**, 2463–2465 (2013)

67. Wang, D., Liang, R.: Simultaneous polarization Mirau interferometer based on pixelated polarization camera. Opt. Lett. **41**, 41–44 (2016)
68. Popescu, G., Ikeda, T., Dasari, R.R., Feld, M.S.: Diffraction phase microscopy for quantifying cell structure and dynamics. Opt. Lett. **31**, 775–777 (2006)
69. Bhaduri, B., Edwards, C., Pham, H., Zhou, R., Nguyen, T.H., Goddard, L.L., Popescu, G.: Diffraction phase microscopy: principles and applications in materials and life sciences. Adv. Opt. Photonics **6**, 57–119 (2014)
70. Bhaduri, B., Pham, H., Mir, M., Popescu, G.: Diffraction phase microscopy with white light. Opt. Lett. **37**, 1094–1096 (2012)
71. Pham, H., Bhaduri, B., Ding, H., Popescu, G.: Spectroscopic diffraction phase microscopy. Opt. Lett. **37**, 3438–3440 (2012)
72. Edwards, C., Bhaduri, B., Griffin, B.G., Goddard, L.L., Popescu, G.: Epi-illumination diffraction phase microscopy with white light. Opt. Lett. **39**, 6162–6165 (2014)
73. Park, Y.K., Popescu, G., Badizadegan, K., Dasari, R.R., Feld, M.S.: Diffraction phase and fluorescence microscopy. Opt. Express **14**, 8263–8268 (2006)
74. Edwards, C., McKeown, S.J., Zhou, J., Popescu, G., Goddard, L.L.: In situ measurements of the axial expansion of palladium microdisks during hydrogen exposure using diffraction phase microscopy. Opt. Mater. Express **4**, 2559–2564 (2014)
75. Hu, C., Zhu, S., Gao, L., Popescu, G.: Endoscopic diffraction phase microscopy. Opt. Lett. **43**, 3373–3376 (2018)
76. Singh, A.S.G., Anand, A., Leitgeb, R.A., Javidi, B.: Lateral shearing digital holographic imaging of small biological specimens. Opt. Express **21**, 23617–23622 (2012)
77. Javidi, B., Markman, A., Rawat, S., O'Connor, T., Anand, A., Andemariam, B.: Sickle cell disease diagnosis based on spatio-temporal cell dynamics analysis using 3D printed shearing digital holographic microscopy. Opt. Express **26**, 13614–13627 (2018)
78. Rawat, S., Komatsu, S., Markman, A., Anand, A., Javidi, B.: Compact and field-portable 3D printed shearing digital holographic microscope for automated cell identification. Appl. Opt. **56**, D127–D133 (2017)
79. O'Connor, T., Anand, A., Andemariam, B., Javidi, B.: Deep learning-based cell identification and disease diagnosis using spatio-temporal cellular dynamics in compact digital holographic microscopy. Biomed. Opt. Express **11**, 4491–4508 (2020)
80. Wang, Z., Millet, L., Mir, M., Ding, H., Unarunotai, S., Rogers, J., Gillette, M.U., Popescu, G.: Spatial light interference microscopy (SLIM). Opt. Express **19**, 1016–1026 (2011)
81. Nguyen, T.H., Popescu, G.: Spatial light interference microscopy (SLIM) using twisted-nematic liquid-crystal modulation. Biomed. Opt. Express **4**, 1571–1583 (2013)
82. Wang, Z., Chun, I.S., Li, X., Ong, Z.-Y., Pop, E., Millet, L., Gillette, M., Popescu, G.: Topography and refractometry of nanostructures using spatial light interference microscopy. Opt. Lett. **35**, 208–210 (2010)
83. Ding, H., Popescu, G.: Instantaneous spatial light interference microscopy. Opt. Express **18**, 1569–1575 (2010)
84. Ma, L., Rajshekhar, G., Wang, R., Bhaduri, B., Sridharan, S., Mir, M., Chakraborty, A., Iyer, R., Prasanth, S., Millet, L., Gillette, M.U., Popescu, G.: Phase correlation imaging of unlabeled cell dynamics. Sci. Rep. **6**, 32702 (2016)
85. Nguyen, T.H., Kandel, M., Shakir, H.M., Best-Popescu, C., Arikkath, J., Do, M.N., Popescu, G.: Halo-free phase contrast microscopy. Sci. Rep. **7**, 44034 (2017)
86. Kandel, M.E., Fanous, M., Best-Popescu, C., Popescu, G.: Real-time halo correction in phase contrast imaging. Biomed. Opt. Express **9**, 623–635 (2018)
87. Sridharan, S., Katz, A., Soto-Adames, F., Popescu, G.: Quantitative phase imaging of arthropods. J. Biomed. Opt. **20**, 111212 (2015)
88. Takabayashi, M., Majeed, H., Kajdacsy-Balla, A., Popescu, G.: Tissue spatial correlation as cancer marker. J. Biomed. Opt. **24**, 016502 (2019)
89. Majeed, H., Kandel, M.E., Han, K., Luo, Z., Macias, V., Tangella, K., Balla, A., Popescu, G.: Breast cancer diagnosis using spatial light interference microscopy. J. Biomed. Opt. **20**, 111210 (2015)

90. Majeed, H., Okoro, C., Kajdacsy-Balla, A., Toussaint, K.C., Popescu, G.: Quantifying collagen fiber orientation in breast cancer using quantitative phase imaging. J. Biomed. Opt. **22**, 046004 (2017)
91. Majeed, H., Nguyen, T.H., Kandel, M.E., Kajdacsy-Balla, A., Popescu, G.: Label-free quantitative evaluation of breast tissue using spatial light interference microscopy (SLIM). Sci. Rep. **8**, 6875 (2018)
92. Sridharan, S., Macias, V., Tangella, K., Kajdacsy-Balla, A., Popescu, G.: Prediction of prostate cancer recurrence using quantitative phase imaging. Sci. Rep. **5**, 9976 (2015)
93. Sridharan, S., Macias, V., Tangella, K., Melamed, J., Dube, E., Kong, M.X., Kajdacsy-Balla, A., Popescu, G.: Prediction of prostate cancer recurrence using quantitative phase imaging: validation on a general population. Sci. Rep. **6**, 33818 (2016)
94. Fanous, M., Keikhosravi, A., Kajdacsy-Balla, A., Eliceiri, K.W., Popescu, G.: Quantitative phase imaging of stromal prognostic markers in pancreatic ductal adenocarcinoma. Biomed. Opt. Express **11**, 1354–1364 (2020)
95. Bhaduri, B., Wickland, D., Wang, R., Chan, V., Bashir, R., Popescu, G.: Cardiomyocyte imaging using real-time spatial light interference microscopy (SLIM). PLoS ONE **8**, e56930 (2013)
96. Gannavarpu, R., Bhaduri, B., Tangella, K., Popescu, G.: Spatiotemporal characterization of a fibrin clot using quantitative phase imaging. PLoS ONE **9**, e111381 (2014)
97. Bhaduri, B., Kandel, M., Brugnara, C., Tangella, K., Popescu, G.: Optical assay of erythrocyte function in banked blood. Sci. Rep. **4**, 6211 (2014)
98. Ceballos, S., Kandel, M., Sridharan, S., Majeed, H., Monroy, F., Popescu, G.: Active intracellular transport in metastatic cells studied by spatial light interference microscopy. J. Biomed. Opt. **20**, 111209 (2015)
99. Lee, Y.J., Cintora, P., Arikkath, J., Akinsola, O., Kandel, M., Popescu, G., Best-Popescu, C.: Quantitative assessment of neural outgrowth using spatial light interference microscopy. J. Biomed. Opt. **22**, 066015 (2017)
100. Kandel, M.E., Teng, K.W., Selvin, P.R., Popescu, G.: Label-free imaging of single microtubule dynamics using spatial light interference microscopy. ACS Nano **11**, 647–655 (2017)
101. Hartman, J.S., Gordon, R.L., Lessor, D.L.: Development of Nomarski microscopy for quantitative determination of surface topography. Proc. SPIE **192**, 223–230 (1979)
102. Shimada, W., Sato, T., Yatagai, T.: Optical surface microtopography using phase-shifting Nomarski microscope. Proc. SPIE **1332**, 525–529 (1990)
103. Hariharan, P., Roy, M.: Achromatic phase-shifting for two-wavelength phase-stepping interferometry. Opt. Commun. **126**, 220–222 (1996)
104. Cogswell, C.J., Smith, N.I., Larkin, K.G., Hariharan, P.: Quantitative DIC microscopy using a geometric phase shifter. Proc. SPIE **2984**, 72–81 (1997)
105. van Munster, E.B., van Vliet, L.J., Aten, J.A.: Reconstruction of optical pathlength distributions from images obtained by a wide-field differential interference contrast microscope. J. Microsc. **188**, 149–157 (1997)
106. van Munster, E.B., Winter, E.K., Aten, J.A.: Measurement-based evaluation of optical pathlength distributions reconstructed from simulated differential interference contrast images. J. Microsc. **191**, 170–176 (1998)
107. Arnison, M.R., Larkin, K.G., Sheppard, C.J.R., Smith, N.I., Cogswell, C.J.: Linear phase imaging using differential interference contrast microscopy. J. Microsc. **214**, 7–12 (2004)
108. McIntyre, T.J., Maurer, C., Fassl, S., Khan, S., Bernet, S., Ritsch-Marte, M.: Quantitative SLM-based differential interference contrast imaging. Opt. Express **18**, 14063–14078 (2010)
109. Baffou, G.: Quantitative phase microscopy using quadriwave lateral shearing interferometry (QLSI): principle, terminology, algorithm and grating shadow description. J. Phys. D: Appl. Phys. **54**, 294002 (2021)
110. Bon, P., Maucort, G., Wattellier, B., Monneret, S.: Quadriwave lateral shearing interferometry for quantitative phase microscopy of living cells. Opt. Express **17**, 13080–13094 (2009)
111. Bon, P., Aknoun, S., Monneret, S., Wattellier, B.: Enhanced 3D spatial resolution in quantitative phase microscopy using spatially incoherent illumination. Opt. Express **22**, 8654–8671 (2014)

112. Aknoun, S., Bon, P., Savatier, J., Wattellier, B., Monneret, S.: Quantitative retardance imaging of biological samples using quadriwave lateral shearing interferometry. Opt. Express **23**, 16383–16406 (2015)
113. Aknoun, S., Savatier J., Bon, P., Galland, F., Abdeladim, L., Wattellier, B., Monneret, S.: Living cell dry mass measurement using quantitative phase imaging with quadriwave lateral shearing interferometry: an accuracy and sensitivity discussion. J. Biomed. Opt. **20**, 126009 (2015)
114. Aknoun, S., Aurrand-Lions, M., Wattellier, B., Monneret, S.: Quantitative retardance imaging by means of quadri-wave lateral shearing interferometry for label-free fiber imaging in tissues. Opt. Commun. **422**, 17–27 (2018)
115. Bellanger, C., Toulon, B., Primot, J., Lombard, L., Bourderionnet, J., Brignon, A.: Collective phase measurement of an array of fiber lasers by quadriwave lateral shearing interferometry for coherent beam combining. Opt. Lett. **35**, 3931–3933 (2010)
116. Han, Z.G., Meng, L.Q., Huang, Z.Q., Shen, H., Chen, L., Zhu, R.H.: Determination of the laser beam quality factor (M2) by stitching quadriwave lateral shearing interferograms with different exposures. Appl. Opt. **56**, 7596–7603 (2017)
117. Zhang, R., Yang, Y., Zhao, H., Liang, Z., Liu, S., Bai, J.: Non-null testing of the aspheric surface using a quadriwave lateral shearing interferometer. Appl. Opt. **59**, 5447–5456 (2020)
118. Khadir, S., Bon, P., Vignaud, D., Galopin, E., McEvoy, N., McCloskey, D., Monneret, S., Baffou, G.: Optical imaging and characterization of graphene and other 2D materials using quantitative phase microscopy. ACS Photonics **4**, 3130–3139 (2017)
119. Khadir, S., Andren, D., Chaumet, P.C., Monneret, S., Bonod, N., Kall, M., Sentenac, A., Baffou, G: Full optical characterization of single nanoparticles using quantitative phase imaging. Optica **7**, 243–248 (2020)
120. Khadir, S., Andrén, D., Verre, R., Song, Q., Monneret, S., Genevet, P., Käll, M., Baffou, G.: Metasurface optical characterization using quadriwave lateral shearing interferometry. ACS Photonics **8**, 603–613 (2021)
121. Poon, T.-C., Korpel, A.: Optical transfer function of an acousto-optic heterodyning image processor. Opt. Lett. **4**, 317–319 (1979)
122. Poon, T.-C.: Scanning holography and two-dimensional image processing by acousto-optic two-pupil synthesis. J. Opt. Soc. Am. A **2**, 521–527 (1985)
123. Duncan, B.D., Poon, T.-C.: Gaussian beam analysis of optical scanning holography. J. Opt. Soc. Am. A **9**, 229–236 (1992)
124. Poon, T.-C.: Optical Scanning Holography with Matlab. Springer, Berlin (2007)
125. Poon, T.-C., Liu, J.P.: Introduction to Modern Digital Holography. Cambridge University Press, Cambridge (2014)
126. Indebetouw, G., Kim, T., Poon, T.-C., Schilling, B.W.: Three-dimensional location of fluorescent inhomogeneities in turbid media by scanning heterodyne holography. Opt. Lett. **23**, 135–137 (1998)
127. Lam, E.Y., Zhang, X., Vo, H., Poon, T.-C., Indebetouw, G.: Three-dimensional microscopy and sectional image reconstruction using optical scanning holography. Appl. Opt. **48**, H113–H119 (2009)
128. Zhang, X., Lam, E.Y., Kim, T., Kim, Y.S., Poon, T.-C.: Blind sectional image reconstruction for optical scanning holography. Opt. Lett. **34**, 3098–3100 (2009)
129. Xin, Z., Dobson, K., Shinoda, Y., Poon, T.-C.: Sectional image reconstruction in optical scanning holography using a random-phase pupil. Opt. Lett. **35**, 2934–2936 (2010)
130. Ke, J., Poon, T.-C., Lam, E.Y.: Depth resolution enhancement in optical scanning holography with a dual-wavelength laser source. Appl. Opt. **50**, H285–H296 (2011)
131. Ou, H., Poon, T.-C., Wong, K.K.Y., Lam, E.Y.: Depth resolution enhancement in double-detection optical scanning holography. Appl. Opt. **52**, 3079–3087 (2013)
132. Zhang, Y., Wang, R., Tsang, P., Poon, T.-C.: Sectioning with edge extraction in optical incoherent imaging processing. OSA Continuum **3**, 698–708 (2020)
133. Indebetouw, G., Klysubun, P., Kim, T., Poon, T.-C.: Imaging properties of scanning holographic microscopy. J. Opt. Soc. Am. A **17**, 380–390 (2000)

134. Poon, T.-C., Indebetouw, G.: Three-dimensional point spread functions of an optical heterodyne scanning image processor. Appl. Opt. **42**, 1485–1492 (2003)
135. Liu, J.P.: Spatial coherence analysis for optical scanning holography. Appl. Opt. **54**, A59–A66 (2015)
136. Indebetouw, G., Tada, Y., Leacock, J.: Quantitative phase imaging with scanning holographic microscopy: an experimental assessment. Biomed. Eng. Online **5**, 63 (2006)
137. Rosen, J., Brooker, G.: Digital spatially incoherent Fresnel holography. Opt. Lett. **32**, 912–914 (2007)
138. Katz, B., Rosen, J., Kelner, R., Brooker, G.: Enhanced resolution and throughput of Fresnel incoherent correlation holography (FINCH) using dual diffractive lenses on a spatial light modulator (SLM). Opt. Express **20**, 9109–9121 (2012)
139. Rosen, J., Brooker, G.: Non-scanning motionless fluorescence three-dimensional holographic microscopy. Nat. Photonics **2**, 190–195 (2008)
140. Brooker, G., Siegel, N., Wang, V., Rosen, J.: Optimal resolution in Fresnel incoherent correlation holographic fluorescence microscopy. Opt. Express **19**, 5047–5062 (2011)
141. Rosen, J., Siegel, N., Brooker, G.: Theoretical and experimental demonstration of resolution beyond the Rayleigh limit by FINCH fluorescence microscopic imaging. Opt. Express **19**, 26249–26268 (2011)
142. Siegel, N., Rosen, J., Brooker, G.: Reconstruction of objects above and below the objective focal plane with dimensional fidelity by FINCH fluorescence microscopy. Opt. Express **20**, 19822–19835 (2012)
143. Brooker, G., Siegel, N., Rosen, J., Hashimoto, N., Kurihara, M., Tanabe, A.: In-line FINCH super resolution digital holographic fluorescence microscopy using a high efficiency transmission liquid crystal GRIN lens. Opt. Lett. **38**, 5264–5267 (2013)
144. Kelner, R., Katz, B., Rosen, J.: Optical sectioning using a digital Fresnel incoherent-holography-based confocal imaging system. Optica **1**, 70–74 (2014)
145. Kim, M.K.: Adaptive optics by incoherent digital holography. Opt. Lett. **37**, 2694–2696 (2012)
146. Kim, M.K.: Incoherent digital holographic adaptive optics. Appl. Opt. **52**, A117–A130 (2013)
147. Siegel, N., Rosen, J., Brooker, G.: Faithful reconstruction of digital holograms captured by FINCH using a Hamming window function in the Fresnel propagation. Opt. Lett. **38**, 3922–3925 (2013)
148. Kashter, Y., Vijayakumar, A., Miyamoto, Y., Rosen, J.: Enhanced super resolution using Fresnel incoherent correlation holography with structured illumination. Opt. Lett. **41**, 1558–1561 (2016)
149. Jeon, P., Kim, J., Lee, H., Kwon, H.S., Kim, D.Y.: Comparative study on resolution enhancements in fluorescence-structured illumination Fresnel incoherent correlation holography. Opt. Express **29**, 9231–9241 (2021)
150. Katz, B., Wulich, D., Rosen, J.: Optimal noise suppression in Fresnel incoherent correlation holography (FINCH) configured for maximum imaging resolution. Appl. Opt. **49**, 5757–5763 (2010)
151. Kelner, R., Rosen, J.: Parallel-mode scanning optical sectioning using digital Fresnel holography with three-wave interference phase-shifting. Opt. Express **24**, 2200–2214 (2016)
152. Rosen, J., Katz, B., Brooker, G.: Fresnel incoherent correlation hologram-a review. Chin. Opt. Lett. **7**, 1134–1141 (2009)
153. Anand, V., Katkus, T., Ng, S.H., Juodkazis, S.: Review of Fresnel incoherent correlation holography with linear and non-linear correlations. Chin. Opt. Lett. **19**, 020501 (2021)
154. Rosen, J., Alford, S., Anand, V., Art, J., Bouchal, P., Bouchal, Z., Erdenebat, M.U., Huang, L., Ishii, A., Juodkazis, S., Kim, N., Kner, P., Koujin, T., Kozawa, Y., Liang, D., Liu, J., Mann, C., Marar, A., Matsuda, A., Nobukawa, T., Nomura, T., Oi, R., Potcoava, M., Tahara, T., Thanh, B.L., Zhou, H.: Roadmap on recent progress in FINCH technology. J. Imaging **7**, 197 (2021)
155. Rosen, J., Vijayakumar, A., Kumar, M., Rai, M.R., Kelner, R., Kashter, Y., Bulbul, A., Mukherjee, S.: Recent advances in self-interference incoherent digital holography. Adv. Opt. Photonics **11**, 1–66 (2019)

156. Vijayakumar, A., Kashter, Y., Kelner, R., Rosen, J.: Coded aperture correlation holography–a new type of incoherent digital holograms. Opt. Express **24**, 12430–12441 (2016)
157. Vijayakumar, A., Rosen, J.: Interferenceless coded aperture correlation holography–a new technique for recording incoherent digital holograms without two-wave interference. Opt. Express **25**, 13883–13896 (2017)
158. Hai, N., Rosen, J.: Interferenceless and motionless method for recording digital holograms of coherently illuminated 3D objects by coded aperture correlation holography system. Opt. Express **27**, 24324–24339 (2019)
159. Vijayakumar, A., Rosen, J.: Spectrum and space resolved 4D imaging by coded aperture correlation holography (COACH) with diffractive objective lens. Opt. Lett. **42**, 947–950 (2017)
160. Dubey, N., Rosen, J., Gannot, I.: High-resolution imaging system with an annular aperture of coded phase masks for endoscopic applications. Opt. Express **28**, 15122–15137 (2020)
161. Hai, N., Rosen, J.: Coded aperture correlation holographic microscope for single-shot quantitative phase and amplitude imaging with extended field of view. Opt. Express **28**, 27372–27386 (2020)
162. Bulbul, A., Rosen, J.: Super-resolution imaging by optical incoherent synthetic aperture with one channel at a time. Photonics Res. **9**, 1172–1181 (2021)
163. Rai, M.R., Vijayakumar, A., Ogura, Y., Rosen, J.: Resolution enhancement in nonlinear interferenceless COACH with point response of subdiffraction limit patterns. Opt. Express **27**, 391–403 (2019)
164. Rai, M.R., Vijayakumar, A., Rosen, J.: Non-linear adaptive three-dimensional imaging with interferenceless coded aperture correlation holography (I-COACH). Opt. Express **26**, 18143–18154 (2018)
165. Rai, M.R., Rosen, J.: Noise suppression by controlling the sparsity of the point spread function in interferenceless coded aperture correlation holography (I-COACH). Opt. Express **27**, 24311–24323 (2019)
166. Wan, Y., Liu, C., Ma, T., Qin, Y., lv, S.: Incoherent coded aperture correlation holographic imaging with fast adaptive and noise-suppressed reconstruction. Opt. Express **29**, 8064–8075 (2021)
167. Rai, M.R., Vijayakumar, A., Rosen, J.: Single camera shot interferenceless coded aperture correlation holography. Opt. Lett. **42**, 3992–3995 (2017)
168. Hai, N., Rosen, J.: Doubling the acquisition rate by spatial multiplexing of holograms in coherent sparse coded aperture correlation holography. Opt. Lett. **45**, 3439–3442 (2020)
169. Kumar, M., Vijayakumar, A., Rosen, J., Matoba, O.: Interferenceless coded aperture correlation holography with synthetic point spread holograms. Appl. Opt. **59**, 7321–7329 (2020)
170. Rai, M.R., Vijayakumar, A., Rosen, J.: Extending the field of view by a scattering window in an I-COACH system. Opt. Lett. **43**, 1043–1046 (2018)
171. Vijayakumar, A., Kashter, Y., Kelner, R., Rosen, J.: Coded aperture correlation holography system with improved performance. Appl. Opt. **56**, F67–F77 (2017)
172. Jiang, Q., Jin, G., Cao, L.: When metasurface meets hologram: principle and advances. Adv. Opt. Photonics **11**, 518–576 (2019)
173. Matsushima, K.: Introduction to Computer Holography. Springer, Berlin (2020)
174. Tricoles, G.: Computer generated holograms: an historical review. Appl. Opt. **26**, 4351–4360 (1987)
175. Tsang, P.W.M., Poon, T.-C., Wu, Y.M.: Review of fast methods for point-based computer-generated holography. Photonics Res. **6**, 837–846 (2018)
176. Zhang, S.: High-Speed 3D Imaging with Digital Fringe Projection Techniques. CRC Press, Boca Raton (2019)
177. Geng, J.: Structured-light 3D surface imaging: a tutorial. Adv. Opt. Photonics **3**, 128–160 (2011)
178. Xu, J., Zhang, S.: Status, challenges, and future perspectives of fringe projection profilometry. Opt. Lasers Eng. **135**, 106193 (2020)

179. Marrugo, A.G., Gao, F., Zhang, S.: State-of-the-art active optical techniques for three-dimensional surface metrology: a review Invited. J. Opt. Soc. Am. A **37**, B60–B77 (2020)
180. Zhang, S.: Absolute phase retrieval methods for digital fringe projection profilometry: a review. Opt. Lasers Eng. **107**, 28–37 (2018)
181. Zhang, S.: High-speed 3D shape measurement with structured light methods: a review. Opt. Lasers Eng. **106**, 119–131 (2018)
182. https://www.lynceetec.com
183. https://www.tomocube.com
184. https://phioptics.com
185. https://www.phasics.com/en/
186. http://celloptic.com
187. Poon, T.C., Liu, J.P.: Introduction to Modern Digital Holography: With Matlab. Cambridge University Press, Cambridge (2014)
188. Kim, M.K.: Digital Holographic Microscopy: Principles, Techniques, and Applications. Springer, Berlin (2011)
189. Malacara, D., Servín, M., Malacara, Z.: Interferogram Analysis for Optical Testing, 2nd edn. CRC Press, Boca Raton (2016)
190. Popescu, G.: Quantitative phase imaging of cells and tissues. McGraw-Hill Education, New York City (2011)

Chapter 4
Non-interferometric Quantitative Optical Phase Imaging

Interferometry-based quantitative phase imaging methods support precise phase measurements and are widely used for optical metrology and biological imaging. However, these methods rely on the high degree of coherence of light beams to achieve interference, and their accuracy heavily depends on the quality of optical elements adopted and the stability of working environment. Thus, it is difficult to apply interferometric methods for imaging with short wavelength including X-ray and high-energy electron beam, where high quality optical elements are not available, and the coherence of radiation beam is much lower than that of common laser. Furthermore, the high requirement of interferometry on the mechanical stability of instrument strictly limits their applications in lots of circumstances. Non-interferometric methods, including coherent diffraction imaging, phase diversity and transport of intensity equation (TIE), etc., can provide a powerful alternative solution to quantitative phase imaging problems without using complex optical alignment and highly coherent illumination. This chapter outlines the principle of such non-interferometric quantitative phase imaging techniques systematically followed by several numerical simulation in MATLAB®.

4.1 Coherent Diffraction Imaging

The quest for ever-higher spatial resolution has prompted imaging technologies to adopt shorter wavelength illumination with X-rays and high-energy electron beams. Despite the fact that most phase measurements are performed using interferometry [1], the lack of high quality optical elements in short wavelength ranges makes it difficult to generate high quality reference beams for short wavelength interferometry. For short wavelengths, high quality optical components such as mirrors, lenses, and beam splitters are still unavailable. Zone platers are still used extensively for X-ray imaging [2], and spherical aberration is still a big problem in electron microscopes [3]. New lensless microscopy methods combined with advanced numerical methods

© The Author(s), under exclusive license to Springer Nature Singapore Pte Ltd. 2022   109
C. Liu et al., *Computational Optical Phase Imaging*, Progress in Optical Science
and Photonics 21, https://doi.org/10.1007/978-981-19-1641-0_4

for image processing and raster scanning can overcome several drawbacks of conventional imaging. In recent years, coherent diffraction imaging (CDI) and its extension Ptychography have made significant technological progress for achieving high resolution lensless imaging for the characterization of a wide range of samples. This section discusses several algorithms and imaging methodologies that have enabled important scientific breakthroughs across numerous disciplines.

Since the phase distribution $\varphi(x_0, y_0)$ of a light field $U(x_0, y_0)$ on x_0–y_0 plane is encoded into its intensity distribution $I(x, y)$ on x–y plane at a distance Δz, there exists a possibility to retrieve the complex light field $U(x_0, y_0)$ including $\varphi(x_0, y_0)$ from the recorded intensity $I(x, y)$. This section will introduce coherent diffraction imaging (CDI) algorithms used for this purpose [4–8]. Since the phase information is totally lost in x–y plane, the reconstruction of complex amplitude cannot be realized analytically, and iterative computations are required to obtain a converging solution. Classical coherent diffraction imaging algorithms such as G-S (Gerchburg-Saxton) algorithm [4], ER (Error Reduction) algorithm, and HIO (Hybrid Input–Output) algorithm [5] have the advantages of simple experimental setup and faster data acquisition. However, their convergence reliability is low, especially in imaging object with complex structures. To improve the performance of these classical CDI algorithms, Ptychographic Iterative Engine (PIE) algorithm has been proposed by combining the principles of classical ptychography and classical coherent diffraction imaging [6]. Compared to ER and HIO algorithms of Fienup, PIE has both faster convergence and higher reconstruction quality. At present, PIE has been successfully realized in the regimes of visible light [9], X-ray [10], high energy electron beam [11], E-UV [12], and Terahertz [13]. PIE in the Fourier domain has been realized recently and achieved a higher resolution imaging with wide field of view by using objectives of lower numerical aperture [14].

4.1.1 G-S Algorithm

Figure 4.1a schematically shows the optical alignment of classical G-S algorithm [4]. The sample $O(x_0, y_0)$ is illuminated by parallel laser beam with uniform intensity

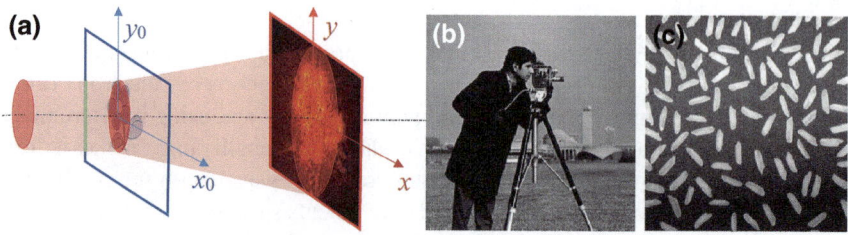

Fig. 4.1 Principle of G-S algorithm **a** G-S scheme; **b** transmitting modulus and **c** phase of the sample used for numerical simulation

and plane wave-front. The transmitted light leaving x_0–y_0 plane is $U_0(x_0, y_0)$ and it forms a diffracted field in x–y plane $U(x, y)$ at a distance z. If the intensities $|U_0(x_0, y_0)|^2$ and $|U_1(x, y)|^2$ are recorded on x_0–y_0 plane and x–y plane, respectively, the phase $\varphi(x_0, y_0)$ of $U_0(x_0, y_0)$ and the phase $\varphi_1(x, y)$ of $U_1(x, y)$ can be iteratively reconstructed with the following steps, starting with an random initial guess for $\varphi_0(x_0, y_0)$.

(1) Propagate the complex amplitude $U_0(x_0, y_0)$ from x_0-y_0 plane by a distance of z to x–y plane to get a the complex amplitude of $\left|U_1'(x, y)\right|e^{i\varphi_1'(x,y)}$.

(2) Replace $\left|U_1'(x, y)\right|$ with the square root of recorded intensity $|U_1(x, y)|^2$ and keep the phase term $e^{i\varphi_1'(x,y)}$ unchanged.

(3) Back propagate the obtained light field $|U_1(x, y)|e^{i\varphi_1'(x,y)}$ from x–y plane to x_0–y_0 plane by a distance of $-z$ to get a complex amplitude of $\left|U_0'(x_0, y_0)\right|e^{i\varphi_0'(x_0,y_0)}$.

(4) Calculate the reconstruction error with $\varepsilon = \dfrac{\iint \left|\left|U_0'(x_0,y_0)\right|-|U_0(x_0,y_0)|\right| dx_0 dy_0}{\iint |U_0(x_0,y_0)| dx_0 dy_0}$. If ε is smaller than target reconstruction error, the iterative reconstruction stops, else go to Step (5).

(5) Replacing $\left|U_0'(x_0, y_0)\right|$ with the square root of recorded intensity $|U_0(x_0, y_0)|^2$ and keep the phase term $e^{i\varphi_1'(x_0,y_0)}$ unchanged to get an improved $U_0(x_0, y_0)$ of x_0–y_0 plane.

(6) Jump to step (1) to start another round of iterative computation.

In above iterative computations, light fields including $U_0(x_0, y_0)$ and $U_1(x, y)$ are discrete matrices of complex value and can be written as $U_0(m\Delta x_0, n\Delta y_0)|_{m=1...M,n=1...N}$ and $U_1(m\Delta x, n\Delta y)|_{m=1...M,n=1...N}$, respectively. Thus their propagation between x_0–y_0 plane and x–y plane is computed with discrete Fresnel diffraction formula. For $\Delta x = \Delta x_0$ and $\Delta y = \Delta y_0$, the Fresnel diffraction formula using transfer function are

$$U_0(m\Delta x_0, n\Delta y_0)$$
$$= F^{-1}\left[F[U_1(m\Delta x, n\Delta y)]e^{-i\pi\lambda z[(m_1\Delta k_x)^2+(n_1\Delta k_y)^2]}\right]\Big|_{m,m_1=1...M,n,n_1=1...N}$$

$$U_1(m\Delta x, n\Delta y)$$
$$= F^{-1}\left[F[U_0(m\Delta x_0, n\Delta y_0)]e^{i\pi\lambda z[(m_1\Delta k_{x_0})^2+(n_1\Delta k_{y_0})^2]}\right]\Big|_{m,m_1=1...M,n,n_1=1\cdots N}$$

$$(4.1)$$

where F and F^{-1} are Fourier and inverse Fourier transforms, Δx and Δy and sampling intervals at x–y plane, and Δx_0 and Δy_0 are sampling intervals at x_0–y_0 plane. $e^{-i\pi\lambda z[(m_1\Delta k_x)^2+(n_1\Delta k_y)^2]}$ and $e^{i\pi\lambda z[(m_1\Delta k_{x_0})^2+(n_1\Delta k_{y_0})^2]}$ are transfer functions for forward and backward propagations of light between x–y plane and x_0–y_0 plane, respectively. $\Delta k_x = \frac{1}{N\Delta x}$, $\Delta k_y = \frac{1}{M\Delta y}$, $\Delta k_{x_0} = \frac{1}{N\Delta x_0}$, $\Delta k_y = \frac{1}{M\Delta y_0}$. In computations with Fresnel diffraction formula using transfer function, Δx_0 and Δy_0 are equal to Δx and Δy, respectively.

Fig. 4.2 Reconstructed modulus image and phase. **a** Diffraction pattern; **b** reconstructed transmitting modulus; **c** reconstructed phase; **d** reconstruction error

The Fresnel diffraction using spatial integral is

$$U_1(m\Delta x, n\Delta y) = \frac{1}{i\lambda z}e^{i\frac{1}{\lambda z}[(m\Delta x)^2+(n\Delta y)^2]}$$
$$F\left[U_0(m\Delta x_0, n\Delta y_0)e^{i\frac{1}{\lambda z}[(m\Delta x_0)^2+(n\Delta y_0)^2]}\right]\Bigg|_{m,m_1=1...M,n,n_1=1...N}$$
$$U_0(m\Delta x_0, n\Delta y_0) = \frac{1}{i\lambda z}e^{-i\frac{1}{\lambda z}[(m\Delta x_0)^2+(n\Delta y_0)^2]}$$
$$F^{-1}\left[U_1(m\Delta x, n\Delta y)e^{-i\frac{1}{\lambda z}[(m\Delta x_0)^2+(n\Delta y_0)^2]}\right]\Bigg|_{m,m_1=1...M,n,n_1=1...N}$$

$$(4.2)$$

where $\Delta x_0 = \frac{\lambda z}{M\Delta x}$, $\Delta y_0 = \frac{\lambda z}{N\Delta y}$, accordingly sampling intervals on x–y plane and x_0–y_0 plane are not equal to each other in most of the cases when Fresnel diffraction formula in the form of spatial integral is used.

To show the properties of G-S algorithm, a set of simulations are carried out by using an object with a transmitting phase and amplitude shown in Fig. 4.1b and c, respectively. The size of the sample is 1792 μm × 1792 μm and is sampled into a matrix of 256 × 256 at an interval of 7 μm. Figure 4.2a shows the diffraction patterns formed at a plane with distance of 3 cm from the object.

The reconstruction follows the above procedure with the following reconstruction code in Matlab. The reconstructed transmitting modulus and phase are shown in Fig. 4.2b and c, respectively. The reconstruction error with respect to the number of iteration is shown in Fig. 4.2d, where we can find that the reconstruction error has reduced to 1% after 400 iterations, showing a quite good convergence speed.

Matlab Code for G-S Algorithm

```
1    % computational phase imaging
2    % G-S algorithm
3    clc;clear;
4    Phase=zeros(512);
5    Modul=zeros(512);
6    [x y]=meshgrid(-256:255,-256:255);
7    PixelSize=7; %um
8    Distance=7e4;%um
9    Lambda=0.6;%um
10   k=2*pi/Lambda;% wave vector
11   Delt_k=1/512/PixelSize;
12   temp=double(imread('cameraman.tif'));
13   temp=temp/max(max(temp))*pi  ;%make phase between 0-pi
14   Phase(128:383,128:383)=temp; % Put phase into center of matrix
15   temp=double(imread('rice.png'));
16   temp=temp/max(max(temp))+0.1; %make module between 0.1-1.1
17   Modul(128:383,128:383)=temp; % Put module into center of matrix
18   Trans=@(z)exp(1i*k*z)*exp(-
     1i*pi*Lambda*z*((x*Delt_k).^2+(y*Delt_k).^2)); %Transfer
     function
19   U=Modul.*exp(1i*Phase);          %U is the transmitted light
20   I1=abs(U).^2;                   % Intensity on sample plane
21   U_ft=fftshift(fft2(fftshift(U)));
22   I2=abs(ifftshift(ifft2(ifftshift(U_ft.*Trans(Distance))))).^2;
     % Intensity on diffraction plane.
23   U1=sqrt(I1).*exp(1i*rand(512));              % initial
     guess to the phase of transmitted light;
24   n=1;
25   while 1
26       U1_ft=fftshift(fft2(fftshift(U1))); %Fourier transform of
     computed transmitted light
27
     U2=ifftshift(ifft2(ifftshift(U1_ft.*Trans(Distance)))); %field
     on the second plane
28       U2=sqrt(I2).*exp(1i*angle(U2)); % Intensity constraint on
     the second plane
29       U2_ft=fftshift(fft2(fftshift(U2)));% Fourier transform of
     light on second plane
30       U1=ifftshift(ifft2(ifftshift(U2_ft.*Trans(-Distance)))); %
     Back propagate to the first plane
31       error(n)=sum(sum(abs(abs(U1)-
     abs(U))))/sum(sum(abs(U))); %error computation
32       U1=sqrt(I1).*exp(1i*angle(U1)); % Intensity constraint on
     the first plane and
33       if (error(n)<0.001)
34           break
35       end
36       drawnow;
37       figure(1);
38       imagesc(angle(U1(128:383,128:383)));
39       colormap('gray');
40       n=n+1;
41   end
42       figure(2);
43       plot(error);
```

4.1.2 ER and HIO Algorithms of Fienup

G-S algorithm has a faster convergence, however, it requires two frames of intensities that are recorded at two parallel planes separated by a distance. This makes the data acquisition relatively complex and makes the G-S algorithm work only for imaging static object. Furthermore, for imaging with short wavelengths such as X-ray and high electron beam, diffraction patterns recorded in both planes will have similar intensity distributions except a magnification coefficient since both detectors are in Fraunhofer region in most of cases. Similar intensity patterns in both planes result in a lower convergence speed. To simplify G-S algorithm, Fienup proposed improved CDI method using Error Reduction (ER) algorithm and Hybrid Input and Output (HIO) algorithm. Both ER and HIO algorithms use the same optical alignment as shown in Fig. 4.3, where the field of view of the sample under inspection is limited by a tiny aperture $A(x_0, y_0)$. $A(x_0, y_0)$ takes values of 1.0 and 0.0 when (x_0, y_0) is within and outside the aperture, respectively. A detector is placed at a proper distance behind the object to record the diffraction intensity. An outstanding advantage of Fienup's method is that only a single frame of diffraction intensity is required and hence the data acquisition and optical alignment become simple. After the diffraction intensity $|U(x, y)|^2$ is recorded, an initial random guess is given to the complex amplitude $U_0(x_0, y_0)$ of the object to begin the iterative reconstruction following the ER algorithm.

(1) Multiply the complex amplitude $U_0(x_0, y_0)$ with the aperture function $A(x_0, y_0)$ and propagate the obtained product from x_0–y_0 plane by a distance of z to x–y plane to get the complex amplitude of $|U_1(x, y)|e^{i\varphi_1(x,y)}$.

(2) Calculate the reconstruction error with $E_{rror} = \frac{\iint ||U_R(x,y)| - |U_1(x,y)||dxdy}{\iint |U_1(x,y)|dxdy}$. If E_{rror} is smaller than the target reconstruction error, the iterative reconstruction stops, else go to step (3).

(3) Replace the $|U_1(x, y)|$ with the recorded $|U_R(x, y)|$ and keep the phase term $e^{i\varphi_1(x,y)}$ unchanged to get improved $|U_R(x, y)|e^{i\varphi_1(x,y)}$.

(4) Propagate the obtained light field $U_1'(x, y) = |U_R(x, y)|e^{i\varphi_1(x,y)}$ from x–y plane to x_0–y_0 plane by a distance of $-z$ to get an updated complex amplitude of $U_0'(x_0, y_0) = |U_0(x_0, y_0)|e^{i\varphi_0'(x_0, y_0)}$

(5) Jump to Step (1) to start another round of iterative computation.

Fig. 4.3 Principle of Fienup's algorithm

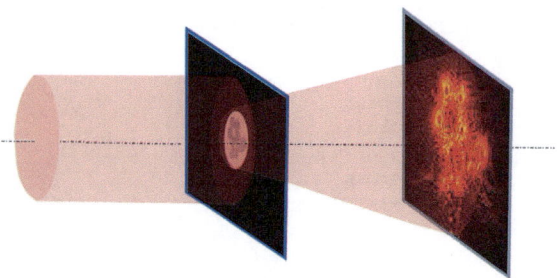

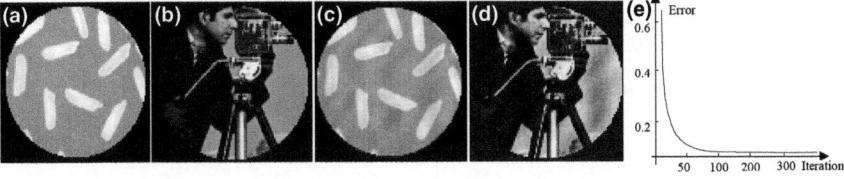

Fig. 4.4 Numerical simulation on Fienup's ER algorithm. **a** Transmitting modulus and **b** phase; **c** reconstructed modulus and **d** phase; **e** reconstruction error

To show the performance of ER algorithm, a set of simulations were carried out by using an object with a transmitting modulus and phase as shown in Fig. 4.4a and b, respectively, where the field of view of the object is limited by a circular aperture of 700 μm in diameter. The reconstructed modulus and phase of the object are shown in Fig. 4.4c and d, respectively, where obvious reconstruction errors can be found in the back ground of two images. The reconstruction with respect to number of iteration is shown in Fig. 4.4e, where we can find that the reconstruction error has reduced to 15% after 150 iterations, and the final residual reconstruction error is about 13%. It is worthy to point out that, in the simulations shown Fig. 4.4, the range of transmission phase of object was set to $[0, 0.25\pi]$. If the transmitting phase varies too drastically, no satisfying reconstruction can be obtained. Experimentally this would mean that the performance of ER algorithm is not good for imaging samples with complex structures.

Matlab code for ER algorithm of Fienup

```
1   % computational phase imaging
2   % ER algorithm of Fienup
3     clc;clear;
4   Phase=zeros(1024);
5   Modul=zeros(1024);
6   [x y]=meshgrid(-512:511,-512:511);
7   PixelSize=7; % Pixel size of detector in μm
8   Distance=30e4;% distance between object to detector in μm
9   Lambda=0.6;% wavelength of light in μm
10  k=2*pi/Lambda;% wave vector
11  Delt_k=1/1024/PixelSize; % sampling interval in K-space
12  temp=double(imread('cameraman.tif'));
13  temp=temp/max(max(temp))*pi/4; % Make phase between 0-pi/4
14  Phase(385:640,385:640)=temp; % Move phase into the center of
    matrix
15  temp=double(imread('rice.png'));
16  temp=temp/max(max(temp))+0.5; % Make module between 0.5-1.5
17  Modul(385:640,385:640)=temp; % Move module into the center of
    matrix
18  Trans=@(z)exp(1i*k*z)*exp(-
    1i*pi*Lambda*z*((x*Delt_k).^2+(y*Delt_k).^2)); %Transfer
    function
```

```
19  Aperture=(((x).^2+(y+25).^2)<2500); % Aperture for generating
    diffraction pattern
20  U=Modul.*exp(1i*Phase).*Aperture;   % U is the transmitted light
21  U_ft=fftshift(fft2(fftshift(U)));
22  I2=abs(ifftshift(ifft2(ifftshift(U_ft.*Trans(Distance))))).^2;
    % Intensity on diffraction plane.
23  U1=rand(1024);                      % Initial guess of phase of
    transmitted light;
24  n=1;
25  while 1
26      U1_ft=fftshift(fft2(fftshift(U1))); %Fourier transform of
    computed transmitted light
27
    U2=ifftshift(ifft2(ifftshift(U1_ft.*Trans(Distance))));  %Field
    on the second plane
28      error(n)=sum(sum(abs(abs(U2)-
    sqrt(I2))))/sum(sum(sqrt(I2)))  % Error computation
29      if (error(n)<0.012)
30          break
31      end
32      U2=sqrt(I2).*exp(1i*angle(U2)); % Intensity constraint on
    the second plane
33      U2_ft=fftshift(fft2(fftshift(U2)));% Fourier transform of
    light on second plane
34      U=ifftshift(ifft2(ifftshift(U2_ft.*Trans(-Distance)))); %
    Back propagate to the first plane
35      U1=(U).*Aperture;
36      drawnow;
37      figure(1);
38      imagesc(angle(U1(385:640,385:640)));
39      colormap('gray');
40      n=n+1;
41  end
42      figure(2);
43      plot(error);
```

In practice, the edge of the aperture $A(x_0, y_0)$ that is limiting the field of view cannot be exactly known in advance and hence the aperture used in the iterative computation is always a little larger than the actual one in most of the cases. In the above simulations, the convergent speed of ER algorithm is much lower than that of G-S algorithm. Generally speaking, larger aperture always means lower reconstruction speed. To improve the convergence speed of ER algorithm, Fienup proposed another modified reconstruction method called Hybrid-Input–Output algorithm, which uses the same optical alignment to record the diffraction patterns, however, in iterative computation the aperture function $A'(x_0, y_0)$ in Eq. (4.3) is applied in the first step of iteration to replace the real aperture $A(x_0, y_0)$, where ε is a real constant that is much smaller than 1.0.

$$A'(x_0, y_0) = \begin{cases} 1.0 \text{ when } (x_0, y_0) \in A(x_0, y_0) \\ \varepsilon \text{ when } (x_0, y_0) \notin A(x_0, y_0) \end{cases} \tag{4.3}$$

The reconstruction algorithm using modified aperture function $A'(x_0, y_0)$ has much faster convergence than ER algorithm and is called Hybrid-Input–Output algorithm (HIO) [5]. To show this advantage of HIO algorithm, the reconstruction was carried out by using HIO algorithm to deal with the diffraction patterns used in the

Fig. 4.5 Simulation with HIO algorithm. **a** Reconstructed modulus and **b** phase; **c** reconstruction error

simulations in Fig. 4.4, and the reconstructed images and reconstruction error are shown in Fig. 4.5. We can find that the reconstruction error of HIO is reduced to 12% after 60 iterations. The reason why HIO algorithm can have a faster convergence is explained by Fienup in one of his works [5].

Matlab code for ER algorithm of Fienup

```
1   % computational phase imaging
2   % ER algorithm of Fienup
3   clc;clear;
4   Phase=zeros(1024);
5   Modul=zeros(1024);
6   [x y]=meshgrid(-512:511,-512:511);
7   PixelSize=7; %um
8   Distance=30e4;%um
9   Lambda=0.6;%um
10  k=2*pi/Lambda;% wave vector
11  Delt_k=1/1024/PixelSize;
12  temp=double(imread('cameraman.tif'));
13  temp=temp/max(max(temp))*pi/4;  % Make phase between 0-pi/4
14  Phase(385:640,385:640)=temp;  % Move phase to the center of
    matrix
15  temp=double(imread('rice.png'));
16  temp=temp/max(max(temp))+0.5; % Make module between 0.5-1.5
17  Modul(385:640,385:640)=temp;  % Move module to the center of
    matrix
18  Trans=@(z)exp(1i*k*z)*exp(-
    1i*pi*Lambda*z*((x*Delt_k).^2+(y*Delt_k).^2)); %Transfer
    function
19  Aperture=(((x).^2+(y+25).^2)<2500); % Aperture for generating
    diffraction pattern
20  Aperture2=(((x).^2+(y+25).^2)<3600); % A little larger
    aperture for reconstruction
```

```
21   Aperture2=(Aperture2-Aperture)/10+Aperture; % ε takes a value
     of 0.1
22   U=Modul.*exp(1i*Phase).*Aperture;  %U is the transmitted light
23   U_ft=fftshift(fft2(fftshift(U)));
24   I2=abs(ifftshift(ifft2(ifftshift(U_ft.*Trans(Distance))))).^2;
     % Intensity on recording plane.
25   U1=rand(1024);                      % Initial guess of the phase
     of transmitted light;
26   n=1;
27   while 1
28       U1_ft=fftshift(fft2(fftshift(U1))); %Fourier transform of
     computed transmitted light
29
     U2=ifftshift(ifft2(ifftshift(U1_ft.*Trans(Distance)))); %Fiel
     d on the recording plane
30       error(n)=sum(sum(abs(abs(U2)-
     sqrt(I2))))/sum(sum(sqrt(I2))) % Error computation
31       if (error(n)<0.012)
32           break
33       end
34       U2=sqrt(I2).*exp(1i*angle(U2)); % Intensity constraint on
     the recording plane
35       U2_ft=fftshift(fft2(fftshift(U2)));% Fourier transform of
     light on recording plane
36       U=ifftshift(ifft2(ifftshift(U2_ft.*Trans(-Distance)))); %
     Back propagate to the first plane
37       U1=(U).*Aperture; % spatial constraint on object plane
38       drawnow;
39       figure(1);
40       imagesc(angle(U1(385:640,385:640)));
41       colormap('gray');
42       n=n+1;
43   end
44       figure(2);
45       plot(error);
```

The reconstruction accuracy of both HIO and ER algorithms is inversely proportional to the size of the aperture $A(x_0, y_0)$. To show this, a series of simulations were carried out by using the object transmitting phase and amplitude shown in Fig. 4.1. The simulated result is shown in Fig. 4.6, where Fig. 4.6a–c are the reconstructed images with aperture diameters of 500 μm, 600 μm, and 700 μm, respectively. We can find that with the increasing size of the aperture, the reconstruction quality becomes worse.

For a given sample, the reconstruction quality not only depends on the aperture size but also depends on the shape of aperture. Figure 4.6d shows reconstructed modulus and phase of the same sample with a larger aperture of quite irregular shape. We can find that two reconstructed images are much clearer than that of Fig. 4.6b and c. This phenomenon has been proved experimentally with X-ray [15], however the underlying physics is yet to be explored.

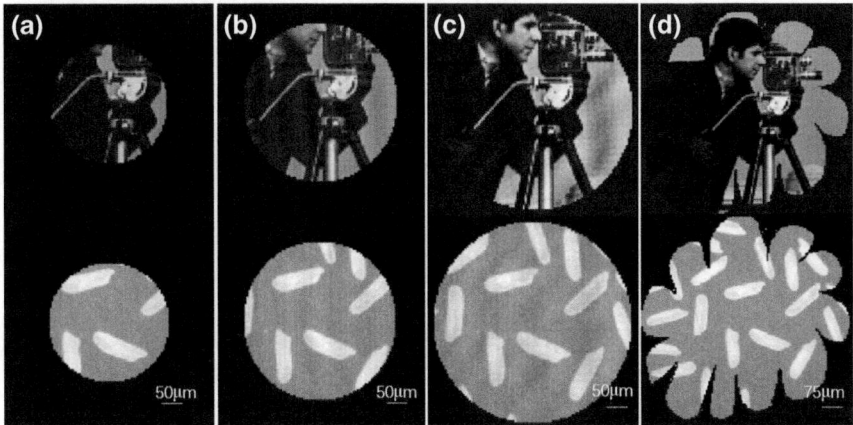

Fig. 4.6 Simulations with apertures of different size. Aperture diameter of **a** 500 μm; **b** 600 μm; **c** 700 μm; **d** aperture with irregular shape

4.1.3 Existence of Convergence in Iterative CDI Algorithms

In coherent diffraction imaging, only the intensity of diffraction patterns is recorded, and its phase information is lost. At the first glance, there are countless phase distributions that can match with the intensity recorded. However, with classic CDI algorithms, including G-S algorithm and ER algorithm, a convergent numerical solution can always be generated. Fienup explained the reason why classic CDI algorithm can generate a convergent numerical solution mathematically [5], and Heinz H. Bauschke explained this from a view of convex optimization [16]. Let's assume that $\mathbf{u}_1(x_1, y_1)$ is the complex amplitude on the diffraction plane, and it is assumed to be the Fourier transform of the complex amplitude $\mathbf{u}_0(x_0, y_0)$ on the object plane. Their relationships in discrete form are

$$\mathbf{u}_1(m_1, n_1) = \sum_{m_0=0}^{M} \sum_{n_0=0}^{N} \mathbf{u}_0(m_0, n_0) e^{-i2\pi \left(\frac{m_0 m_1}{M} + \frac{n_0 n_1}{N} \right)}$$

$$\mathbf{u}_0(m_0, n_0) = \frac{1}{MN} \sum_{m_1=0}^{M} \sum_{n_1=0}^{N} \mathbf{u}_1(m_1, n_1) e^{i2\pi \left(\frac{m_0 m_1}{M} + \frac{n_0 n_1}{N} \right)}$$

(4.4)

Since $\mathbf{u}_1(m_1, n_1)$ and $\mathbf{u}_0(m_0, n_0)$ are complex valued, they can be written as $\mathbf{u}_1(m_1, n_1) = |\mathbf{u}_1(m_1, n_1)| e^{i\varphi_1(m_1,n_1)}$ and $\mathbf{u}_0(m_0, n_0) = |\mathbf{u}_0(m_0, n_0)| e^{i\varphi_0(m_0,n_0)}$, respectively. In the kth iteration, the complex amplitude on object plane is $\mathbf{u}_{0k}(m_0, n_0) = |\mathbf{u}_0(m_0, n_0)| e^{i\varphi_{0k}(m_0,n_0)}$, its Fourier transform $\mathbf{u}_{1k}(m_1, n_1) = |\mathbf{u}_{1k}(m_1, n_1)| e^{i\varphi_{1k}(m_1,n_1)}$ can be updated as $\mathbf{u}'_{1k}(m_1, n_1) = |\mathbf{u}_1^R(m_1, n_1)| e^{i\varphi_{1k}(m_1,n_1)}$, and then the complex amplitude on object plane becomes $\mathbf{u}'_{0k}(m_0, n_0) = |\mathbf{u}'_{0k}(m_0, n_0)| e^{i\varphi'_{0k}(m_0,n_0)}$, which is the Fourier transform of $\mathbf{u}'_{1k}(m_1, n_1)$. Finally, this complex amplitude is updated as $\mathbf{u}_{0k+1}(m_0, n_0) = |\mathbf{u}_{0k+1}(m_0, n_0)| e^{i\varphi'_{0k}(m_0,n_0)}$, where $|u_{0k+1}(m_0, n_0)|$ is the recorded intensity $\left|u_0^R(m_0, n_0)\right|$ in G-S algorithm or $\left|u_{0k}'(m_0, n_0)\right|$ limited by aperture $A(x_0, y_0)$ in ER algorithm.

The squared error of the kth iteration on object plane is

$$E_{0k}^2 = \sum_{m_0, n_0} \left| \mathbf{u}_{0k+1}(m_0, n_0) - \mathbf{u}_{0k}'(m_0, n_0) \right|^2 \tag{4.5}$$

By Parseval's theorem, Eq. (4.5) can be written as [17]

$$E_{0k}^2 = (NM)^{-1} \sum_{m_1, n_1} \left| \mathbf{u}_{1k+1}(m_1, n_1) - \mathbf{u}_{1k}'(m_1, n_1) \right|^2 \tag{4.6}$$

The squared error of the kth iteration on diffraction plane is

$$E_{1k}^2 = (NM)^{-1} \sum_{m_1, n_1} \left| \mathbf{u}_{1k}(m_1, n_1) - \mathbf{u}_{1k}'(m_1, n_1) \right|^2$$

$$= \sum_{m_0, n_0} \left| \mathbf{u}_{0k}(m_0, n_0) - \mathbf{u}_{0k}'(m_0, n_0) \right|^2 \tag{4.7}$$

Since at any coordinate the value of $\mathbf{u}_{0k+1}(m_0, n_0)$ is averagely closer to $\mathbf{u}_{0k}'(m_0, n_0)$ than that of $\mathbf{u}_{0k}(m_0, n_0)$, thus

$$\left| \mathbf{u}_{0k+1}(m_0, n_0) - \mathbf{u}_{0k}'(m_0, n_0) \right| \leq \left| \mathbf{u}_{0k}(m_0, n_0) - \mathbf{u}_{0k}'(m_0, n_0) \right| \tag{4.8}$$

From Eq. (4.8), Eq. (4.7), and Eq. (4.5),

$$E_{0k}^2 \leq E_{1k}^2 \tag{4.9}$$

Since, and the value of $\mathbf{u}_{1k+1}(m_1, n_1)$ is closer to $\mathbf{u}_{1k+1}'(m_1, n_1)$ than to $\mathbf{u}_{1k}'(m_1, n_1)$.

$$\left| \mathbf{u}_{1k+1}(m_1, n_1) - \mathbf{u}_{1k+1}'(m_1, n_1) \right| < \left| \mathbf{u}_{1k+1}(m_1, n_1) - \mathbf{u}_{1k}'(m_1, n_1) \right| \tag{4.10}$$

$$E_{1k+1}^2 \leq E_{0k}^2 \tag{4.11}$$

By combing Eq. (4.11) and Eq. (4.9) we can find

$$E_{1k+1}^2 \leq E_{0k}^2 \leq E_{1k}^2 \tag{4.12}$$

The reconstruction error becomes smaller and smaller as the number of iterations increases, eventually leading to a convergent solution.

4.1.4 Equivalence of ER Algorithm and Steepest-Decent Method

Previous section has explained the capability of CDI algorithm to generate a converging solution. What makes us believe that the obtained convergent numerical solution is the correct complex amplitude? This reason is also explained by Fienup in 1982 [5]. For simplicity in analysis, we can assume the square error on diffraction plane as

$$B_k \equiv E_{1k}^2 = (NM)^{-1} \sum_{m_1,n_1} \left[|\mathbf{u}_{1k}(m_1, n_1)| - |\mathbf{u}_1^R(m_1, n_1)| \right]^2 \tag{4.13}$$

From the point of view of optimization, the objective of iterative reconstruction of ER algorithm is to search for a $\mathbf{u}_0(m_0, n_0)$ step by step to get a minimum B from an initial guess. In the kth iteration, the optimum step size to be taken in the direction of gradient can be determined by forming a first-order Taylor series expansion of B about $\mathbf{u}_0(m_0, n_0)$ at the point $\mathbf{u}_{0k}(m_0, n_0)$.

$$B \approx B_k + \sum_{m_0,n_0} \frac{\partial B_k}{\partial \mathbf{u}_0(m_0, n_0)} [\mathbf{u}_0(m_0, n_0) - \mathbf{u}_{0k}(m_0, n_0)] \tag{4.14}$$

This first-order expansion of B is equal to zero at

$$\mathbf{u}_0(m_0, n_0) = \mathbf{u}_0^{\text{opt}}(m_0, n_0)\mathbf{u}_0^{\text{opt}}(m_0, n_0) - \mathbf{u}_{0k}(m_0, n_0)$$
$$= -\frac{B_k}{\sum_{m_0,n_0} \left(\frac{\partial B_k}{\partial \mathbf{u}_0(m_0,n_0)} \right)^2} \left(\frac{\partial B_k}{\partial \mathbf{u}_0(m_0, n_0)} \right) \tag{4.15}$$

Equation (4.15) can be verified by inserting it into Eq. (4.14). The partial derivative of B_k with respect to a value at a given point $\mathbf{u}_0(m_0, n_0)$ is

$$\frac{\partial B}{\partial \mathbf{u}_0(m_0, n_0)} \equiv 2(NM)^{-1} \sum_{m_1,n_1} \left[|\mathbf{u}_{1k}(m_1, n_1)| - |\mathbf{u}_1^R(m_1, n_1)| \right] \frac{\partial |\mathbf{u}_{1k}(m_1, n_1)|}{\partial \mathbf{u}_0(m_0, n_0)} \tag{4.16}$$

$$\frac{\partial \mathbf{u}_{1k}(m_1, n_1)}{\partial \mathbf{u}_0(m_0, n_0)} = \frac{\partial}{\partial \mathbf{u}_0(m_0, n_0)} \sum_{n_0,m_0=0}^{N} \mathbf{u}_0(m_0, n_0) e^{-i2\pi \left(\frac{m_0 m_1}{M} + \frac{n_0 n_1}{N} \right)} = e^{-i2\pi \left(\frac{m_0 m_1}{M} + \frac{n_0 n_1}{N} \right)} \tag{4.17}$$

since

$$
\frac{\partial |\mathbf{u}_{1k}(m_1, n_1)|}{\partial \mathbf{u}_0(m_0, n_0)} = \frac{\partial \sqrt{|\mathbf{u}_{1k}(m_1, n_1)|^2}}{\partial \mathbf{u}_0(m_0, n_0)} = \frac{1}{2|\mathbf{u}_{1k}(m_1, n_1)|} \frac{\partial |\mathbf{u}_{1k}(m_1, n_1)|^2}{\partial \mathbf{u}_0(m_0, n_0)}
$$
$$
= \frac{\mathbf{u}_{1k}(m_1, n_1)e^{i2\pi\left(\frac{m_0 m_1}{M} + \frac{n_0 n_1}{N}\right)} + \mathbf{u}_{1k}^*(m_1, n_1)e^{-i2\pi\left(\frac{m_0 m_1}{M} + \frac{n_0 n_1}{N}\right)}}{2|\mathbf{u}_{1k}(m_1, n_1)|}
$$

$$(4.18)$$

Equation (4.16) becomes

$$
\frac{\partial B}{\partial \mathbf{u}_0(m_0, n_0)}
$$
$$
\equiv (NM)^{-1} \sum_{m_1, n_1} \left[\mathbf{u}_{1k}(m_1, n_1) - |\mathbf{u}_1^R(m_1, n_1)| \frac{\mathbf{u}_{1k}(m_1, n_1)}{|\mathbf{u}_{1k}(m_1, n_1)|} \right] e^{i2\pi\left(\frac{m_0 m_1}{M} + \frac{n_0 n_1}{N}\right)}
$$
$$
+ (NM)^{-1} \sum_{m_1, n_1} \left[\mathbf{u}_{1k}^*(m_1, n_1) - |\mathbf{u}_1^R(m_1, n_1)| \frac{\mathbf{u}_{1k}^*(m_1, n_1)}{|\mathbf{u}_{1k}(m_1, n_1)|} \right] e^{-i2\pi\left(\frac{m_0 m_1}{M} + \frac{n_0 n_1}{N}\right)}
$$

$$(4.19)$$

Since $|\mathbf{u}_1^R(m_1, n_1)| \frac{\mathbf{u}_{1k}(m_1, n_1)}{|\mathbf{u}_{1k}(m_1, n_1)|} = \mathbf{u}'_{1k}(m_1, n_1)$, Eq. (4.19) is in the form of a discrete Fourier transform, it can be reduced to Eq. (4.20) under the assumption that $\mathbf{u}_0(m_0, n_0)$ has real value.

$$
\frac{\partial B_k}{\partial \mathbf{u}_0(m_0, n_0)} = 2\left[\mathbf{u}_{0k}(m_0, n_0) - \mathbf{u}'_{0k}(m_0, n_0) \right] \tag{4.20}
$$

Under the assumption of $\mathbf{u}_0(m_0, n_0)$ takes no-negative and real value

$$
\sum_{m_0, n_0} \left(\frac{\partial B_k}{\partial \mathbf{u}_0(m_0, n_0)} \right)^2 = 4\left[\mathbf{u}_{0k}(m_0, n_0) - \mathbf{u}'_{0k}(m_0, n_0) \right]^2 = 4B_k \tag{4.21}
$$

Then Eq. (4.15) can be written as

$$
\mathbf{u}_0^{\text{opt}}(m_0, n_0) - \mathbf{u}_{0k}(m_0, n_0) = \frac{1}{4}\frac{\partial B_k}{\partial \mathbf{u}_0(m_0, n_0)} = -\frac{1}{2}\left[\mathbf{u}_{0k}(m_0, n_0) - \mathbf{u}'_{0k}(m_0, n_0) \right] \tag{4.22}
$$

Above searching step in obtained with the first-order linear expansion, however since B is quadratic in $\mathbf{u}_0(m_0, n_0)$, the linear approximation above can be expected to predict a step size half as large as the optimum. Therefore, we should use the double-length step obtained in Eq. (4.22), that is

$$
\mathbf{u}_0^{\text{opt}}(m_0, n_0) - \mathbf{u}_{0k}(m_0, n_0) = \mathbf{u}'_{0k}(m_0, n_0) - \mathbf{u}_{0k}(m_0, n_0) \tag{4.23}
$$

Equation (4.23) means that the updated complex amplitude $\mathbf{u}'_{0k}(m_0, n_0)$ in object domain in each iteration of ER algorithm is the searched value $\mathbf{u}_0^{opt}(m_0, n_0)$ of each step in the steepest-decent method. In other words, ER algorithm is mathematically the steepest-decent method, and this is the reason why the convergent numerical solution represents the correct light field.

4.1.5 Ptychographic Iterative Engine

Classical CDI imaging algorithms followed a simple optical alignment for a faster data acquisition and were widely used in the field of material science and biomedical engineering to determine the microscopic structure of samples. Modern applications in emerging fields of science require imaging of more complex structures. Slower convergence of most of the GS and Fienup algorithms along with restricted field of view limited its applications in several fields and were inadequate to meet the new challenges in the field of phase retrieval imaging.

Ptychographic Iterative Engine (PIE) was proposed as an alternative to conventional CDI methods [6]. PIE scans the sample under inspection using a localized illuminating beam at various raster positions and records the corresponding diffraction patterns that are formed by each sample positions. The transmission function of the sample, including both the phase and modulus, is faithfully reconstructed from these recorded diffraction patterns using an updating formula developed for this purpose. Since more than a hundred frames of diffraction patterns are recorded in PIE, the data redundancy is quite high, which led to the development of ePIE algorithm that is capable of reconstructing both the sample and illumination simultaneously [18]. An automatic position correction is introduced using a quenching algorithm [19] and maximum correlation searching algorithm [20]. The multi-mode algorithm [21, 22] is applied to overcome the incoherency of illumination beam and vibration of the imaging system. By now, PIE has been realized with X-ray, visible light, high energy electron beam, and deep ultraviolet.

The concept of ptychography was proposed by late Walter Hoppe between 1968 and 1973 [23–25], and its basic idea is illustrated in Fig. 4.7. A sample with a

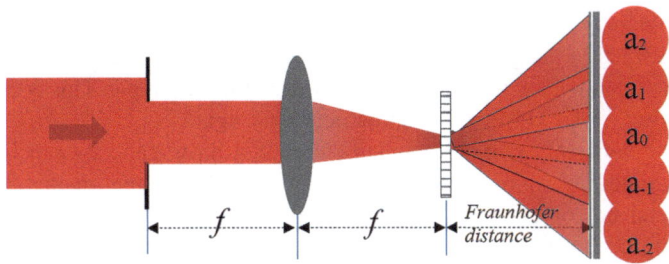

Fig. 4.7 Principle of classical ptychography

roughly periodic structure $\mathbf{O}(x_0, y_0)$ is illuminated by a focused laser beam $\mathbf{illu}_0(x_0, y_0)$, the diffraction pattern appears in the Fraunhofer diffraction region is the Fourier transform of the exiting field of the sample. When the divergent angle of $\mathbf{illu}_0(x_0, y_0)$ is small enough, a grid of small circular diffraction disks can be expressed in the form $\mathbf{U}(k_x, k_y) = \sum_{m,n} a_{mn} \exp(i\varphi_{mn}) \delta(k_x - m\Delta k_x, k_y - m\Delta k_y) \text{cir}\left(\alpha\sqrt{k_x^2 + k_y^2}\right)$. The intensity a_{mn} can be directly recorded by the detector, however, the phase $\exp(i\varphi_{mn})$ is lost during the recording restricting a direct calculation of the complex amplitude of the exiting light that can reveal the structure of sample. It is possible to adjust the amount of overlap in the diffraction pattern by increasing the divergent angle of the illuminating beam $\mathbf{illu}_0(x_0, y_0)$. The intensity within the overlapping region of two neighboring diffraction disks can be written as

$$I_{m\_n,m+1\_n} = \left|a_{m,n}\right|^2 + \left|a_{m+1,n}\right|^2 + 2\left|a_{m,n}\right|\left|a_{m+1,n}\right|\cos\left(\varphi_{m,n} - \varphi_{m+1,n}\right) \quad (4.24)$$

Though $I_{m\_n,m+1\_n}$ depends on the phase difference $\Delta\varphi = \varphi_{m,n} - \varphi_{m+1,n}$, we can not determine $\Delta\varphi$ directly from $I_{m\_n,m+1\_n}$, $\left|a_{m,n}\right|$ or $\left|a_{m+1,n}\right|$, which are recorded in Fig. 4.7. Since $\cos\Delta\varphi$ is a symmetric function of $\Delta\varphi$ in the range of $[-\pi, \pi]$, we cannot distinguish between $\Delta\varphi$ and $-\Delta\varphi$ from the value of $\cos\Delta\varphi$. If we shift the sample by a distance of Δx, a corresponding shift appears in the phase φ_{mn} of the mnth disk that is equal to $\Delta m\Delta k_x \Delta x$ and the new phase difference between neighboring disks becomes

$$I'_{m\_n,m+1\_n} = \left|a_{m,n}\right|^2 + \left|a_{m+1,n}\right|^2 + 2\left|a_{m,n}\right|\left|a_{m+1,n}\right|\cos\left(\varphi_{m,n} - \varphi_{m+1,n} + \Delta k_x \Delta x\right)$$
$$(4.25)$$

From Eqs. (4.24) and (4.25), we obtain the phase difference between the neighboring disks which is further used to calculate the phase $\varphi_{m,n}$ of all diffraction patterns in sequence. Finally, the structure of the periodic sample can be determined by calculating the inverse Fourier transform of $\sum_{m,n} a_{mn} \exp(i\varphi_{mn}) \delta(k_x - m\Delta k_x, k_y - m\Delta k_y)$. For a periodic sample, the phase change in diffraction patterns can be analytically calculated from change in diffraction intensity occurring while shifting the sample. However, this is not the case with samples with no periodic structures where we will have several equations from numerous overlapping diffraction disks on the recording plane that is to be solved to obtain $\varphi_{m,n}$. Only an iterative reconstruction approach similar to classic coherent diffraction imaging would be capable of retrieving the phase information from diffraction intensity variation caused by moving the sample. This is essentially the underlying physics of Ptychographic Iterative Engine.

A sample with only non-periodic structures will have many Δk_x in Eq. (4.25) to be determined by using illumination of very small divergent angles, and these equations can be solved by shifting the sample to many different distances and by analyzing each diffraction pattern formed. As a practical matter, however, the analytical method using Eqs. (4.24) and (4.25) is not applicable to non-periodic samples, since all diffraction disks overlap. As phase information of all diffraction

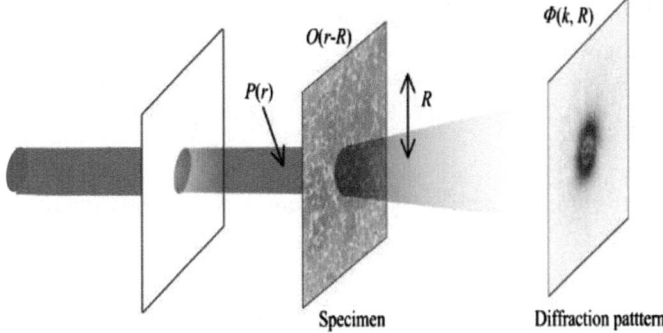

Fig. 4.8 Basic principle of PIE

disks is still involved in the change in diffraction intensity when sample is shifted, iterative reconstruction can be used to find the sample's transmission function from all recorded diffraction disks. Essentially, this is how the Ptychographic Iterative Engine works.

The setup of PIE is schematically shown in Fig. 4.8, where the object $O\left(\vec{r} - \vec{R_n}\right)$ is fixed on a translation stage and can scan through a known localized radiation beam of $P(x, y)$ to many positions, where $\vec{R_n}$ is the nth position of the object. The diffraction patterns I_n are recorded when the object is at the position $\vec{R_n}$. During the shifting of sample, neighboring illuminated regions should be partially overlapped during the scanning of the object. After the data acquisition is finished, iterative calculations are carried out between object and recording planes to retrieve the complex transmittance of the object by following the procedure below. It starts with an initial random guess given to $O\left(\vec{r} - \vec{R_n}\right)$ [6].

(1)   The exit field $U\left(\vec{r}, \vec{R_n}\right)$ of the object at the position $\vec{R_n}$ is calculated

$$U\left(\vec{r}, \vec{R_n}\right) = P(\vec{r}) \cdot O\left(\vec{r} - \vec{R_n}\right) \tag{4.26}$$

(2)   Light field $\Psi\left(\vec{r}, \vec{R_n}\right)$ on the data recording plane is calculated using the Fresnel diffraction formula as

$$\Psi\left(\vec{r}, \vec{R_n}\right) = \left[U\left(\vec{r}, \vec{R_n}\right)\right] = \left|\Psi\left(\vec{r}, \vec{R_n}\right)\right| e^{i\varphi\left(\vec{r}, \vec{R_n}\right)} \tag{4.27}$$

(3)   The amplitude of $\Psi\left(\vec{r}, \vec{R_n}\right)$ is replaced by the square root of I_n:

$$\Psi\left(\vec{r}, \vec{R_n}\right) = \sqrt{I_n}\, e^{i\varphi\left(\vec{r}, \vec{R_n}\right)} \tag{4.28}$$

(4)   Update the exit field $U\left(\vec{r}, \vec{R_n}\right)$ of the object by propagating $\Psi\left(\vec{r}, \vec{R_n}\right)$ back to the object plane

$$U'\left(\vec{r}, \vec{R_n}\right) = F^{-1}\left[\Psi\left(\vec{r}, \vec{R_n}\right)\right] \tag{4.29}$$

(5)   Update the object function $O\left(\vec{r}, \vec{R_n}\right)$ with the following formula

$$O'\left(\vec{r} - \vec{R_n}\right) = O\left(\vec{r} - \vec{R_n}\right) + \frac{|P(\vec{r})|}{|P(\vec{r})|_{\max}} \frac{P^*(\vec{r})}{\left[|P(\vec{r})|^2 + \alpha\right]}$$
$$\left[U'\left(\vec{r}, \vec{R_n}\right) - P(\vec{r}) \cdot O\left(\vec{r} - \vec{R_n}\right)\right] \tag{4.30}$$

(6)   Change to $(n+1)$th position and repeat the above Steps 1–5 until all diffraction patterns are addressed.
(7)   Calculate the reconstruction error

$$\varepsilon = \int \left|O'\left(\vec{r} - \vec{R_n}\right) - O\left(\vec{r} - \vec{R_n}\right)\right|d\vec{r} \Big/ \int \left|O\left(\vec{r} - \vec{R_n}\right)\right|d\vec{r} \tag{4.31}$$

(8)   If ε is smaller than the required value, the reconstruction stops, else, start another round of reconstruction form the first scanning position.

In the above PIE reconstruction algorithm, the illumination function $P(\vec{r})$ should be known accurately in advance, however, this is quite difficult in experiments. To solve this problem, e-PIE algorithm was proposed to reconstruct both the illumination and object simultaneously [26]. In e-PIE algorithm, the above Eq. (4.30) is composed of two counterpart formula as

$$O'\left(\vec{r} - \vec{R_n}\right) = O\left(\vec{r} - \vec{R_n}\right) + \frac{|P(\vec{r})|}{|P(\vec{r})|_{\max}} \frac{P^*(\vec{r})}{\left[|P(\vec{r})|^2 + \alpha\right]}$$
$$\left[U'\left(\vec{r}, \vec{R_n}\right) - P(\vec{r}) \cdot O\left(\vec{r} - \vec{R_n}\right)\right]$$
$$P(\vec{r}) = P(\vec{r}) + \frac{\left|O\left(\vec{r} - \vec{R_n}\right)\right|}{\left|O\left(\vec{r} - \vec{R_n}\right)\right|_{\max}} \frac{O^*\left(\vec{r} - \vec{R_n}\right)}{\left[\left|O\left(\vec{r} - \vec{R_n}\right)\right|^2 + \alpha\right]}$$
$$\left[U'\left(\vec{r}, \vec{R_n}\right) - P(\vec{r}) \cdot O\left(\vec{r} - \vec{R_n}\right)\right] \tag{4.32}$$

Since $P(\vec{r})$ is not required to be known in e-PIE algorithm, this greatly improves the accuracy of reconstructed image and the convenience of the experiment, thus, in most of the works, e-PIE algorithm is adopted to replace standard PIE algorithm to perform iterative reconstruction.

Since the sample is transversely scanned in PIE, the field of view is much larger than that of common microscopy and that of classical coherent diffractive imaging methods. On the other hand, since tens or hundreds of diffraction patterns are recorded in PIE, the information redundancy of PIE is quite high, and thus the imaging quality of PIE is very good in most of the cases when compared to other single shot phase imaging methods, including holography.

A set of simulations were carried out to show the properties of PIE. The light probe incident on the sample was formed by illuminating a pinhole of 420 μm in diameter with a parallel laser beam of 0.7 μm in wavelength. The distance of the scanning sample to pinhole and that to detector are 2 cm and 18 cm, respectively. Figures 4.9a and b show the modulus and phase of the light probe incident on the scanning sample. Figure 4.9c shows a single frame of diffraction patterns formed when the sample with transmitting phase and modulus shown in Fig. 4.2 were scanned to 7 × 7 positions. Figure 4.9d and e show the reconstructed phase and modulus of reconstructed light probe, respectively. Figure 4.9f and g show the reconstructed phase and modulus of reconstructed sample, respectively. Figure 4.9h shows the reconstruction error with number of iterations. The reconstruction code of ePIE is listed as follows.

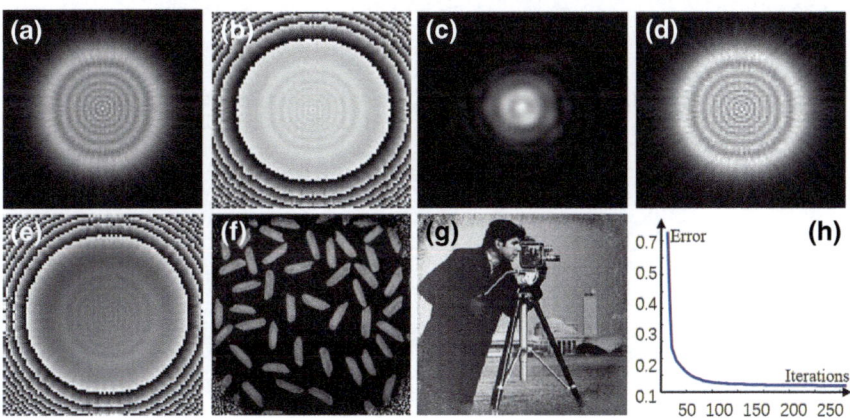

Fig. 4.9 Simulations on PIE. **a** Setting probe modulus and **b** phase; **c** example of a single frame of diffraction pattern; **d** reconstructed probe modulus and **e** phase; **f** reconstructed sample modulus and **g** phase; **h** reconstruction error

Matlab Code for ePIE algorithm

```
1    % computational phase imaging
2    % ePIE algorithm
3    clc;clear;
4    Phase=zeros(512);
5    Modul=zeros(512);
6    [x y]=meshgrid(-256:255,-256:255);
7    PixelSize=7; %um
8     Distance=30e4; %um
9    Lambda=0.6;%um
10   k=2*pi/Lambda;% wave vector
11   Delt_k=1/1024/PixelSize;
12   temp=double(imread('cameraman.tif'));
13   temp=temp/max(max(temp))*pi/4;  % Make phase between 0-pi
14   Phase(128:383,128:383)=temp;  % Put phase into center of matrix
15   temp=double(imread('rice.png'));
16   temp=temp/max(max(temp))+0.5;  % Make modul between 0.1-1.1
17   Modul(128:383,128:383)=temp;  % Put modul into center of matrix
18   Trans=@(z)exp(1i*k*z)*exp(-
     1i*pi*Lambda*z*((x*Delt_k).^2+(y*Delt_k).^2)); %Transfer
     function
19   Aperture=(((x).^2+(y).^2)<900);% Aperture to generate
     illuminating probe
20   probe=ifftshift(ifft2(fftshift(ifftshift(fft2(fftshift(Aperture
     ))).*Trans(2e4))));%Generaring the illuminating light probe
21   sample=Modul.*exp(1i*Phase); % forming sample
22   I(1:11,1:11,1:512,1:512)=0; % recorded diffraction intensity
23   imagesc(abs(probe));
24   for m=-5:5
25      for n=-5:5
26          S=zeros(512);
27          S(256-100:256+100,256-100:256+100)=sample(256+m*10-
     100:256+m*10+100,256+n*10-100:256+n*10+100);% scanne the sample
28          T_ft=fftshift(fft2(fftshift(S.*probe))); % Fourier
     transform of transmitted light
29
     I(m+6,n+6,1:512,1:512)=abs(ifftshift(ifft2(ifftshift(T_ft.*Tran
     s(20e4))))).^2; % compute the diffraction intensity
30      end
31   end
32
33   sample=rand(512);  % initial guess of smaple
34   probe=rand(512);   % initial guess of illumination
35   error(1:500)=0;    % computation error
36
37   for k=1:500
38      t=sample;
39   for m=-5:5
40      for n=-5:5
41
42          S=S.*0;
```

```
43        S(256-100:256+100,256-100:256+100)=sample(256+m*10-
     100:256+m*10+100,256+n*10-100:256+n*10+100);%scanning of the
     sample
44        T0=S.*probe; % transmited light
45        T_ft=fftshift(fft2(fftshift(S.*probe))); % Fourier
     transform of transmited light
46        diff=ifftshift(ifft2(ifftshift(T_ft.*Trans(20e4))));%
     diffraction on recording plane
47        temp(1:512,1:512)=sqrt(I(m+6,n+6,1:512,1:512));
48        diff=temp.*exp(1i*angle(diff)); % undate the light on
     recording plane
49        diff_ft=fftshift(fft2(fftshift(diff))); % Fourier
     transform of the undated diffraction
50        T=ifftshift(ifft2(ifftshift(diff_ft.*Trans(-20e4)))); %
     Transmited light copmuted by back propagation
51        probe2=probe+(T-
     T0)./(abs(S).^2+0.1).*conj(S).*abs(S)/max(max(abs(S))); %
     updating the illuminating probe
52        S=S+(T-
     T0)./(abs(probe).^2+0.1).*conj(probe).*abs(probe)/max(max(abs(p
     robe))); % updating the sample
53        probe=probe2;
54        sample(256+m*10-100:256+m*10+100,256+n*10-
     100:256+n*10+100)=S(256-100:256+100,256-100:256+100);% put the
     undated part to its original position
55        drawnow;
56        figure(1);
57        imagesc(abs(sample(185:300,185:330)));
58      end
59   end
60   error(k)=sum(sum(abs(t-sample)))/sum(sum(abs(sample)));
61   figure(2)
62   drawnow;
63   plot(error);
64   end
```

4.1.6 Fourier Ptychographic Microscopy

Ptychographic Iterative Engine (PIE) can also be realized in Fourier domain with the optical setup shown in Fig. 4.10, where a sample with a transmission function $O(x, y)$ is imaged by a lens of small numerical aperture and with a parallel laser beam of the form $A_0 \exp[ik_0(x_0 \cos \alpha + y_0 \cos \beta)]$. By changing the incident angle of illumination beam to a series of $(\alpha_n, \beta_n)|_{n=1:N}$, a series of images $I_n (x_2, y_2)|_{n=1:N}$ are recorded.

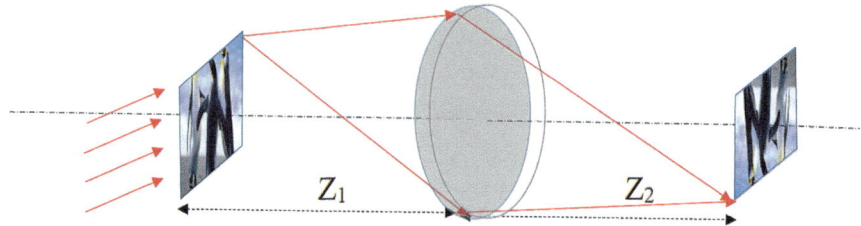

Fig. 4.10 Principle of FPM

By assuming $k_0 \cos \alpha_n$ and $k_0 \cos \beta_n$ as k_x^n and k_y^n, respectively, the transmitted light of the sample $O(x, y)$ under the illumination of $A_0 \exp[ik_0(x_0 \cos \alpha_n + y_0 \cos \beta_n)]$ is $A_0 O(x, y) \exp[i(k_x^n x + k_y^n y)]$, and the light field incident on the front surface of lens is

$$U_L(x_1, y_1) \propto \exp\left(ik_0 \frac{x_1^2 + y_1^2}{2z_1}\right) \iint O(x_0, y_0) \exp\left[i\left(k_x^n x_0 + k_y^n y_0\right)\right]$$
$$\exp\left(ik_0 \frac{x_0^2 + y_0^2}{2z_1}\right) \exp\left(ik_0 \frac{x_1 x_0 + y_1 y_0}{z_1}\right) dx_0 dy_0$$

$$U_L(x_1, y_1) \propto \exp\left(ik_0 \frac{x_1^2 + y_1^2}{2z_1}\right) \iint O(x_0, y_0) \exp\left[i\left(k_x^n x_0 + k_y^n y_0\right)\right]$$
$$\exp\left(ik_0 \frac{x_0^2 + y_0^2}{2z_1}\right) \exp\left(ik_0 \frac{x_1 x_0 + y_1 y_0}{z_1}\right) dx_0 dy_0$$

$$= \exp\left(ik_0 \frac{x_1^2 + y_1^2}{2z_1}\right) F\left\{O(x_0, y_0) \exp\left[i\left(k_x^n x_0 + k_y^n y_0\right)\right] \exp\left(ik_0 \frac{x_0^2 + y_0^2}{2z_1}\right)\right\}$$
$$\tag{4.33}$$

Assuming $z_1 = z_2 = 2f$, where f is the focal length of the imaging lens, the light field on the image plane is

$$U^n(x_1, y_1) \propto \exp\left(ik_0 \frac{x_1^2 + y_1^2}{2z_1}\right)$$
$$F\left\{P(x, y) F\left\{O(x_0, y_0) \exp\left(ik_0 \frac{x_0^2 + y_0^2}{4f}\right) \exp[ik_0(\cos \alpha_n x_0 + \cos \beta_n y_0)]\right\}\right\}$$
$$\tag{4.34}$$

Assuming $O(x_0, y_0)\exp\left(ik_0\frac{x_0^2+y_0^2}{4f}\right)$ as $O'(x_0, y_0)$,

$$U^n(x_2, y_2) \propto \exp\left(ik_0\frac{x_2^2+y_2^2}{2z_1}\right)F\{P(x_1, y_1)F\{O'(x_0, y_0)\exp[ik_0(\cos\alpha_n x_0 + os\beta_n y_0)]\}\}$$

$$= \exp\left(ik_0\frac{x_0^2+y_0^2}{4f}\right)F\left\{P(x_1, y_1)\widetilde{O'}\left(\frac{x_1}{2\lambda f} - k_0\cos\alpha_n, \frac{y_1}{2\lambda f} - k_0\cos\beta_n\right)\right\} \tag{4.35}$$

Thus, with a change of incident angle $(\alpha_n, \beta_n)|_{n=1:N}$ of the illumination beam, the spatial frequency components of $\tilde{O}'\left(\frac{x_1}{2\lambda f} - k_0\cos\alpha_n, \frac{y_1}{2\lambda f} - k_0\cos\beta_n\right)$ will transversely scan through the aperture $P(x_1, y_1)$ of the imaging lens. When the aperture of $P(x_1, y_1)$ is small, recorded images $I_n(x_2, y_2) = |U^n(x_2, y_2)|^2$ are of lower resolution. Ptychographic iterative engine can be realized in Fourier domain to reconstruct a high resolution image $O(x, y)$ of the sample from these recorded low resolution images of $I_n(x_2, y_2)|_{n=1:N}$ by following the computational steps as below:

(1) Give an initial guess of the transmission function of the sample to reconstruct as $O(x, y)$ and compute its Fourier transform as $\tilde{O}(f_x, f_y)$, and give an initial uniform guess to the aperture function as $P(f_x, f_y)$.

(2) Shift $\tilde{O}(f_x, f_y)$ distances of $k_0\cos\alpha_n$ and $k_0\cos\beta_n$ in directions of f_x and f_y, respectively, obtaining $\tilde{O}(f_x - k_0\cos\alpha_n, f_y - k_0\cos\beta_{ny})$.

(3) Multiply $P(f_x, f_y)$ with $\tilde{O}(f_x - k_0\cos\alpha_n, f_y - k_0\cos\beta_{ny})$ and generate an image of $O'_n(x, y)$ as $O'_n(x, y) = F\left(P(f_x, f_y)\tilde{O}(f_x - k_0\cos\alpha_n, f_y - k_0\cos\beta_{ny})\right) = |O'_n(x, y)|e^{i\varphi(x,y)}$.

(4) Update $O'_n(x, y)$ by replacing its modulus with the square root of the recorded intensity $I_n(x, y)$ and keep its phase unchanged $O'_n(x, y) = \sqrt{I_n(x, y)}e^{i\varphi(x,y)}$.

(5) Make Inverse Fourier transform on $O'_n(x, y)$ to obtain $\tilde{O}'_n(f_x, f_y) = F^{-1}\left(\sqrt{I_n(x, y)}e^{i\varphi(x,y)}\right)$.

(6) Update $\tilde{O}(f_x, f_y)$ as

$$\tilde{O}(f_x, f_y) = \tilde{O}(f_x, f_y)$$
$$+ \frac{\tilde{O}'_n(f_x, f_y) - P(f_x, f_y)\tilde{O}(f_x - k_0\cos\alpha_n, f_y - k_0\cos\beta_{ny})}{|P(f_x, f_y)|^2 + \alpha}P$$
$$*(f_x, f_y)\frac{|P(f_x, f_y)|}{|P(f_x, f_y)|_{max}}$$

$$P(f_x, f_y) = P(f_x, f_y)$$
$$+ \frac{\tilde{O}'_n(f_x, f_y) - P(f_x, f_y)\tilde{O}(f_x - k_0\cos\alpha_n, f_y - k_0\cos\beta_{ny})}{|\tilde{O}(f_x, f_y)|^2 + \alpha}\tilde{O}$$

$$* \left(f_x, f_y\right) \frac{\left|\tilde{O}\left(f_x, f_y\right)\right|}{\left|\tilde{O}\left(f_x, f_y\right)\right|_{\max}} \qquad (4.36)$$

(7)   Jump to step 2 to repeat the above computations with regard to another illumi-
       nation angle corresponding to another recorded intensity until all $I_n(x, y)$ are
       addressed.

(8)   Compute reconstruction ε with Eq. (4.36). If E_{rror} becomes smaller than target
       value, the iterative reconstruction is completed, else jump to Step 2 to start
       another round of iteration.

$$\varepsilon = \sum_{x,y} \left|\left|\tilde{O}_n\left(f_x, f_y\right)\right| - \left|\tilde{O}_{n-1}\left(f_x, f_y\right)\right|\right| \Big/ \sum_{x,y} \left|\tilde{O}_{n-1}\left(f_x, f_y\right)\right| \qquad (4.37)$$

The numerical aperture of imaging lens is 0.1 and two images in Fig. 4.2b are
used as modulus and phase of sample. 11×11 frames of low resolution images were
recorded by changing illumination angle in the range $[-65°, 65°]$ with respect to
both x and y axis at an interval of $5°$. Figure 4.11a shows some of the generated low
resolution images under different illuminating angles, Fig. 4.11b shows reconstructed
modulus and phase of pupil function of $P(x, y)$, Fig. 4.11c shows the reconstructed
transmitting modulus and phase of sample, and Fig. 4.11d shows the reconstructed
error with number of iterations.

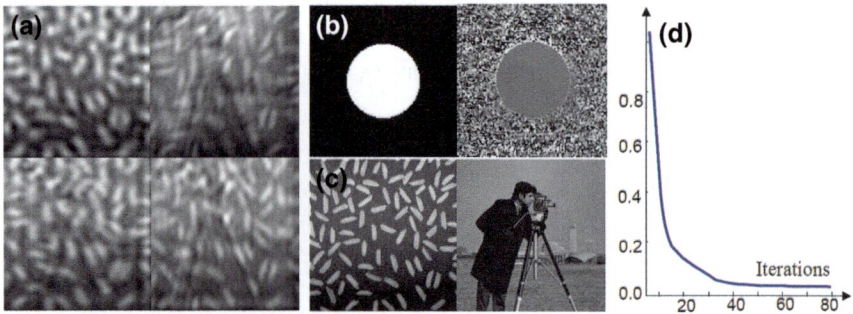

Fig. 4.11 Simulations of FPM. **a** Generated low resolution images under different illuminating
angles; **b** reconstructed modulus and phase of pupil function; **c** reconstructed transmitting modulus
and phase of sample; **d** reconstruction error

Matlab Code for FPM algorithm

```
1    % computational phase imaging
2    % FPM algorithm
3    clc;clear;
4    Phase=zeros(512);
5    Modul=zeros(512);
6    sample=zeros(512);
7    [x y]=meshgrid(-256:255,-256:255);
8    PixelSize=7; %um
9    Distance=30e4;%um
10   Lambda=0.6;%um
11   k=2*pi/Lambda;% wave vector
12   temp=double(imread('rice.png'));
13   temp=temp/max(max(temp))+0.5; % Make modulus between 0.5-1.5
14   sample(128:383,128:383)=temp;
15   temp=double(imread('cameraman.tif'));
16   temp=temp/max(max(temp))*pi/4; % Make phase between 0-pi/4
17   sample(128:383,128:383)=sample(128:383,128:383).*exp(1i*temp);
     % Put modulus into center of matrix
18
19   Trans=@(z)exp(1i*k*z)*exp(-
     1i*pi*Lambda*z*((x*Delt_k).^2+(y*Delt_k).^2)); %Transfer
     function
20   Aperture=(((x).^2+(y).^2)<1000);% Aperture to generate
     illuminating probe
21   I(1:21,1:21,1:512,1:512)=0; % Recorded diffraction intensity
22   Delt_k=fix(sin(2.5/180*pi)*256); % Shifting interval in K-space
     corresponding to 2.5 degrees
23   sample_ft=fftshift(fft2(fftshift(sample))); % Fourier transform
     of sample
24   S_sample_ft=sample_ft(162:35,162:350); % Cut central part of
     Fourier transform of sample
25   M=[0,-1.5,2.7,-3.4,3.2,-1.5,3.3,-3.2,5.5,-3.2,4.7,-5.3,6.2,-
     6.7,8.3,-7.7,6.5,-8.4,7.7,-9.3,10];
26   N=[0,-2.1,1.7,-2.3,1.4,-3.3,2.3,-4.3,4.7,-5.3,6.5,-4.7,3.5,-
     8.5,5.4,-5.7,7.5,-9.5,8.5,-8.7,9.2];
27    % M and N are random for shifting distance of S_sample_ft in K-
     space, using random number can avoid periodical structure in
     reconstructed image and aperture.
28   for m=1:21
29      for n=1:21
30         S=zeros(512);
31         deltx=round(M(m)*Delt_k); %shifting distance in k_x-
     direction
32         delty=round(N(n)*Delt_k); % shifting distance in k_y-
     direction
33         S(256+deltx-144:256+deltx+144,256+delty-
     144:256+delty+144)=sample_ft(256-144:256+144,256-
     144:256+144); % shifting sample_ft
34         temp=abs(ifftshift(ifft2(ifftshift(S.*Aperture)))).^2; %
     compute the diffraction intensity
35         I(m,n,1:512,1:512)=temp; % store the diffraction
     intensity
36      end
```

```
37  end
38  sample(1:512,1:512)=sqrt(I(11,11,1:512,1:512));  % initial guess
    of sample
39  Aperture=double(((x).^2+(y).^2)<1600);  % initial guess of
    aperture larger than the real one
40  sample_ft=fftshift(fft2(fftshift(sample)));  % Compute the
    Fourier frequency of guessed sample
41  error(1:100)=0;    % computation error
42
43  for k=1:100
44      t=sample;  % store the sample before each round of iterative
    computation
45      t0=max(max(abs(Aperture)));
46      Aperture=Aperture/t0;  % Normalization of aperture
47      sample_ft=sample_ft*t0;  % keep filtered energy unchanged
    after normalization on aperture
48    for m=1:21
49      for n=1:21
50        S=S.*0;
51        deltx=round(M(m)*Delt_k);  %shifting distance in k_x-
    direction
52        delty=round(N(n)*Delt_k);  %shifting distance in k_y-
    direction
53        S(256+deltx-144:256+deltx+144,256+delty-
    144:256+delty+144)=sample_ft(256-144:256+144,256-
    144:256+144);  % shifting sample_ft
54        I2=ifftshift(ifft2(ifftshift(S.*Aperture)));% diffraction
    on recording plane
55        temp(1:512,1:512)=sqrt(I(m,n,1:512,1:512));  % intensity
    constraint
56        I2=temp.*exp(1i*angle(I2));  % undate the light on
    recording plane
57        sample_ft2=fftshift(fft2(fftshift(I2)));% Fourier
    transform of the undated diffraction
58        S2=S+(sample_ft2-
    S.*Aperture)./(abs(Aperture).^2+0.01).*conj(Aperture).*abs(Aper
    ture)/max(max(abs(Aperture)));
59        Aperture=Aperture+(sample_ft2-
    S.*Aperture)./(abs(S).^2+0.01).*conj(S).*abs(S)/max(max(abs(S))
    );
60        sample_ft=sample_ft.*0;
61        sample_ft(256-144:256+144,256-144:256+144)=S2(256+deltx-
    144:256+deltx+144,256+delty-144:256+delty+144);% put the
    undated part to its original position
62        sample=ifftshift(ifft2(ifftshift(sample_ft)));  %
    computing the sample
63        drawnow;
64        figure(1);
65        imagesc(abs(sample));
66      % from line 67 to line 83 is another round of computation by
    exchanging deltx and delty
67        S=S.*0;
68        deltx=round(M(n)*Delt_k);
69        delty=round(N(m)*Delt_k);
```

```
70      S(256+deltx-144:256+deltx+144,256+delty-
   144:256+delty+144)=sample_ft(256-144:256+144,256-144:256+144);
71         I2=ifftshift(ifft2(ifftshift(S.*Aperture)));
72         temp(1:512,1:512)=sqrt(I(n,m,1:512,1:512));
73         I2=temp.*exp(1i*angle(I2));
74         sample_ft2=fftshift(fft2(fftshift(I2)));
75         S2=S+(sample_ft2-
   S.*Aperture)./(abs(Aperture).^2+0.01).*conj(Aperture).*abs(Aper
   ture)/max(max(abs(Aperture)));
76         Aperture=Aperture+(sample_ft2-
   S.*Aperture)./(abs(S).^2+0.01).*conj(S).*abs(S)/max(max(abs(S))
   );
77         sample_ft=sample_ft.*0;
78         sample_ft(256-144:256+144,256-144:256+144)=S2(256+deltx-
   144:256+deltx+144,256+delty-144:256+delty+144);
79         sample=ifftshift(ifft2(ifftshift(sample_ft)));
80         drawnow;
81         figure(1);
82         imagesc(abs(sample));
83      end
83   end
85      error(k)=sum(sum(abs(t-sample)))/sum(sum(abs(sample)));  %
   computing reconstruction error
86      figure(2)
87      drawnow;
88      plot(error);
89      figure(3)
90      drawnow;
91      imagesc(abs(Aperture));
92   end
```

4.1.7 Coherent Modulation Imaging

Coherent modulation imaging (CMI) is a newly developed single shot CDI technique, which adopts a highly random phase plate to diffract the light field to be observed into a speckle pattern [27]. When modulation and intensity constraints are used together, CMI reaches rapid convergence, as results conflicting with these two constraints are rapidly eliminated in iterative reconstruction. Since only a single frame of diffraction pattern is required, CMI has higher temporal resolution than PIE and is a promising tool for the observation of dynamical samples.

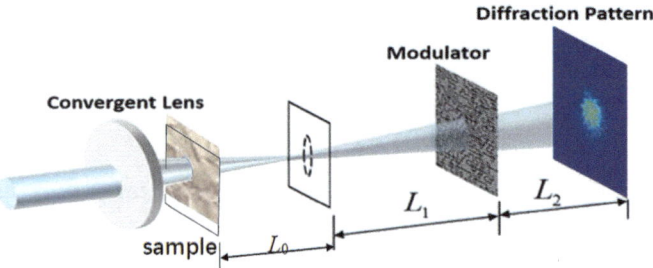

Fig. 4.12 Principle of CMI

Figure 4.12 shows the schematic diagram of standard CMI, where the specimen under detection was illuminated by a convergent radiation beam, and for most of weak specimens, the transmitted radiation will focus around the original focal spot. A highly random phase plate with known structure of $\mathbf{P}(x,y)$ is placed after the focal spot, forming a diffracted speckle patterns $\mathbf{I}(x, y)$ on the detector. Similar to PIE, the complex amplitude $\mathbf{E}(x,y)$ of the illumination beam on the phase plate surface can be reconstructed iteratively with the following procedure with an initial guess given to the light field $U_I(x, y)$ incident on the phase plate $\varphi(x, y)$.

(1) Calculate the transmitted light field of phase plate as $P(x, y)U_I^n(x, y)$ and numerically propagate it to the detector plane as $U_D^n(x, y) = \mathfrak{F}_{L_2}^+\{U_I^n(x, y)\}$. $F_{L_2}^{+1}$ indicates the numerical propagation of light with Fresnel formula by a distance of Z_2 along with the propagation direction of light, and n is the iteration number.

(2) Replace the modulus of $U_D^n(x, y)$ with the square root of recorded intensity $\sqrt{I(x, y)}$ and keep its phase unchanged. $U_D^n(x, y) = \sqrt{I(x, y)}\exp[i\varphi_n(x, y)]$.

(3) Numerically propagate $U_D^n(x, y)$ to the phase plate plane to get updated transmitted light field as $U_T^n(x, y) = F_{L_2}^-\{\sqrt{I(x, y)}\exp[i\varphi_n(x, y)]\}$, where $F_{L_2}^-$ indicates the numerical propagation of light with Fresnel formula by a distance of Z_2 along a direction opposite to the propagating light.

(4) The illumination $U_I^n(x, y)$ incident on phase plate is updated as

$$U_I^n(x, y) = U_I^{n-1}(x, y) + \left[U_T^n(x, y) - U_T^{n-1}(x, y)\right]\frac{P^*(x, y)}{|P(x, y)|^2 + \alpha} \quad (4.38)$$

(5) Propagate $U_I^n(x, y)$ to the focal spot plane and using a small aperture with radius of r to confine it.

$$F_I^n(x, y) = \begin{cases} F_{L_1}^{-1}\{U_I^n(x, y)\} & \sqrt{x^2 + y^2} < r \\ 0 & \sqrt{x^2 + y^2} > r \end{cases} \qquad (4.39)$$

(6) Propagate $F_I^n(x, y)$ to phase plate plane to get a new illumination beam $U_I^{n+1}(x, y)$, and computer the convergence as

$$C = \frac{\sum_{x,y} ||U_I^{n+1}(x, y)| - |U_I^n(x, y)||}{\sum_{x,y} |U_I^{n+1}(x, y)|} \qquad (4.40)$$

(7) Jump to Step (1) to start another round of iteration when C is larger than given value C_0, else, stop the iterative computation and propagate $U_I^{n+1}(x, y)$ the sample plane to get sample's transmitted light $U_0(x, y)$, which can indicate the structure of sample.

$$U_0(x, y) = F_{L0}^-\{U_I^{n+1}(x, y)\} \qquad (4.41)$$

The distance L_0 between sample and focal plane is 8.0 cm, the distance L_1 from focal plane to phase plate is 4.0 cm, the distance L_2 from phase plate to the detector is 2.0 cm. The initial amplitude and phase chosen are the same as in previous simulations. The diffraction patterns on the detector have a structure shown in Fig. 4.13a. Phase and modulus of the transmitted light of sample computed with above illustrated iterative computation procedure from recorded intensity are shown in Figs. 4.13b and c which are roughly the same as original modulus and phase.

Fig. 4.13 Simulations of CMI. **a** Diffraction pattern; **b** reconstructed sample phase; **c** reconstructed sample modulus; **d** reconstruction error

Matlab Code for CMI algorithm

```
1   clc;clear;
2   Modul=zeros(512); Phase=zeros(512);
3   [x y]=meshgrid(-256:255,-256:255);
4   PixelSize=7; %um
5   L0=8e4; %um
6   L1=4e4; %um
7   L2=2e4; %um
8   Lambda=0.6; %um
9   k=2*pi/Lambda;% wave vector
10  Delt_k=1/512/PixelSize;
11  Trans=@(z)exp(1i*k*z)*exp(-
12  1i*pi*Lambda*z*((x*Delt_k).^2+(y*Delt_k).^2));
13  %Transfer function
14  plate=kron(ones(4),fix(rand(128)*2))*pi;
15  plate=exp(1i*plate); %genration of binary pure phae plate of
    pi~0
16  temp=double(imread('cameraman.tif'));
17  temp=temp/max(max(temp))*pi/2; % make phase between 0-pi/4
18  Phase(128:383,128:383)=temp; % Put phase into center of matrix
19  temp=double(imread('rice.png'));
20  temp=temp/max(max(temp))+0.1; %make modul between 0.1-1.1
21  Modul(128:383,128:383)=temp; % Put modul into center of matrix
22  Sample=Modul.*exp(1i*Phase);
23  illu=exp(-1i*sqrt(49*x.^2+49*y.^2+L0.^2)*k).*exp(-
24  (x.^2+y.^2)/40000);
25  T0=illu.*Sample;     % T0 is the transmitted light
26  T0_ft=fftshift(fft2(fftshift(T0)));
27  U1=ifftshift(ifft2(ifftshift(T0_ft.*Trans(L0+L1))));   %  light
28  field on phase plate plane.
29  T1=U1.*plate;       % T1 is transmitted light of phase plate
30  T1_ft=fftshift(fft2(fftshift(T1)));
31  I=abs(ifftshift(ifft2(ifftshift(T1_ft.*Trans(L2))))).^2;     %
    Intensity
    recorded
32  U1=rand(512); % initial guess of light field incident on of
    phase plate;
33  n=1;
34  while 1
35      T1_ft=fftshift(fft2(fftshift(U1.*plate)));         %Fourier
    transform of computed transmited light of phae plate
36      U2=ifftshift(ifft2(ifftshift(T1_ft.*Trans(L2))));  % field
    on the detector plane
37      U2=sqrt(I).*exp(1i*angle(U2)); % Intensity constraint on the
    detector plane
38      U2_ft=fftshift(fft2(fftshift(U2)));% Fourier transform of
    light on detect plane
39      T1=ifftshift(ifft2(ifftshift(U2_ft.*Trans(-L2)))); % Back
    propagate to plate plane
40      U12=U1+(T1-U1.*plate)./(abs(plate)+0.001).*conj(plate); %
    updating the light field incident on phase plate
```

```
41      U12_ft=fftshift(fft2(fftshift(U12)));  %  Fourier  transform
   of the updated light field
42      U_f=ifftshift(ifft2(ifftshift(U12_ft.*Trans(-L1)))); %
   Fourier
   transform of incident light field
43      U_f=U_f.*((x.^2+y.^2)<(100+n/100).^2); % filtering on the
   focal plane
44      U12_ft=fftshift(fft2(fftshift(U_f))); %Fourier transform of
   light on focal plane
45      U12=ifftshift(ifft2(ifftshift(U12_ft.*Trans(L1)))); %
   computating the light incident on phase plate plane
46   error(n)=sum(sum(abs(abs(U12)-abs(U1))))/sum(sum(abs(U1))); %
   computation errof
47   U1=U12; %put updated field into U1 for next iterntion
48   if (error(n)<0.0001)
49       break
50     end
51   U0=ifftshift(ifft2(ifftshift(U12_ft.*Trans(-
   L0)))); %computation of the light flield leaving the
   sample
52      temp=U0./(illu); % computating the sample's transmitting
   function.
53      figure(1);
54      drawnow;
55      imagesc(abs(temp(128:383,128:383)));
56      colormap('gray');
57      figure(2);
58      drawnow;
59      imagesc(angle(temp(128:383,128:383)));
60      colormap('gray');
61      n=n+1;
62   end
63      figure(3);
64      plot(error);
```

4.2 Transport of Intensity Equation Method

Many optical phase imaging methods such as interferometry and coherent diffraction imaging are successful in operating in the optical band using coherent light. However, they can hardly be used in short-wavelength imaging by electron beams and X-rays which are not completely coherent. Transport of intensity equation (TIE) provides a non-interferometric approach to optical phase imaging problems. Though TIE method was initially designed [28] and adopted for adaptive optics in astronomy [29, 30], it soon became a useful tool in short-wavelength imaging such as X-ray imaging [31], transmission electron microscope [32], quantum mechanical wave measurement [33], neutron radiography [34], and Lorentz microscopy [35]. TIE is also compatible with commercial microscopes which do not rely on coherent sources or reference beam. TIE can retrieve the continuous phase without phase wrapping and provide an alternative solution for optical phase imaging. Principle of TIE is discussed in this part along with a MATLAB code that proposes a classical solution. Further, improvements made in phase retrieval algorithms and optical systems are reviewed from both software and hardware perspectives. This part ends with a discussion on pros and cons of TIE.

4.2.1   Theory of Transport of Intensity Equation and Its Classical Solution

Existing optoelectronic sensors are unable to record the frequency of light and hence we cannot directly measure the phase. Encoding the phase into the intensity is a possible solution to extract the phase. Phase information in the form of sinusoidal fringes in interferometry is such an example. In TIE, the specimen phase distorts the incident wavefront and thus induces changes in the intensity distribution. Figure 4.14 schematically reveals the principle of TIE in both geometrical and physical optics perspectives. The intensity distribution hardly changes during propagation when the wavefront is a plane wave, even considering diffraction. A plane wave maintains a unique intensity distribution during its propagation as shown in Fig. 4.14a, where the arrows indicate the rays and the dotted lines represent the wavefront, representations of both geometrical and physical optics. When a specimen with phase gradient is introduced, it induces a phase change and inevitably distorts the wavefront and thus generates an inhomogeneity in the intensity distribution as shown in Fig. 4.14b. It is very similar to the aberration induced intensity distortions. As an example, in high power lasers, the aberration caused by the defects of the system and the thermal stress may induce local foci responsible for laser damages. Inhomogeneity in the intensity

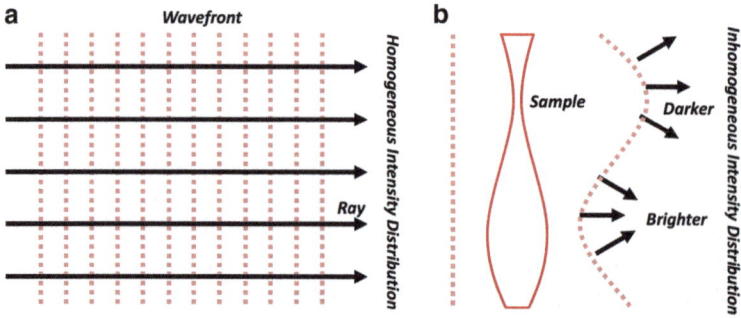

Fig. 4.14 Schematic Description of the TIE in both geometrical and physical optics perspectives. **a** Plane wave generates the unique intensity distribution; **b** distorted wave induces an inhomogeneity in the intensity distribution

distribution in Fig. 4.14b is due to the distorted phase, while the intensity of the plane wave is almost unique as shown in Fig. 4.14a. Therefore, it can be concluded that the phase can affect the intensity, which is the origin of the TIE.

Theoretical background of TIE was first discussed by Teague [28] to extract the phase in a non-interferometric way [28, 36]. It is generally known that the wave propagation in the free space satisfies the Helmholtz equation in Eq. (4.42), in which, $\Psi(r)$ represents a wavefront propagating along the propagation axis to a distance z, r indicates the Cartesian coordinate, ∇^2 is $\partial_x^2 + \partial_y^2 + \partial_z^2$, k is the wave number computed as $2\pi/\lambda$, and λ is the wavelength.

$$\left(\nabla^2 + k^2\right)\Psi(r) = 0 \tag{4.42}$$

$\Psi(r)$ can also be described as Eq. (4.43), in which $\psi(r_\perp, z)$ represents the wavefront distribution and $r_\perp$ indicates the x–y plane of the Cartesian coordinate.

$$\Psi(r_\perp, z) = \exp(ikz)\psi(r_\perp, z) \tag{4.43}$$

Equation (4.44) can be obtained by substituting Eq. (4.43) into Eq. (4.42), and the second partial derivative of $\psi(r_\perp, z)$ with respect to z as $\partial_z^2\psi(r_\perp, z)$ can be approximated to 0 according to paraxial approximation. It should be noted that Eq. (4.44) is actually a parabolic equation.

$$\left(\nabla_\perp^2 + 2ik\partial_z\right)\psi(r_\perp, z) = 0 \tag{4.44}$$

In terms of intensity and phase distributions, I and φ, we can also express $\psi(r)$ as

$$\psi(r_\perp, z) = \sqrt{I(r_\perp, z)}exp[i\varphi(r_\perp, z)] \tag{4.45}$$

We derive TIE from the parabolic equation and Eqs. (4.46) and (4.47) which are conjugates of Eqs. (4.44) and (4.45).

$$\left(\nabla_\perp^2 - 2ik\partial_z\right)\psi^*(r_\perp, z) = 0 \tag{4.46}$$

$$\psi^*(r_\perp, z) = \sqrt{I(r_\perp, z)}\exp[-i\varphi(r_\perp, z)] \tag{4.47}$$

Multiplying Eq. (4.44) and Eq. (4.47), we have

$$ik\partial_z I - 2Ik\partial_z\varphi - \frac{1}{4}I^{-1}(\partial_x I)^2 + \frac{1}{2}\partial_x^2 I + i\partial_x I\partial_x\varphi - I(\partial_x\varphi)^2$$
$$+ iI\partial_x^2\varphi - \frac{1}{4}I^{-1}(\partial_y I)^2 + \frac{1}{2}\partial_y^2 I + i\partial_y I\partial_y\varphi - I(\partial_y\varphi)^2 + iI\partial_y^2\varphi = 0 \tag{4.48}$$

Multiplying Eqs. (4.46) and (4.45), we have

$$- ik\partial_z I - 2Ik\partial_z\varphi - \frac{1}{4}I^{-1}(\partial_x I)^2 + \frac{1}{2}\partial_x^2 I - i\partial_x I\partial_x\varphi - I(\partial_x\varphi)^2 - iI\partial_x^2\varphi$$

$$- \frac{1}{4}I^{-1}(\partial_y I)^2 + \frac{1}{2}\partial_y^2 I - i\partial_y I\partial_y\varphi - I(\partial_y\varphi)^2 - iI\partial_y^2\varphi = 0 \qquad (4.49)$$

Subtracting Eq. (4.48) and Eq. (4.49),

$$2ik\partial_z I + 2i\partial_x I\partial_x\varphi + 2iI\partial_x^2\varphi + 2i\partial_y I\partial_y\varphi + 2iI\partial_y^2\varphi = 0 \qquad (4.50)$$

Simplifying further,

$$k\partial_z I + (\partial_x I\partial_x\varphi + \partial_y I\partial_y\varphi) + I(\partial_x^2\varphi + \partial_y^2\varphi) = 0 \qquad (4.51)$$

Equation (4.51) can be rewritten as Eq. (4.52) and then further simplified to Eq. (4.53).

$$-k\partial_z I = \nabla_\perp I\nabla_\perp\varphi + I\Delta_\perp\varphi \qquad (4.52)$$

$$-k\partial_z I = \nabla_\perp \cdot (I\nabla_\perp\varphi) \qquad (4.53)$$

Equation (4.53) is defined as the TIE and it is the bridge between intensity derivative and phase. It means that the phase can be extracted from intensity derivative.

Several methods are available to solve TIE to obtain the phase from intensity derivative. We provide a very basic and classical way that relies on auxiliary function proposed by Teague [28] and the derivative feature of Fourier transform. According to the definition, the vector field satisfies Eq. (4.54), in which A is the vector potential.

$$I\nabla_\perp\varphi = \nabla_\perp\phi + \nabla \times A \qquad (4.54)$$

As the curl term in Eq. (4.54) can be ignored in the approximation in [28], the auxiliary function is defined as Eq. (4.55).

$$\nabla_\perp\phi = I\nabla_\perp\varphi \qquad (4.55)$$

Using the auxiliary function, the TIE in Eq. (4.53) can be rewritten into two Poisson equations as in Eqs. (4.56) and (4.57).

$$\nabla_\perp^2\phi = -k\partial_z I \qquad (4.56)$$

$$\nabla_\perp^2\varphi = \nabla_\perp \cdot (I^{-1}\nabla_\perp\phi) \qquad (4.57)$$

Equation (4.58) describes the derivative feature of the Fourier transform, in which F represents Fourier transform, and k_x is the spatial frequency.

$$F[g^n(x)] = (-ik_x)^n F[g(x)] \tag{4.58}$$

Similarly, Eq. (4.59) is from Eq. (4.58) but in two-dimensional condition, in which $\overline{x}$ and $\overline{y}$ are unit vectors, and k_x and k_y are spatial frequencies and satisfying Eq. (4.60).

$$F[\nabla_\perp g(x, y)] = i k_x F[g(x, y)]\overline{x} + i k_y F[g(x, y)]\overline{y} \tag{4.59}$$

$$k_\perp = \sqrt{k_x^2 + k_y^2} \tag{4.60}$$

Using inverse Fourier transform F^{-1}, Eq. (4.61) can be derived from Eq. (4.59).

$$\nabla_\perp g(x, y) = -i F^{-1}\{k_x F[g(x, y)]\}\overline{x} - i F^{-1}\{k_y F[g(x, y)]\}\overline{y} \tag{4.61}$$

Moreover, Eq. (4.62) can be obtained according to Eqs. (4.58) and (4.61).

$$\nabla_\perp^2 g(x, y) = -F^{-1}\{(k_x^2 + k_y^2) F[g(x, y)]\} = -F^{-1}\{k_\perp^2 F[g(x, y)]\} \tag{4.62}$$

Therefore, the solution to Eq. (4.56) can be obtained as Eq. (4.63) according to Eq. (4.62).

$$\phi = F^{-1}\left[k_\perp^{-2} F(k\partial_z I)\right] \tag{4.63}$$

Similarly, the solution to Eq. (4.57) can be obtained as Eq. (4.64).

$$\varphi = -F^{-1}\left\{k_\perp^{-2} F\left[\nabla_\perp \cdot (I^{-1}\nabla_\perp \phi)\right]\right\} \tag{4.64}$$

Therefore, the phase can be finally obtained by substituting Eq. (4.63) into Eq. (4.64), where I is actual in-focus intensity as I_0, and $\partial_z I$ can be computed from multi-focal images such as Eq. (4.65), where $I_{-\Delta}$ and I_Δ are two symmetric under- and over-focus intensities at a defocus distance of Δ.

$$\partial_z I = (I_\Delta - I_{-\Delta})/(2\Delta) \tag{4.65}$$

It is worth noting that Eqs. (4.63) and (4.64) provide a solution to TIE based on fast Fourier transform, and Eq. (4.65) is an estimation of intensity derivative based on center finite difference approach. According to the principle of the TIE methods, under-, in- and, over-focus images should be captured by correspondingly shifting the image recorder in practical applications, as shown in Fig. 4.15, and specimen phase

Fig. 4.15 Scheme of TIE method

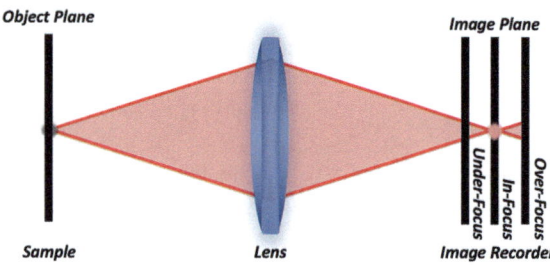

can be reconstructed from these multi-focal images. Besides this approach, many improved methods have also been designed to improve the phase retrieval accuracy and noise suppression capacity. The following numerical simulation is based on basic model for TIE solution.

4.2.2 Numerical Simulations on TIE Method

TIE solution is illustrated using a numerical simulation based on classical fast Fourier transform method. Two important steps involved in this process are multi-focal image generation and phase retrieval. The simulation is done in MATLAB and the corresponding code is provided here.

The first step is to numerically compute all the under-, in-, and over-focus intensities. Two images are used to represent the amplitude and phase distributions of the wavefront. The in-focus wavefront is directly constructed by combining these two images according to Eq. (4.43). The under- and over-focus wavefronts are computed from the in-focus wavefront using the angular spectrum-based numerical wavefront propagation. All multi-focal intensities are computed from their corresponding wavefronts. We can also introduce noise in multi-focal image generation. Multi-focal intensities at any desired distance can be obtained through a numerical wavefront propagation to that distance. We use multi-focal image generation to simulate multi-focal image recording in practical applications.

The second step in numerical simulation is to retrieve the phase from computed multi-focal intensities. In practical situations, we use directly captured intensities in place of computed intensities. Classical fast Fourier transform-based method is adopted here for the phase retrieval. Intensity derivative is computed from multi-focal images and the phase is reconstructed by solving TIE. This method is actually a high-pass filter, therefore, many low-frequency components such as aberrations are lost in the process.

In this numerical simulation, the wavelength is 632.8 nm, the pixel number is 256 × 256, the pixel size is 10 μm, and the numerical wavefront propagation is based on the classical angular spectrum method. Figure 4.16a shows the amplitude and phase distributions of the specimen. According to the numerical wavefront propagation, both the over- and under-focus wavefronts can be computed from the in-focus wavefront with a defocus distance of ±1 mm. Figure 4.16b lists the under-, in-, and over-focus intensities, respectively. Finally, the phase is retrieved from these multi-focal intensities by solving TIE. Figure 4.16c reveals the reconstructed amplitude and phase distributions.

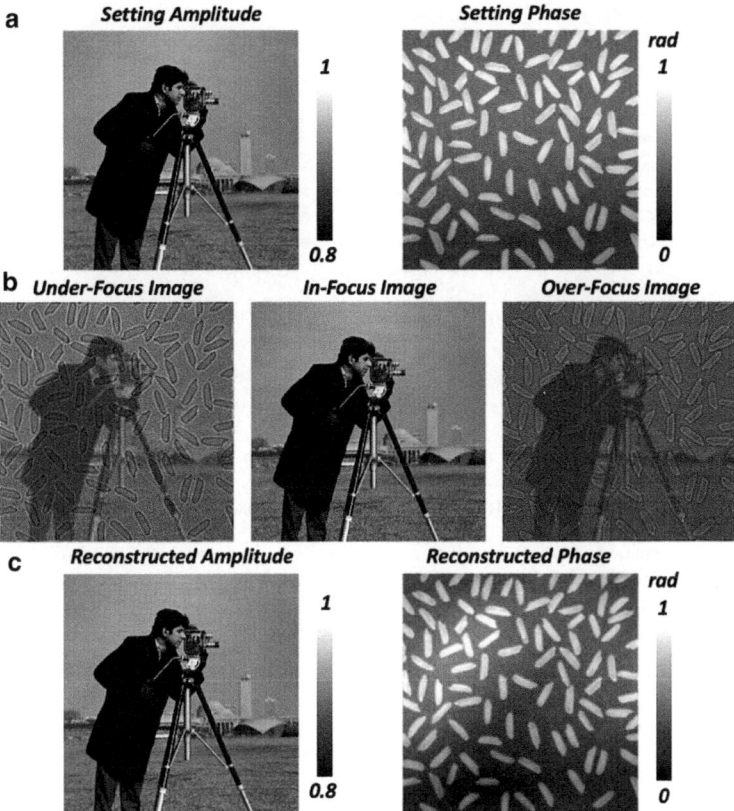

Fig. 4.16 Numerical simulation on TIE method. **a** Setting object amplitude and phase distributions; **b** numerically computed under-, in-, and over-focus intensities; **c** reconstructed object amplitude and phase distributions

Matlab Code for Transport of Intensity Equation Method

```
1    % Computational Optical Phase Imaging
2    % Transport of Intensity Equation Method for Phase Imaging
3
4    %% CCC
5    clear all;
6    close all;
7    clc;
8
9    %% Input Constants
10   lambda = 632.8*10^(-9); % Wavelength
11   PixelNum = 2^8; % Pixel Number
12   PixelSize = 10*10^(-6); % Pixel Size
13   DeltaDistance = 1*10^(-3); % Defocusing Distance
14   RI = 1; % Refractive Index
15
16   %% Other Constants
17   k = 2*pi/lambda; % Wave number
18   Freqency = 1/PixelSize; % Frequency
19   Fxvector = linspace(-Freqency/2,Freqency/2,PixelNum);
20   Fyvector = linspace(-Freqency/2,Freqency/2,PixelNum);
21   [FxMat, FyMat] = meshgrid(Fxvector,Fyvector); % Spectral
     coordinate
22   Hz = exp((1i*2*pi*DeltaDistance*RI/lambda)*(1-
     (lambda.*FxMat/RI).^2-(lambda.*FyMat/RI).^2).^0.5); % Transfer
     function-Propagation
23   Hmz = exp((1i*2*pi*(-DeltaDistance)*RI/lambda)*(1-
     (lambda.*FxMat/RI).^2-(lambda.*FyMat/RI).^2).^0.5); % Transfer
     function-Back Propagation
24
25   %% Sample Wavefront
26   AmplitudeSample = double(imread('cameraman.tif')); % Sample
     amplitude
27   PhaseSample = double(imread('rice.png')); % Sample phase
28   AmplitudeSample =
     (AmplitudeSample/max(max(AmplitudeSample))+4)/5; % For high
     transmittance
29   PhaseSample = PhaseSample/max(max(PhaseSample));
30
31   %% Multi-Focal Images
32   WavefrontSampleInFocus =
     AmplitudeSample.*exp(1i*PhaseSample); % In-focus sample
     wavefront
33   WavefrontSampleOverFocus =
     fftshift(ifft2(ifftshift(fftshift(fft2(ifftshift(WavefrontSamp
     leInFocus)))
     .*Hz))); % Over-focus sample wavefront
34   WavefrontSampleUnderFocus =
     fftshift(ifft2(ifftshift(fftshift(fft2(ifftshift(WavefrontSamp
     leInFocus)))
     .*Hmz))); % Under-focus sample wavefront
35   IntensityInFocus =
     WavefrontSampleInFocus.*conj(WavefrontSampleInFocus); % In-
     focus intensity
```

```
36   IntensityOverFocus =
     WavefrontSampleOverFocus.*conj(WavefrontSampleOverFocus);
     % Over-focus intensity
37   IntensityUnderFocus =
     WavefrontSampleUnderFocus.*conj(WavefrontSampleUnderFocus);
     % Under-focus intensity
38
39   %% Solve Poisson Equation
40   Epsilon = 10^3;  % Avoiding dividing 0
41   FxMat = 2*pi*FxMat;
42   FyMat = 2*pi*FyMat;
43   FMatSqure = 1./(FxMat.^2+FyMat.^2+Epsilon);
44   Derivative = k*(IntensityOverFocus-
     IntensityUnderFocus)/(2*DeltaDistance);
     % Intensity Derivative
45   XPart =
     fftshift(fft2(ifftshift(fftshift(ifft2(ifftshift(FxMat.*FMatSq
     ure.*fftshift(fft2(ifftshift(
     Derivative)))))))./IntensityInFocus))).*FxMat;
46   YPart =
     fftshift(fft2(ifftshift(fftshift(ifft2(ifftshift(FyMat.*FMatSq
     ure.*fftshift(fft2(ifftshift(
     Derivative)))))))./IntensityInFocus))).*FyMat;
47   PhaseSolve =
     real(fftshift(ifft2(ifftshift((XPart+YPart).*FMatSqure))));
48   PhaseSolve = PhaseSolve-min(min(PhaseSolve));  % Retrieved
     Phase
49   AmplitudeSolve = sqrt(IntensityInFocus);  % Retrieved Amplitude
50
51   %% Plot
52   figure
53   subplot(3,3,1); imagesc(AmplitudeSample); axis image; axis
     off; colormap(gray);
     title('Setting Sample Amplitude');
54   subplot(3,3,2); imagesc(PhaseSample); axis image; axis off;
     colormap(gray);
     title('Setting Sample Phase');
55   subplot(3,3,4); imagesc(IntensityUnderFocus); axis image; axis
     off; colormap(gray);
     title('Under-Focus Image');
56   subplot(3,3,5); imagesc(IntensityInFocus); axis image; axis
     off; colormap(gray); title('In-Focus Image');
57   subplot(3,3,6); imagesc(IntensityOverFocus); axis image; axis
     off; colormap(gray);
     title('Over-Focus Image');
58   subplot(3,3,7); imagesc(AmplitudeSolve); axis image; axis off;
     colormap(gray);
     title('Retrieved Amplitude');
59   subplot(3,3,8); imagesc(PhaseSolve); axis image; axis off;
     colormap(gray); title('Retrieved Phase');
```

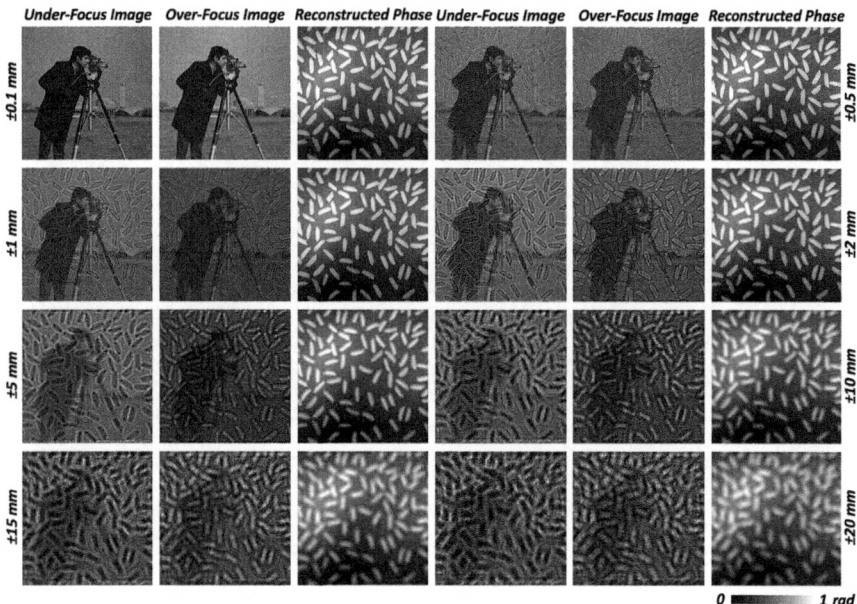

Fig. 4.17 Numerical simulation on TIE method with different defocus distances

The defocus distance significantly affects the phase retrieval accuracy. Figure 4.17 lists the under- and over-focus images computed from the in-focus wavefront in Fig. 4.16a but with different defocus distances of ±0.1 mm, ±0.5 mm, ±1 mm, ±2 mm, ±5 mm, ±10 mm, ±15 mm, and ±20 mm, and also lists the corresponding reconstructed phases. The calculated values of correlation coefficients between actual and reconstructed phases are 0.9633, 0.9585, 0.9490, 0.9388, 0.9158, 0.8935, 0.8679, and 0.8377, respectively, proving that smaller defocus distances maintain higher phase retrieval accuracy.

Presence of noise affects the quality of phase retrieval, particularly with smaller defocus distances. Figure 4.18 lists multi-focal intensities computed from the in-focus wavefront in Fig. 4.16a with different defocus distances as ±0.1 mm, ±2 mm, and ±5 mm but with different signal to noise ratios (SNRs) as 40 dB and 20 dB representing high and low SNR conditions, respectively. Figure 4.18 reveals the corresponding phase retrievals. When the defocus distance is small, many artifacts are left in the retrieved phase distributions, especially in the low SNR condition. It is because of the fact that the adopted TIE solution is actually a high-pass filter and noise significantly influences the phase retrieval accuracy. However, when the defocus distance is large, the noise effects are suppressed. Therefore, a large defocus distance is preferred to suppress the noise.

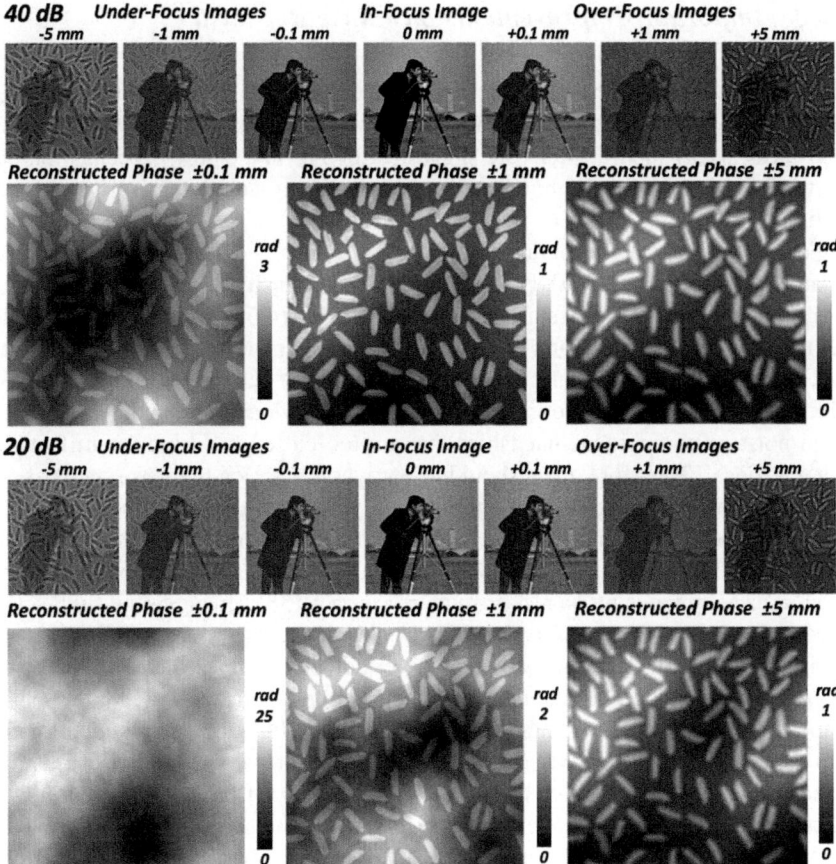

Fig. 4.18 Numerical simulation of TIE method in different SNRs

This classical method of solving TIE still does not address many issues affecting the accuracy of phase reconstruction. For example, besides the defocus distance, the phase retrieval accuracy is also related to the number of multi-focal images, the boundary condition, as well as the imaging quality. Moreover, the multi-focal captures inevitably limit the real-time imaging. In order to illustrate these issues, many important techniques in both software and hardware perspectives are introduced in the following sections.

4.2.3 Important Improvements on Phase Retrieval

Though phase retrieval by classical method of solving TIE is satisfactory to some extent, it needs significant improvement to pursue higher accuracy and better noise robustness. Many modifications are suggested for TIE in the past, especially on the software perspective, and a brief summary of different techniques is given in this section.

For phase retrieval via solving Eq. (4.53), the intensity derivative should be first obtained. In the earliest work by Teague [28], the intensity derivative is computed using center finite difference approach according to Eq. (4.65), which is rather simple and thus has been used in many TIE works. According to the definition of the derivative, when the defocus distance tends to 0, the intensity derivative can be estimated with a high precision. However, according to the previous numerical simulations, when noise exists, defocus intensities that are too close would induce artifacts in phase retrieval. Therefore, there should be a tradeoff between the noise suppression and the intensity derivative estimation accuracy. Many works have been reported on such tradeoff. Nugent group analyzed the suited defocus distances in different SNR conditions [37]. Allen group discussed the phase retrieval based on focus variation, especially focusing on reconstruction accuracy [38]. Huang group studied the effects of non-linearity and noise on different frequency components [39]. All these works provide a similar conclusion that larger defocus distance improves the noise suppression but reduces the intensity derivative computation accuracy, while smaller defocus distance maintains high intensity derivative estimation precision but easily introduces artifacts in phase retrieval. In short, these intensity derivative estimation tactics still suffer from limitations. Though phase retrieval from a few defocus intensities simplifies the optical system, the accuracy of the intensity derivative is limited. Additionally, it is very difficult to obtain high-quality phase reconstruction from noisy intensity distributions.

In order to improve the accuracy of intensity derivative, Ishizuka et al. proposed higher order intensity derivative from massive multi-focal intensities [40]. However, less discussion is provided on the noise suppression; and more importantly, the work did not provide an optimized selection of coefficients corresponding to different orders for intensity derivative estimation. Therefore, many following works employed the same idea but provided different coefficient selection tactics. Soto et al. ignored partial higher order coefficients but enhanced lower order coefficients focusing on noise suppression [41]. Waller et al. designed higher order TIE method using polynomial fitting for coefficient determination, aiming at improving phase retrieval accuracy and correcting non-linearities [42]. Yuan group extends the above works, trying to balance the intensity derivative estimation accuracy and the noise influence [43]. Besides these works using equally-spaced multi-focal intensities, Xue group [44, 45] and Waller group [46] adopted unequally-spaced multi-focal intensities for intensity derivative estimation to obtain better noise suppression capability. Additionally, Zuo group used Savitzky-Golay differentiation filter for phase retrieval in the perspective of frequency maintenance and noise suppression, and tried to provide a general solution for higher order TIE methods [47]. Martinez-Carranza et al. proposed optimized plane selection and band-pass filters to obtain optimal intensity derivative and phase retrieval [40–50]. To pursue an even simpler method, Wang group designed noise adaptive tactic which selects the proper intensity derivative estimation according to the measured SNR [51].

With the intensity derivative, the phase can be extracted by solving TIE. Teague provided Green's function-based analytic method for phase extraction [28]. In this method, the auxiliary function in Eq. (4.55) was introduced, not only in this method but also in other approaches. For example, besides the first application for phase retrieval in interferometry [52], Takeda group also first employed the Fourier transform for TIE solution, still using the auxiliary function [53]; and subsequently, Nugent group adopted the fast Fourier transform-based method for numerical phase retrieval [54]. The fast Fourier transform-based method is the most widely used tactic for phase retrieval in TIE methods due to its fast processing speed. Besides, other classical approaches for TIE solution have also been proposed. Nugent group proposed the propagation-based technique [55] and the Zernike polynomial-based method [56, 57], and Allen group also successfully adopted multi-grid method and iteration method [36]. In astronomy, since the amplitude of the wavefront under detection can be treated as unique, Roddier further simplified the TIE to a Poisson equation relying on the constant intensity but without using the auxiliary function [29, 30].

The phase retrieval accuracy is still limited due to several issues, such as the boundary condition, the applicable condition of auxiliary function, and the light coherence. When TIE was first designed, the Dirichlet boundary condition for phase retrieval was applied [28], however, it requires previously measured data which adds more complexity to the TIE method. Woods and Greenaway solved the TIE relying on the Neumann boundary condition only requiring matrix multiplication, much simpler than the tactic using Dirichlet boundary condition [58]. In addition, Zuo group adopted the discrete cosine transform to solve the TIE using additional stop to satisfy the Neumann boundary condition [59, 60]. Volkov et al. solved the TIE via fast Fourier transform using periodic extension to satisfy periodic boundary condition [61]. Parvizi et al. solved the TIE based on the finite element multi-grid method with the prior knowledge of flat phase in empty areas of the sample, and no periodic boundary conditions are assumed [62]. All these methods reduce the errors and artifacts during the TIE solution, thus improving the phase retrieval accuracy.

Many methods solve the TIE using the auxiliary function as listed in Eq. (4.55), which is obtained by ignoring the curl term in Eq. (4.54). When the intensity distribution of the in-focus image is unique, the error induced by the curl term omission can be ignored; otherwise, the error is still considerable as analyzed by Pavlov group [63] and Zuo group [64]. It seems that iteration can reduce the phase retrieval error [64, 65], however, long-time iteration sacrifices the advantages of the TIE methods especially the high speed phase retrieval.

Though TIE is derived according to the coherent light condition [28], Streibl presented a theoretical explanation for partially coherent illumination in TIE method [66]. Nugent group proved that the TIE has a unique solution only if the intensity distribution has no zeros in both coherent and partially coherent illumination conditions [67], and proved that the TIE tactic still works even when the light source is not completely coherent in experiments [55]. Both theoretical and experimental works extend the application scope of TIE methods from coherent condition to partially coherent illumination condition. Therefore, the TIE methods can be considered as non-interferometric phase imaging tactic that is different from classical holography, interferometry, and coherent diffraction imaging. Since the TIE methods still work even though the light is not completely coherent, many phase retrieval methods for TIE solution ignored the influence of the partially coherent illumination. However, some of the works still discussed such issue. Gureyev group obtained quantitative phase retrieval for TIE solution using the broadband polychromatic radiation but with a prior knowledge on spectral distribution [68–70]. Anastasio group demonstrated the transport of spectrum equation which is the spectral version of TIE, as well as designed the phase retrieval method by taking measurements at multiple frequencies, solving a system of TIE and reconstructing the phase relying on phase weighted average of each mode [71]. Barbastathis group demonstrated another version of the TIE valid for partially coherent illumination, which can measure the phase of thin samples [72]. Zuo group proposed a phase-space formulation of TIE, also for phase retrieval in partially coherent illumination condition [73]. However, most of these methods require the prior knowledge on spectral distribution, complicating the procedures of the TIE methods. Therefore, according to the results provided by Gureyev group [68–70], the phase retrieval error induced by not considering the spectral information can be ignored when the illumination has high coherence. Accordingly, in practical applications, it is preferred to insert filters and reduce the illumination condenser aperture to respectively improve the temporal and spatial coherence to pursue a high phase retrieval accuracy even without using spectral information during the TIE solution, thus keeping the advantage of the TIE as simple setup for phase retrieval.

One key advantage of TIE methods is the simple phase retrieval process in comparison to holography, interferometry, and coherent diffraction imaging. It not only avoids the time-consuming phase unwrapping often required by holography and interferometry, but also does not rely on iteration used in coherent diffraction imaging. Moreover, the setup of TIE methods is often simpler than those of holography, interferometry, and coherent diffraction imaging, since it does not need an extra reference light as in holography and interferometry or spatial/angular illumination scanning as in coherent diffraction imaging. However, the TIE methods require multi-focal image recording for phase retrieval, inevitably limiting the applications in real-time phase imaging. In order to solve the problem, some important improvements on the hardware perspective of the TIE methods are provided and discussed in the following section.

4.2.4 *Important Improvement on Multi-focal Imaging*

Both phase retrieval accuracy and noise suppression can be improved with the above mentioned updated TIE phase retrieval methods. However, TIE requires multiple acquisitions for phase recovery, thus compromising real-time phase imaging. In order to extend its scope, some tactics have been designed such as using fast mechanical scanning or even electrical-driven focal scanning [74]. However, these techniques only accelerate the focal scanning speed, therefore, they are still not real-time solutions. An effective way for real-time TIE approaches is to capture the multi-focal images simultaneously.

Barbastathis group designed a chromatic aberration induced multi-focal imaging tactic for real-time quantitative phase imaging via TIE approach [75] and Fig. 4.19a shows the scheme of this method. Its key principle is that the mechanical axial defocus is transformed into chromatic aberration. Relying on the dispersion property of the imaging system, beams of different wavelengths are focused at different planes. Therefore, using three wavelengths, for instance red, green, and blue, their corresponding images can be recorded using a color image recorder and these monochromatic images can be treated as under-, in-, and over-focus ones for phase retrieval

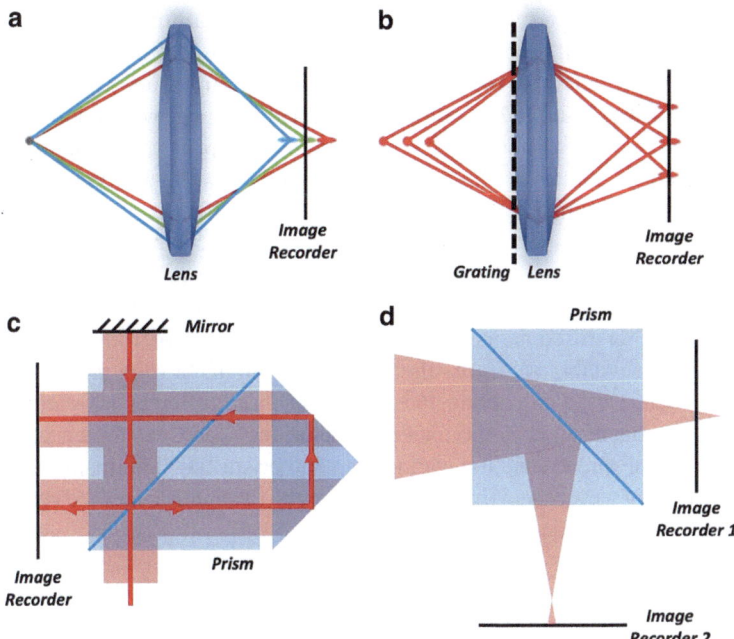

Fig. 4.19 Simultaneous multi-focal image recording tactics. **a** Chromatic aberration-based simultaneous multi-focal imaging; **b** point spread function engineering-based simultaneous multifocal imaging; **c** prism-based simultaneous multi-focal imaging; **d** multi-image recorder-based simultaneous multi-focal imaging

using TIE approach. Using this method, no physical movement is needed for multi-focal image recording and only a color image recorder is required. Therefore, it satisfies the requirement of capturing the multi-focal images simultaneously, thereby supports real-time phase imaging. Though this method is simple, it can be hardly used when the sample exhibits strong dispersive characteristics.

Greenaway group employed a distorted diffraction grating to generate multi-focal images on a single plane which can be simultaneously recorded using only one image recorder [76]. Barbastathis group used a volume holographic grating for simultaneous multi-focal imaging [77]. Both of these systems are rather simple, however, the imaging quality is heavily influenced by the fabrication quality of the distorted diffraction grating and the volume holographic grating. In order to improve multi-focal imaging quality, Wang group adopted the spatial light modulator to generate the multi-focal imaging on a single plane, and still a single image recorder is required for multi-focal image recording [78]. In such a design, a 4-f system is built and the spatial light modulator is set at the frequency plane, and the phase mask can be precisely set for high-quality multi-focal image generation. All these methods can be treated as point spread function engineering tactics as the scheme shown in Fig. 4.19b, and these methods can capture the multi-focal images simultaneously, thus supports real-time phase retrieval using TIE method.

Different from above mentioned techniques, Zuo group adopted both the spatial light modulator and the prism for multi-focal imaging [79]. In order to further simplify the system, spatial light modulator-free setups are proposed. Zhao group designed a flipping imaging module [80, 81] and Nishchal group proposed a Michelson interferometer-like set-up [82], both for multi-focal imaging, which do not require sophisticated fabricated distorted diffraction grating or expensive spatial light modulator. All these techniques are based on a prism, aiming at field of view separation and Fig. 4.19c shows their respective scheme.

All of the above mentioned methods can obtain simultaneous multi-focal image recording, thus support real-time quantitative phase imaging using TIE. Moreover, in these methods, only single image recorder is required, which can guarantee unified image recording for different multi-focal images. However, the chromatic aberration-based method inevitably sacrifices the spatial resolution. In such method, a color image recorder is used to capture the images of three different wavelengths and because of the Bayer layer in the imaging sensor, the spatial resolution is dropped by half. For point spread function engineering and prism-based multi-focal imaging tactics, the field of view is heavily limited since multi-focal images should be captured by a single image recorder.

Besides the above methods relying on only single image recorder, another tactic is also provided to use multiple image recorders for multi-focal image recording, as shown in Fig. 4.19d. As the simplest design, Dai group captured the multi-focal images separated by a beam splitter using two image recorders [83]. Moreover, in order to be suitable for the commercial microscopes, Wang group also designed a dual-view TIE method for real-time quantitative phase imaging [84]: two identical image recorders are set on the binoculars of the microscope, but their image planes are set differently by inserting brass spacer rings. Therefore, both the under- and over-focus images can be simultaneously captured for real-time phase retrieval via TIE. However, it is difficult to maintain the consistent fields of view captured by two image recorders. In order to solve such problem, Wang group also adopted the digital field of view correction algorithm relying on phase correlation to successfully correct these fields of view, even though they represent under- and over-focus images [85]. Therefore, this proposed method can be successfully used in real-time quantitative live cell phase imaging [86]. Moreover, Wang group also designed phase real-time microscope camera (PhaseRMiC), which has a simple and cost-effective configuration only consisting of a beam splitter and a board-level camera with two CMOS imaging chips, and PhaseRMiC could be used by directly connecting it to the C-mount port of the commercial microscope, and was proved for live cell phase imaging [87, 88]. Compared to the single image recorder-based real-time TIE approaches, these multiple image recorders-based real-time TIE approaches often have large fields of view. However, since different image recorders are used, the unity of multi-focal image recording is reduced.

With these proposed simultaneous multi-focal image recording tactics, the TIE approach can be extended to real-time quantitative phase imaging applications. Compared to holography and interferometry techniques that are often used in live cell imaging, the system of TIE approach is much simpler; it not only avoids the reference beam but also more compatible with commercial microscopes. Additionally, compared to the Shack-Hartmann wavefront sensor introduced in the following section, it also can be used for real-time wavefront sensing with much higher spatial resolution. Therefore, the TIE approach is also a solution for real-time quantitative phase imaging, especially in live cell imaging.

4.2.5 Discussion

As an important technique of quantitative phase imaging, TIE methods have been widely used in different wavebands and applications. However, only an outline of this technique is briefly discussed in this section. A series of reviews by Nugent group focusing on both short-wavelength and optical imaging [37, 89–91] and a tutorial on TIE written by Zuo group mostly focusing on optical imaging [92] are available for interested readers. All these works describe many aspects on this approach in detail.

As a non-interferometric method, TIE methods can provide wavefront information only using partially coherent light. Furthermore, it does not rely on the reference beam required by holography and interferometry. It has rather high spatial resolution which can hardly be achieved using Shack-Hartmann wavefront sensor. Use of rapid phase retrieval algorithms avoids time-consuming phase unwrapping often required by holography and interferometry and iterations often needed by coherent diffraction imaging. Furthermore, this method has been successfully used not only in lensless short-wavelength imaging, reducing the need for high-quality elements, but also in optical phase microscopy because it can be used with commercial microscopes. Considering these advantages, TIE works as a solution for optical phase imaging.

TIE methods have certain disadvantages. The key limitation is that they often suffer from lower phase retrieval accuracy compared to holography, interferometry, and coherent diffraction imaging. One reason is that TIE methods can hardly retrieve all the components in the whole frequency band. High frequency components are maintained when the defocus distance is small, but low frequency components are lost or the other way around. Another limitation is that since the phase is directly extracted from the intensity, the extremely low intensity or even zero intensity distributions make these TIE methods invalid. Almost all applications using TIE methods are focused on imaging and hence more suitable for observation rather than measurement purposes.

4.3 Shack-Hartmann Wavefront Sensor

The Shack-Hartmann wavefront sensor is a robust non-interferometric metrology tool for measuring wavefront errors. This is an improved version of Hartmann wavefront sensor [93] where microlens arrays are used for detecting the phase of the wavefront [94, 95]. Though the technique was originally developed for large telescopes to measure wavefront errors induced by atmospheric turbulence, it has evolved to become a versatile tool in optical and laser testing applications. Both Hartmann and Shack-Hartmann wavefront sensors are simple, compact, robust, and relatively insensitive to vibrations. Many astronomical and biological applications currently use Shack-Hartmann wavefront sensors that are coupled with adaptive optical systems to obtain wavefront distortion compensation that leads to greater imaging quality. Principles of both Hartmann and Shack-Hartmann wavefront sensors are given in this section along with numerical simulations.

4.3.1 Theory of Hartmann and Shack-Hartmann Wavefront Sensors

Different from coherent diffraction imaging, holography, interferometry, and transport of intensity equation method for phase retrieval in whole field of view, Hartmann and Shack-Hartmann wavefront sensors measure the wavefront through sampling. Hartmann wavefront sensor adopts aperture array and Shack-Hartmann wavefront sensor adopts microlens array for wavefront sampling as revealed in Fig. 4.20a and b, respectively. Based on aperture array transmitted beam directions and microlens focused beam positions, the sampled phase slope of wavefront can be computed, and the phase of wavefront can be further reconstructed through 2-D integration. Unfortunately, Hartmann and Shack-Hartmann wavefront sensors suffer from low spatial resolution due to their sampling principle.

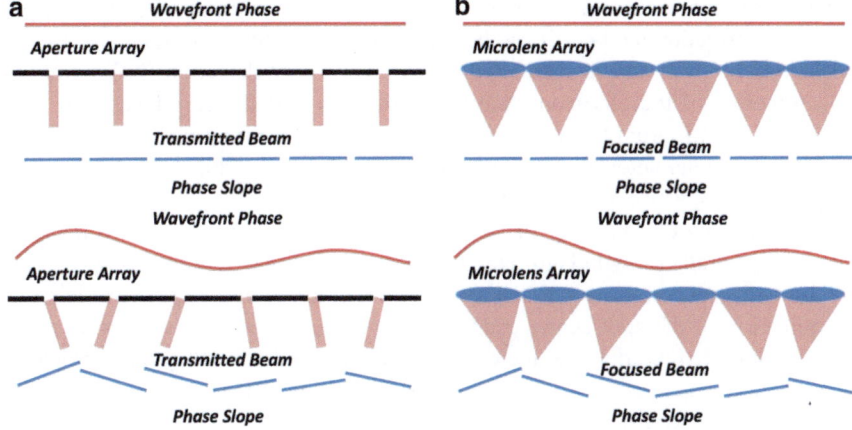

Fig. 4.20 Scheme of **a** Hartmann wavefront sensor and **b** Shack-Hartmann wavefront sensor

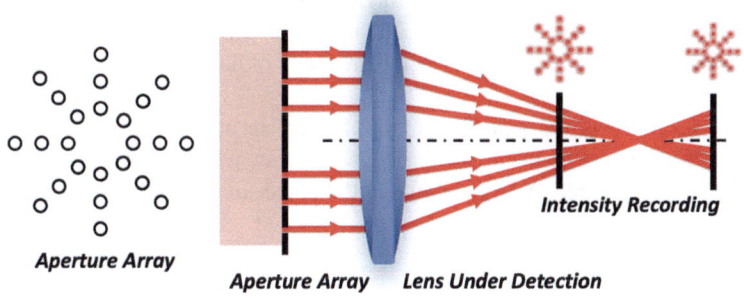

Fig. 4.21 Principle of the Hartmann wavefront sensor

Figure 4.21 shows the principle of the Hartmann wavefront sensor. An aperture array is located in front of an optical element under detection (such as a lens in Fig. 4.21). The beam passing through the aperture array is sampled into many narrow beams which are then modulated by the lens under detection. These modulated narrow beams can generate intensity spots on either side of focal planes, which can be recorded by moving an image recorder along the axis. Based on these recorded spot intensity distributions, their central positions in two imaging planes can be determined. Further, these sampled narrow beam propagation directions can be computed based on rather simple geometric optics. According to Fig. 4.20a, these sampled narrow beam propagation directions actually describe the sampled phase slopes of the wavefront modulated by the lens under detection, and phase of the wavefront can be reconstructed using numerical 2-D integration.

The Hartmann wavefront sensor has a compact configuration and follows a simple phase reconstruction method. It can be used in optical testing to retrieve the optical aberrations [96, 97] and also adopted in laser wavefront detection [98]. Only the main principle of the Hartmann wavefront sensor is briefly introduced here and more details on different Hartmann wavefront sensing configurations, aperture array designs, spot center determination, and phase integration algorithms can be found in many references, especially in Optical Shop Testing [1].

Unfortunately, the intensity of sampled narrow beams used for phase imaging and sensing is rather low in Hartmann wavefront sensor due to the introduced aperture array. Moreover, the generated spots are often large, limiting the precision in determination of the spot center. Additionally, Hartmann wavefront sensing often requires multiple recording for the computation of sampled narrow beam propagation direction. These disadvantages made Hartmann wavefront sensor less useful until Shack improved Hartmann wavefront sensor by using microlens array instead of aperture array in 1971. Ever since Shack-Hartmann wavefront sensor has become a commonly used phase imaging and sensing tool in various fields, especially in adaptive optics since its development. While Hartmann and Shack-Hartmann wavefront sensors share the same concept of detecting phase slopes by sampled beam propagation and phase reconstruction through 2-D integration, the latter only enhances the collected beam intensity and improves the precision of determining the center of the spot.

Figure 4.22 shows the principle of the Shack-Hartmann wavefront sensor. A microlens array is set in front of the image recorder which is at a focal distance from the lens. The focal spots generated at the center of each microlens by a plane

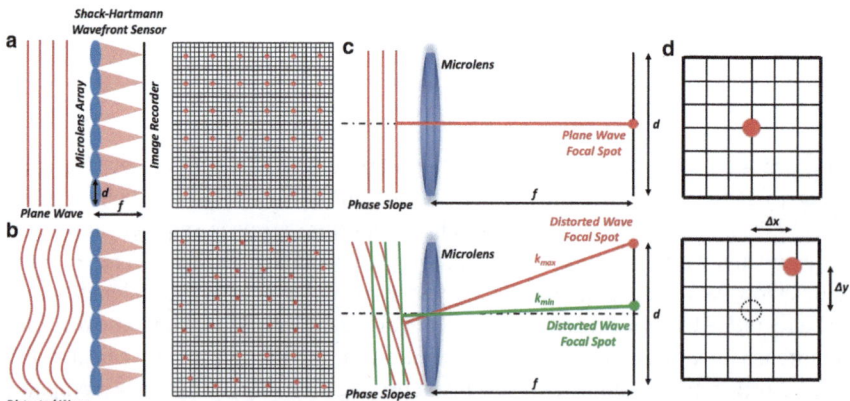

Fig. 4.22 Principle of the Shack-Hartmann wavefront sensor. **a** Plane wave condition; **b** distorted wave condition; **c** Shack-Hartmann wavefront sensor principle in a 1-D perspective of a single microlens; **d** focal spot deviations

wavefront illumination are regarded as a reference focal spot as shown in Fig. 4.22a. When the incident wavefront under detection is distorted, the focal spots are deviated from the centers as shown in Fig. 4.22b. Figure 4.22c depicts the details in a 1-D perspective of a single microlens, in which d and f are the diameter and focal length of the microlens, respectively. According to the focal spot position, the phase slopes corresponding to microlenses can be computed briefly as described in Fig. 4.20b. Limited by the size of the microlens and its occupied pixels in the image recorder, the maximum phase slope that could be detected is $d/2f$, and the minimum slope is mainly determined by the pixel size, the accuracy of the center determination and the signal to noise ratio of recorded focal spots. When the phase slope exceeds $d/2f$, the focal spot corresponding to that microlens cannot be recorded and is missing in the Fig. 4.22b. Therefore, in 2-D condition, when the generated focal spot deviates from the reference point by $(\Delta x, \Delta y)$ as shown in Fig. 4.22d, while less than $d/2f$, the phase slopes of the incident light k_x and k_y in two dimensions are given by Eqs. (4.66) and (4.67) where $\Phi(x, y)$ is the wavefront incident on the Shack-Hartmann wavefront sensor. Similar to Hartmann wavefront sensor, the wavefront is reconstructed through a 2-D integration of the phase slopes that are measured with respect to a reference focal spot.

$$k_x = \frac{\partial \Phi(x, y)}{\partial x} = \frac{\Delta x}{f} \tag{4.66}$$

$$k_y = \frac{\partial \Phi(x, y)}{\partial y} = \frac{\Delta y}{f} \tag{4.67}$$

4.3.2 Numerical Simulations on Shack-Hartmann Wavefront Sensor

Figure 4.23 shows the results of numerical simulations of Shack-Hartmann wavefront sensor obtained with the following Matlab code. Numerical wavefront propagation is based on classical angular spectrum method in the coherent condition, however, Shack-Hartmann wavefront sensor can also be used in incoherent condition. The pixel number is 1024 × 1024 with a pixel size of 10 μm is chosen for simulation. Each microlens in a 32 × 32 array consists of 32 × 32 pixels and has a focal length of 5 mm. The wavelength is 632.8 nm. When the incident wavelength is a plane wave, all the focal spots are generated at the centers of microlenses as shown in Fig. 4.23a. The phase of the distorted wavefront is shown in Fig. 4.23b which is the aberration corresponding to sum of Zernike polynomials from 0th to 2nd orders.

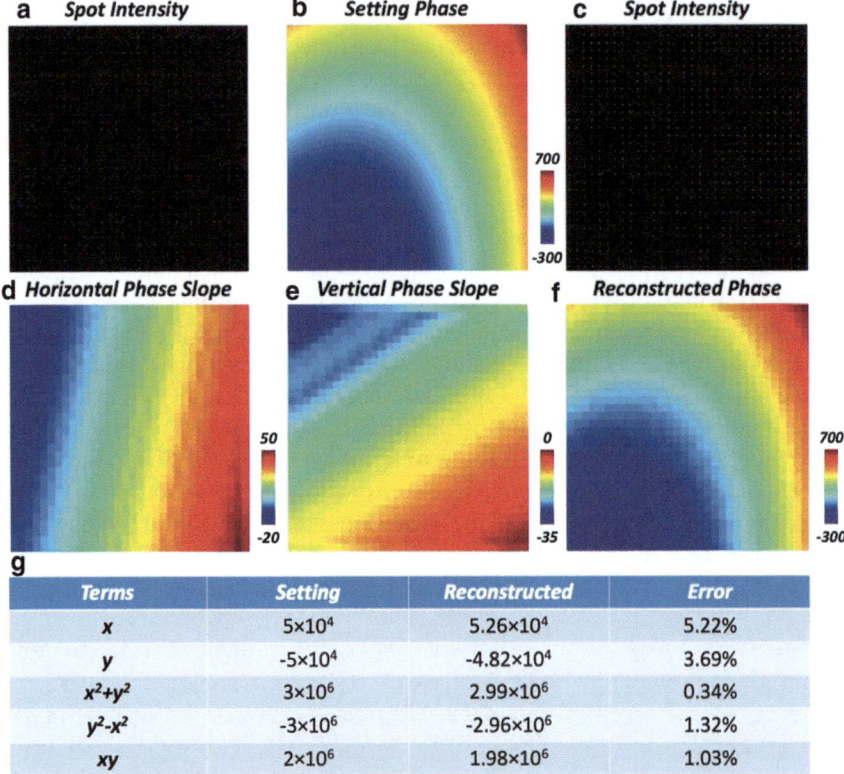

a *Spot Intensity*   **b** *Setting Phase*   **c** *Spot Intensity*

d *Horizontal Phase Slope*   **e** *Vertical Phase Slope*   **f** *Reconstructed Phase*

g

Terms	Setting	Reconstructed	Error
x	$5×10^4$	$5.26×10^4$	5.22%
y	$-5×10^4$	$-4.82×10^4$	3.69%
x^2+y^2	$3×10^6$	$2.99×10^6$	0.34%
y^2-x^2	$-3×10^6$	$-2.96×10^6$	1.32%
xy	$2×10^6$	$1.98×10^6$	1.03%

Fig. 4.23 Numerical Simulations of the Shack-Hartmann wavefront sensor. **a** Captured focal spots with the plane wavefront; **b** setting phase of the wavefront under detection; **c** captured focused spots with the distorted wavefront; **d** phase slopes in x axis; **e** phase slopes in y axis; **f** reconstructed phase; **g** Zernike coefficients of (B) and (F)

Figure 4.23c shows the detected focal spots. Following the determination of each spot center, the deviations are computed, and the wavefront slopes corresponding to different microlenses in two directions are reconstructed as shown in Fig. 4.23d and e. Phase of the distorted wavefront that is retrieved through a 2-D integration of the wavefront slopes in Fig. 4.23d and e. This is shown in Fig. 4.23f. The comparison between the set phase in Fig. 4.23b and the reconstructed one in Fig. 4.23f as well as the error of the extracted Zernike coefficients compared to the setting ones in Fig. 4.23g demonstrate that the Shack-Hartmann wavefront sensor can reconstruct wavefronts with high precision.

Matlab Code for Shack-Hartmann Wavefront Sensor

```
1    % Computational Optical Phase Imaging
2    % Shack-Hartmann Wavefront Sensor
3
4    %% CCC
5    clear all;
6    close all;
7    clc;
8
9    %% Input Constants
10   lambda = 632.8*10^(-9); % Wavelength
11   k = 2*pi/lambda; % Wave number
12   PixelNum = 2^10; % Pixel number
13   PixelSize = 10*10^(-6); % Pixel size
14   RI = 1; % Refractive Index
15
16   %% Microlens Array
17   Focus = 5*10^(-3); % Focal length
18   MicrolensPixelNum = 2^5; % Microlens pixel number
19   MicrolensNum = PixelNum/MicrolensPixelNum; % Microlens number
20   Microlensxvector = linspace(-
     MicrolensPixelNum*PixelSize/2,MicrolensPixelNum*PixelSize/2,Mi
     crolensPixelNum);
21   Microlensyvector = linspace(-
     MicrolensPixelNum*PixelSize/2,MicrolensPixelNum*PixelSize/2,Mi
     crolensPixelNum);
22   [MicrolensxMat, MicrolensyMat] =
     meshgrid(Microlensxvector,Microlensyvector);
     % Microlens coordinate
23   Microlens = exp(-
     1i*k*(MicrolensxMat.^2+MicrolensyMat.^2)/2/Focus);
24   MicrolensArray = zeros(PixelNum,PixelNum);
25   for ii = 1:1:MicrolensNum
26       for jj = 1:1:MicrolensNum
27           MicrolensArray(MicrolensPixelNum*(ii-
     1)+1:MicrolensPixelNum*(ii-1)+
             MicrolensPixelNum,MicrolensPixelNum*(jj-1)+
             1:MicrolensPixelNum*(jj-1)+MicrolensPixelNum) =
     Microlens;
28       end
29   end
30
31   %% Transfer Function
32   Freqency = 1/PixelSize; % Frequency
33   Fxvector = linspace(-Freqency/2,Freqency/2,PixelNum);
34   Fyvector = linspace(-Freqency/2,Freqency/2,PixelNum);
35   [FxMat, FyMat] = meshgrid(Fxvector,Fyvector); % Spectral
     coordinate
36   Hz = exp((1i*2*pi*Focus*RI/lambda)*(1-(lambda.*FxMat/RI).^2-
     (lambda.*FyMat/RI).^2).^0.5);
     % Transfer function-Propagation
37
38   %% Reference
39   %% Reference Wavefront
```

```
40   AmplitudeReference = ones(PixelNum,PixelNum);  % Reference
     amplitude
41   PhaseReference = zeros(PixelNum,PixelNum);  % Reference phase
42   WavefrontReference =
     AmplitudeReference.*exp(1i*PhaseReference);  % Reference
     wavefront
43
44   %% Intensity
45   SpotWavefrontReference =
     fftshift(ifft2(ifftshift(fftshift(fft2(ifftshift(WavefrontRefe
     rence.*MicrolensArray))).*Hz)));
     % Wavefront at image recorder
46   IntensityReference =
     SpotWavefrontReference.*conj(SpotWavefrontReference);  %
     Intensity
47
48   %% Focus Position Determination
49   CenterXReference = zeros(MicrolensNum,MicrolensNum);
50   CenterYReference = zeros(MicrolensNum,MicrolensNum);
51   for ii = 1:1:MicrolensNum
52      for jj = 1:1:MicrolensNum
53          IntensityCutReference =
                IntensityReference(MicrolensPixelNum*(ii-
                1)+1:MicrolensPixelNum*(ii-
                1)+MicrolensPixelNum,MicrolensPixelNum*(jj-
                1)+1:MicrolensPixelNum*(jj-1)+MicrolensPixelNum);
54          CenterXReference(ii,jj) =
                sum(sum(MicrolensxMat.*IntensityCutReference))/sum(s
                um(IntensityCutReference));  % Center position in x
                axis
55          CenterYReference(ii,jj) =
                sum(sum(MicrolensyMat.*IntensityCutReference))/sum(s
                um(IntensityCutReference));
                % Center position in y axis
56      end
57   end
58
59   %% Reference Wavefront Slope Retrieval
60   SlopeXReference =
     CenterXReference/Focus*MicrolensPixelNum*PixelSize*2*pi/lambda
     ;
     % Reference slope phase in x axis
61   SlopeYReference =
     CenterYReference/Focus*MicrolensPixelNum*PixelSize*2*pi/lambda
     ;
     % Reference slope phase in y axis
62   RetrievalPhaseReference = zeros(MicrolensNum,MicrolensNum);  %
     Retrieved reference phase
63   for ii = 1:1:MicrolensNum
64      if ii == 1
65          RetrievalPhaseReference(ii,1) = SlopeYReference(ii,1);
66      else
67          RetrievalPhaseReference(ii,1) =
```

```
       RetrievalPhaseReference(ii-1,1)+SlopeYReference(ii,1);
           % Integration in x axis
68         end
69     end
70     for ii = 1:1:MicrolensNum
71         for jj = 2:1:MicrolensNum
72             RetrievalPhaseReference(ii,jj) =
       RetrievalPhaseReference(ii,jj-1)+SlopeXReference(ii,jj);
           % Integration in y axis
73         end
74     end
75
76     %% Testing
77     %% Wavefront Under Detection
78     Sxvector = linspace(-
       PixelNum*PixelSize/2,PixelNum*PixelSize/2,PixelNum);
79     Syvector = linspace(-
       PixelNum*PixelSize/2,PixelNum*PixelSize/2,PixelNum);
80     [SxMat, SyMat] = meshgrid(Sxvector,Syvector);  % Microlens
       coordinate
81     Z0 = 0;  % Zernike coefficients
82     Z1 = 5*10^4;
83     Z2 = -5*10^4;
84     Z3 = 3*10^6;
85     Z4 = -3*10^6;
86     Z5 = 2*10^6;
87
88     Phase =
       Z0+Z1*SxMat+Z2*SyMat+Z3*2*(SxMat.^2+SyMat.^2)+Z4*(SyMat.^2-
       SxMat.^2)+Z5*2*SxMat.*SyMat;  % Phase
89     Phase = Phase-Phase(PixelNum/2+1,PixelNum/2+1);
90     Amplitude = ones(PixelNum,PixelNum);  % Amplitude
91     Wavefront = Amplitude.*exp(1i*Phase);  % Wavefront
92
93     %% Intensity
94     SpotWavefront =
       fftshift(ifft2(ifftshift(fftshift(fft2(ifftshift(Wavefront.*Mi
       crolensArray))).*Hz)));
           % Wavefront at image recorder
95     Intensity = SpotWavefront.*conj(SpotWavefront);  % Intensity
96
97     %% Focus Position Determination
98     CenterX = zeros(MicrolensNum,MicrolensNum);
99     CenterY = zeros(MicrolensNum,MicrolensNum);
100    for ii = 1:1:MicrolensNum
101        for jj = 1:1:MicrolensNum
102            IntensityCut = Intensity(MicrolensPixelNum*(ii-
       1)+1:MicrolensPixelNum*(ii-
       1)+MicrolensPixelNum,MicrolensPixelNum*(jj-
       1)+1:MicrolensPixelNum*(jj-1)+MicrolensPixelNum);
103            CenterX(ii,jj) =
       sum(sum(MicrolensxMat.*IntensityCut))/sum(sum(IntensityCut));
           % Center position in x axis
```

```
104        CenterY(ii,jj) =
     sum(sum(MicrolensyMat.*IntensityCut))/sum(sum(IntensityCut));
           % Center position in y axis
105      end
106  end
107
108  %% Wavefront Under Detection Slope Retrieval
109  SlopeX =
     CenterX/Focus*MicrolensPixelNum*PixelSize*2*pi/lambda;  % Slope
     phase in x axis
110  SlopeY =
     CenterY/Focus*MicrolensPixelNum*PixelSize*2*pi/lambda;  % Slope
     phase in y axis
111
112  RetrievalPhase = zeros(MicrolensNum,MicrolensNum);  % Retrieved
     phase
113  for ii = 1:1:MicrolensNum
114      if ii == 1
115          RetrievalPhase(ii,1) = SlopeY(ii,1);
116      else
117          RetrievalPhase(ii,1) = RetrievalPhase(ii-
     1,1)+SlopeY(ii,1);  % Integration in x axis
118      end
119  end
120  for ii = 1:1:MicrolensNum
121      for jj = 2:1:MicrolensNum
122          RetrievalPhase(ii,jj) = RetrievalPhase(ii,jj-
     1)+SlopeX(ii,jj);  % Integration in y axis
123      end
124  end
125
126  %% Reconstructed Phase
127  ReconstructedPhase = RetrievalPhase-RetrievalPhaseReference;  %
     Reconstructed phase
128  ReconstructedPhase = ReconstructedPhase-
     ReconstructedPhase(MicrolensNum/2+1,MicrolensNum/2+1);
129
130  %% Regression
131  RxMat = zeros(MicrolensNum,MicrolensNum);  % Regress coordinate
132  RyMat = zeros(MicrolensNum,MicrolensNum);  % Regress coordinate
133  for ii = 1:1:MicrolensNum
134      for jj = 1:1:MicrolensNum
135          PositionX = SxMat(MicrolensPixelNum*(ii-
     1)+1:MicrolensPixelNum*(ii-
     1)+MicrolensPixelNum,MicrolensPixelNum*(jj-
     1)+1:MicrolensPixelNum*(jj-1)+MicrolensPixelNum);
136          PositionY = SyMat(MicrolensPixelNum*(ii-
     1)+1:MicrolensPixelNum*(ii-
     1)+MicrolensPixelNum,MicrolensPixelNum*(jj-
     1)+1:MicrolensPixelNum*(jj-1)+MicrolensPixelNum);
137          RxMat(ii,jj) =
     sum(sum(PositionX))/MicrolensPixelNum/MicrolensPixelNum;
138          RyMat(ii,jj) =
```

```
       sum(sum(PositionY))/MicrolensPixelNum/MicrolensPixelNum;
139       end
140   end
141
142   RxMat1D = zeros(MicrolensNum^2,1);
143   RyMat1D = zeros(MicrolensNum^2,1);
144   ReconstructedPhase1D = zeros(MicrolensNum^2,1);
145   for jj = 1:1:MicrolensNum
146       for ii = 1:1:MicrolensNum
147           RxMat1D(ii+(jj-1)*MicrolensNum,1) = RxMat(ii,jj);
148           RyMat1D(ii+(jj-1)*MicrolensNum,1) = RyMat(ii,jj);
149           ReconstructedPhase1D(ii+(jj-1)*MicrolensNum,1) =
       ReconstructedPhase(ii,jj);
150       end
151   end
152   RCoordinate1D =
       [ones(size(RxMat1D)),RxMat1D,RyMat1D,2*(RxMat1D.^2+RyMat1D.^2)
       ,(RyMat1D.^2-RxMat1D.^2),2*RxMat1D.*RyMat1D];   % 1-D Regress
153   ReconstructedZernike =
       regress(ReconstructedPhase1D,RCoordinate1D);
154
155   %% Plot
156   figure
157   subplot(2,2,1); imagesc(IntensityReference); axis image; axis
       off; colormap(gray); title('Reference Spot Intensity');
158   subplot(2,2,2); imagesc(Phase); axis image; axis off;
       colormap(gray); title('Setting Sample Phase');
159   subplot(2,2,3); imagesc(Intensity); axis image; axis off;
       colormap(gray); title('Spot Intensity');
160   subplot(2,2,4); imagesc(ReconstructedPhase); axis image; axis
       off; colormap(gray); title('Reconstructed ample Phase');
```

The Shack-Hartmann wavefront sensor is compact and uses a simple phase reconstruction method, but its spatial resolution is limited by the microlens array. However, the Shack-Hartmann wavefront sensor has found multiple applications, especially in adaptive optics in astronomical [99] and biological [100] fields. The reference contains more information on Shack-Hartmann wavefront sensors, especially in Optical Shop Testing [1].

4.3.3 Discussion

Both Hartman and Shack-Hartmann wavefront sensors employ the same principle of phase detection by sampling: they sample the wavefront under detection using an aperture array or microlens array, then retrieve the phase slope of each sampled wavefront based on the sampled wavefront propagation direction, and finally reconstruct the phase via 2-D integration. As compared to Hartmann wavefront sensor, Shack-Hartmann wavefront sensor is more widely used because it uses microlens array instead of aperture array, which collects more light and determines propagation direction more accurately. Furthermore, unlike holography and interferometry, neither Hartmann nor Shack-Hartmann wavefront sensors rely on a reference beam or coherent light source. In addition, they can rapidly reconstruct the phase of the sample wavefront from a single shot without iteration compared to coherent diffraction imaging and transport of intensity equation methods. Both sensors are also compact, robust, and vibration-insensitive. In either case, the spatial sampling rates are inevitably reduced when using aperture arrays or microlens arrays, resulting in poor spatial resolutions for Hartmann and Shack-Hartmann wavefront sensors. Thus, only phases in low frequency can be reconstructed. Since aberrations in optical systems and atmospheric fluctuations mainly consist of low-frequency phases, Hartmann and Shack-Hartmann wavefront sensors can still be used in adaptive optics for both astronomy and biology.

4.4 Other Quantitative Computational Optical Phase Imaging Techniques

As well as widely used quantitative computational optical phase imaging techniques such as holography, interferometry, coherent diffraction imaging, transport of intensity equation imaging, and Hartmann and Shack-Hartmann wavefront sensing, there are also other quantitative computational optical phase imaging techniques designed and used in different applications such as differential phase contrast microscopy, pyramid wavefront sensing, Moiré deflectometry, coded aperture phase imaging, and phase diversity. To make the book a good reference for computational optical phase imaging techniques, the following sections provide a brief introduction to these techniques.

4.4.1 Differential Phase Contrast Microscopy

First designed by Hamilton, Sheppard, and Wilson [101–103], differential phase contrast microscopy was adopted to enhance the imaging contrast. The main idea of differential phase contrast microscopy is based on the light deflection due to existing phase gradient. When a transparent specimen has a unique phase distribution without any phase gradient, the intensity of the collected image should be unique. While it has a non-zero phase gradient, the collected image is no more unique due to light deflection. Initially applied in confocal microscopy [101–105], differential phase contrast microscopy was further developed to use in bright field mode [106–108], and finally in quantitative phase imaging [109–118]. Compared to microscopy techniques of phase contrast, DIC and Hoffman contrast, differential phase contrast microscopy has rather simple optical system. Moreover, it does not rely completely on coherent light source. Therefore, it is an ideal tool in both qualitative and quantitative phase imaging applications.

The mathematical model of quantitative differential phase contrast microscopy is introduced in illumination modulation [110, 111], or explained in another perspective as in pupil modulation [112]. Two tactics are similar and illumination modulation is briefly explained in this section. The transmission function of the specimen $o(\boldsymbol{r_o})$ can be explained by Eq. (4.68), in which $\mu(\boldsymbol{r_o})$ represents the absorption, $\varphi(\boldsymbol{r_o})$ represents the phase delay, and $\boldsymbol{r_o}$ represents the specimen coordinates in the spatial domain.

$$o(\boldsymbol{r_o}) = e^{[-\mu(\boldsymbol{r_o})+i\varphi(\boldsymbol{r_o})]} \tag{4.68}$$

When a specimen is illuminated by a single point source, the light passing through the object can be described as $q(\boldsymbol{r_o})o(\boldsymbol{r_o})$, in which $q(\boldsymbol{r_o})$ is the point source illumination. The image of the specimen is explained by Eq. (4.69) or (4.70), in which $P(\boldsymbol{u''})$ is the pupil function in the frequency domain, $\boldsymbol{u''}$ represents the pupil plane, and $\boldsymbol{r_i}$ represents the imaging coordinates in the spatial domain.

$$I(\boldsymbol{r_i}) = \left| F^{-1}\left\{ F[q(\boldsymbol{r_o})o(\boldsymbol{r_o})] \cdot P(\boldsymbol{u''}) \right\} \right|^2 \tag{4.69}$$

$$I(\boldsymbol{r_i}) = \left| \iint \left[\iint q(\boldsymbol{r_o})o(\boldsymbol{r_o})e^{-i2\pi\boldsymbol{r_o}\cdot\boldsymbol{u''}}d^2\boldsymbol{r_o} \right] P(\boldsymbol{u''})e^{-i2\pi\boldsymbol{r_i}\cdot\boldsymbol{u''}}d^2\boldsymbol{u''} \right|^2 \tag{4.70}$$

Equations (4.69) and (4.70) only consider the point source illumination, the image at the camera is the incoherent sum of images corresponding to various point source illumination as explained by Eqs. (4.71) and (4.72), in which $\boldsymbol{u'}$ represents the illumination plane, and the illumination can be explained by Eq. (4.73).

$$I(\boldsymbol{r_i}) = \sum_{q(\boldsymbol{r_o})} \left| F^{-1}\left\{ F[q(\boldsymbol{r_o})o(\boldsymbol{r_o})] \cdot P(\boldsymbol{u''}) \right\} \right|^2 \tag{4.71}$$

$$I(\boldsymbol{r_i}) = \iint \left| \iint \left[\iint q(\boldsymbol{r_o})o(\boldsymbol{r_o})e^{-i2\pi\boldsymbol{r_o}\cdot\boldsymbol{u''}}d^2\boldsymbol{r_o} \right] P(\boldsymbol{u''})e^{-i2\pi\boldsymbol{r_i}\cdot\boldsymbol{u''}}d^2\boldsymbol{u''} \right|^2 d^2\boldsymbol{u'} \tag{4.72}$$

$$q(\boldsymbol{r_o}) = \sqrt{S(\boldsymbol{u'})}e^{-i2\pi\boldsymbol{r_o}\cdot\boldsymbol{u''}} \tag{4.73}$$

According to Eqs. (4.71) and (4.72), different illumination [110, 111] (or different pupil [112]) introduces different imaging intensities. Here, in illumination modulation case, different illumination induces different low-pass filtering, thus generating different images. In order to derive the phase transfer function similar to [109], a simplified model is provided according to the weak object transfer function as Eq. (4.74) used in [110–112]. Here, the effect of the weak object transfer function is to make the complex amplitude of specimen into two separated absorption and phase terms, therefore, the absorption transfer function is only determined by $\mu(\boldsymbol{r_o})$, and the phase transfer function is only determined by $\varphi(\boldsymbol{r_o})$.

$$o(\boldsymbol{r}_o) = e^{[-\mu(\boldsymbol{r}_o)+i\varphi(\boldsymbol{r}_o)]} \approx 1 - \mu(\boldsymbol{r}_o) + i\varphi(\boldsymbol{r}_o) \tag{4.74}$$

According to Eqs. (4.72), (4.73), and (4.74), the image in the frequency domain $I(\boldsymbol{u}_i)$ can be derived to Eq. (4.75) referring to [119].

$$I(\boldsymbol{u}_i) = B(\boldsymbol{u}_i)\delta(\boldsymbol{u}_i) + H_\mu(\boldsymbol{u}_i)\mathrm{F}[\mu(\boldsymbol{r}_o)] + H_\varphi(\boldsymbol{u}_i)\mathrm{F}[\varphi(\boldsymbol{r}_o)] \tag{4.75}$$

In Eq. (4.75), $B(\boldsymbol{u}_i)$ represents the background term explained as Eq. (4.76), $H_\mu(\boldsymbol{u}_i)$ represents the absorption transfer function explained as Eq. (4.77), and $H_\varphi(\boldsymbol{u}_i)$ represents the phase transfer function explained as Eq. (4.78)

$$B(\boldsymbol{u}_i) = \iint S(\boldsymbol{u}')\big|P(\boldsymbol{u}'')\big|^2 \mathrm{d}^2\boldsymbol{u}' \tag{4.76}$$

$$H_\mu(\boldsymbol{u}_i) = -\left[\iint S(\boldsymbol{u}')P^*(\boldsymbol{u}')P(\boldsymbol{u}'+\boldsymbol{u}_i)\mathrm{d}^2\boldsymbol{u}' + \iint S(\boldsymbol{u}')P^*(\boldsymbol{u}')P(\boldsymbol{u}'-\boldsymbol{u}_i)\mathrm{d}^2\boldsymbol{u}'\right] \tag{4.77}$$

$$H_\varphi(\boldsymbol{u}_i) = i\left[\iint S(\boldsymbol{u}')P^*(\boldsymbol{u}')P(\boldsymbol{u}'+\boldsymbol{u}_i)\mathrm{d}^2\boldsymbol{u}' - \iint S(\boldsymbol{u}')P^*(\boldsymbol{u}')P(\boldsymbol{u}'-\boldsymbol{u}_i)\mathrm{d}^2\boldsymbol{u}'\right] \tag{4.78}$$

In differential phase contrast microscopy, $I_T(\boldsymbol{r}_i)$ and $I_B(\boldsymbol{r}_i)$ represent images corresponding to top and bottom illumination, and the differential phase contrast image $I_{\mathrm{DPC}}(\boldsymbol{r}_i)$ can be computed according to Eq. (4.79).

$$I_{\mathrm{DPC}}(\boldsymbol{r}_i) = \frac{I_T(\boldsymbol{r}_i) - I_B(\boldsymbol{r}_i)}{I_T(\boldsymbol{r}_i) + I_B(\boldsymbol{r}_i)} \tag{4.79}$$

Therefore, according to Eq. (4.79), background term in the differential phase contrast image is cancelled. Additionally, since the pupil function is a real and symmetric circular function in the aberration-free condition, the absorption transfer function as $H_\mu(\boldsymbol{u}_i)$ is a symmetric function, therefore, for differential phase contrast imaging, its absorption transfer function $H_\mu^{\mathrm{DPC}}(\boldsymbol{u}_i)$ is zero. While the phase transfer function as $H_\varphi(\boldsymbol{u}_i)$ is an anti-symmetric function, therefore, for differential phase contrast imaging, its phase transfer function $H_\varphi^{\mathrm{DPC}}(\boldsymbol{u}_i)$ is non-zero. Therefore, according to Eqs. (4.76) to (4.79), the differential phase contrast imaging can be explained by Eq. (4.80), in which $H_\varphi^{\mathrm{DPC}}(\boldsymbol{u}_i)$ is $H_\varphi(\boldsymbol{u}_i)/B$.

$$I^{\mathrm{DPC}}(\boldsymbol{u}_i) = H_\varphi^{\mathrm{DPC}}(\boldsymbol{u}_i)F[\varphi(\boldsymbol{r}_o)] \tag{4.80}$$

Specimen phase retrieval in differential phase contrast imaging is very similar to deconvolution according to Eq. (4.81), in which n is the detecting number of differential phase contrast imaging, by adjusting the source, different images can be combined to compute different differential phase contrast images.

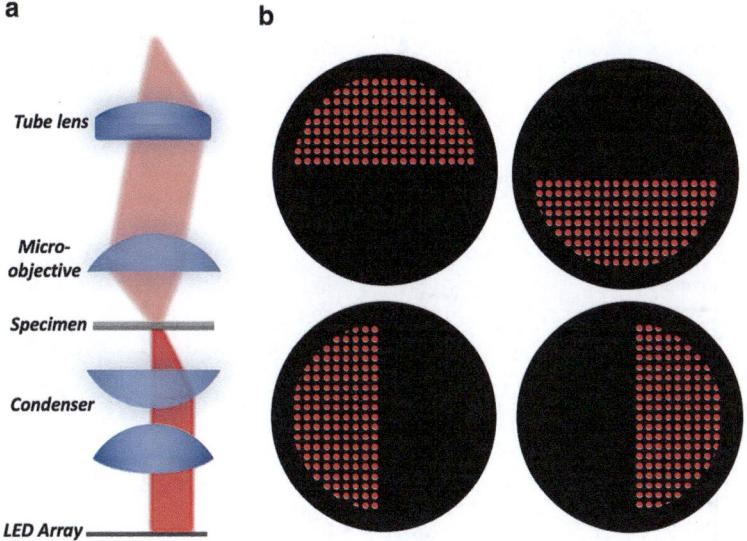

Fig. 4.24 Differential phase contrast microscopy. **a** Optical system; **b** LED array illumination

$$\min \sum_n \left| I_n^{\mathrm{DPC}}(\boldsymbol{u}_i) - H_{\varphi,n}^{\mathrm{DPC}}(\boldsymbol{u}_i) F[\varphi(\boldsymbol{r}_o)] \right|^2 + \alpha |F[\varphi(\boldsymbol{r}_o)]|^2 \qquad (4.81)$$

Using a programmable LED [110, 111, 120] or pupil [112] as shown in Fig. 4.24, various differential phase contrast images can be recorded, therefore, specimen phase can be retrieved according to Eq. (4.81). This method does not rely on coherent light or reference light, moreover, it has simple system and fast phase retrieval speed. Therefore, differential phase contrast is also a promising technique which has been widely used in many applications such as quantitative specimen measurements, aberration reconstruction, and so on.

4.4.2 Pyramid Wavefront Sensing

The pyramid wavefront sensor was first designed by Ragazzoni [121–123] especially for astronomical applications. Figure 4.25a demonstrates the pyramid wavefront sensing system. The pyramid prism at the Fourier plane of the pupil is the key optical element in the system, and its four faces can slightly deflect light into four different directions. All the deflected light can be then re-imaged into four separated pupil images using an extra re-imaging lens. In other words, each point on the entrance pupil corresponds to four points in four separated pupil images, and the wavefront derivative at such a point can be determined according to its four corresponding imaging point intensities.

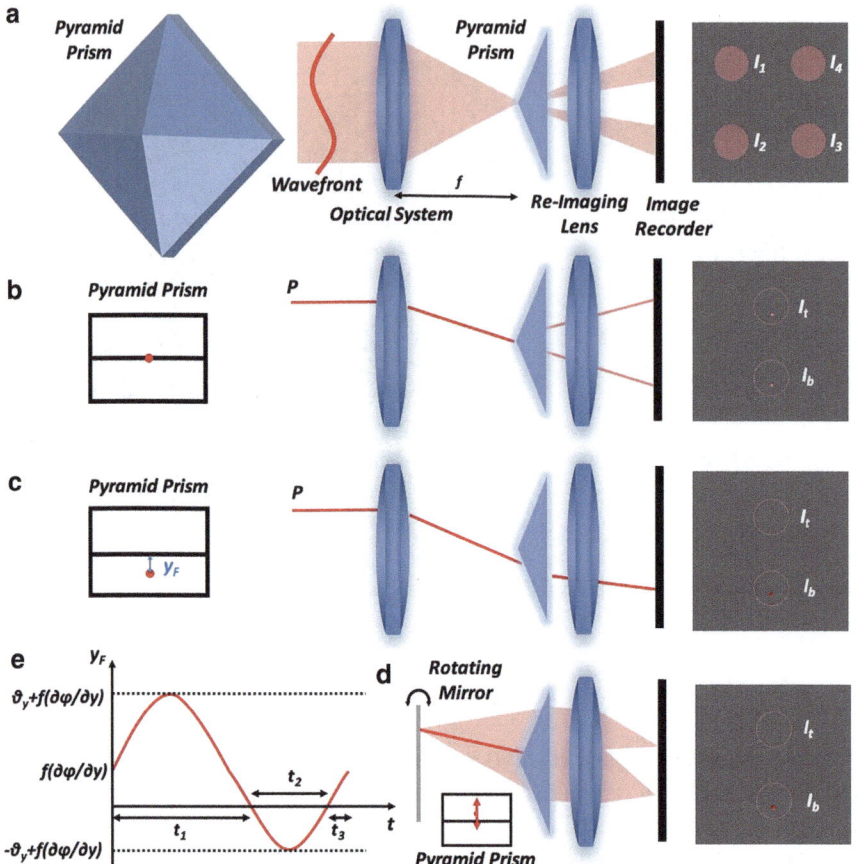

Fig. 4.25 Pyramid wavefront sensing. **b** Pyramid wavefront sensing system; **b** and **c** geometrical optics-based model of planar and non-planar wavefront in static mode, respectively; **d** geometrical optics-based model in dynamic mode; **e** relation between y_F and image intensity in dynamic mode

In order to demonstrate the principle of pyramid wavefront sensing technique, a geometrical optics-based model in one dimension is provided here. It is shown in Fig. 4.25b. A ray emitting from a point P is focused on the Fourier plane. When the wavefront at P is a plane wave, the refracted wave just hits the vertex of the pyramid prism as revealed in Fig. 4.25b, and is evenly separated into two parts and generates two image points with the re-imaging lens. However, if the wavefront phase at P is not planar but with a phase derivative such as $\frac{\partial \varphi}{\partial y_o}$, the emitting ray hitting the Fourier plane has a deviation y_F as shown in Fig. 4.25c, and the deviation satisfies Eq. (4.82), in which f is the equivalent focal distance of the imaging system.

$$y_F = \frac{\partial \varphi}{\partial y_o} f \tag{4.82}$$

However, in such condition, the ray only hits one half of the pyramid prism, while no light reaches the other half as shown in Fig. 4.25c. In such condition, the pyramid wavefront sensing is equivalent to Foucault knife-edge testing [124], which can only determine the direction of the phase derivative, though not its exact value.

In order to obtain the exact values of phase derivatives, dynamic modulations are required in the pyramid wavefront sensing system, such as vibrating the pyramid prism [121] or rotating the mirror [125]. Here, an example using mirror rotation is briefly introduced as revealed in Fig. 4.25d [125]. Note that pyramid prism vibration tactic is rather similar to mirror rotation tactic [121]. In one-dimensional perspective, a mirror with rotating angle ranging from $-\theta_y$ to θ_y is set before the pyramid prism, and $|\theta_y| > \left|\frac{\partial \varphi}{\partial y_o}\right|$. Via rotating such a mirror, the deviation $y_F(t)$ varies according to Eq. (4.83) as described in Fig. 4.25d.

$$y_F(t) = \left[\frac{\partial \varphi}{\partial y_o} + \theta_y(t)\right] f \tag{4.83}$$

In practices, the deviation $y_F(t)$ cannot be measured directly, but it is related to intensity distribution at the imaging plane. According to Fig. 4.25e, during the time interval t_1 and t_3, the ray hits the top half of the pyramid prism, while during the time interval t_2, the ray hits the bottom half of the pyramid prism. Therefore, the bias as $\frac{y_F}{f\theta_y}$ can be computed according to Eq. (4.84). It is worth noting that Eq. (4.84) varies according to different rotation ways [121].

$$\frac{y_F}{f\theta_y} = \sin\left(\frac{\pi}{2} \cdot \frac{t_1 - t_2 + t_3}{t_1 + t_2 + t_3}\right) \tag{4.84}$$

Assuming that the light intensity is evenly distributed during mirror rotation period. During t_1 and t_3, the generated image occurs in the top pupil image; while during t_2, the generated image occurs in the bottom pupil image. Therefore, according to imaging integral within the angular rotation period, the top pupil image intensity $I_t \propto t_1 + t_3$, and the bottom pupil image intensity $I_b \propto t_2$. Therefore, Eq. (4.85) is equivalent to Eq. (4.84), but $\frac{I_t - I_b}{I_t + I_b}$ can be directly measured according to the dynamic pyramid wavefront sensing system.

$$\frac{y_F}{f\theta_y} = \sin\left(\frac{\pi}{2} \cdot \frac{I_t - I_b}{I_t + I_b}\right) \tag{4.85}$$

Therefore, according to both Eqs. (4.82) and (4.85), the phase derivative along vertical direction can be obtained as Eq. (4.86).

$$\frac{\partial \varphi}{\partial y_o} = \theta_y \cdot \sin\left(\frac{\pi}{2} \cdot \frac{I_t - I_b}{I_t + I_b}\right) \tag{4.86}$$

While considering the two-dimensional condition as shown in Fig. 4.25a, phase derivatives in both horizontal and vertical directions can be demonstrated by Eqs. (4.87) and (4.88), in which, I_1, I_2, I_3, and I_4 are all marked in Fig. 4.25a.

$$\frac{\partial \varphi(x_0, y_0)}{\partial x_o} = \theta_x \cdot \sin\left(\frac{\pi}{2} \cdot \frac{(I_1 + I_2) - (I_3 + I_4)}{I_1 + I_2 + I_3 + I_4}\right) \tag{4.87}$$

$$\frac{\partial \varphi(x_0, y_0)}{\partial y_o} = \theta_y \cdot \sin\left(\frac{\pi}{2} \cdot \frac{(I_1 + I_4) - (I_2 + I_3)}{I_1 + I_2 + I_3 + I_4}\right) \tag{4.88}$$

Finally, the phase at the entrance pupil can be reconstructed via 2-D integration from $\frac{\partial \varphi(x_0, y_0)}{\partial x_o}$ and $\frac{\partial \varphi(x_0, y_0)}{\partial y_o}$.

Above demonstrations on pyramid wavefront sensing are based on geometrical theory [121, 125], while diffraction theory has also been employed to describe pyramid wavefront sensing rigorously, as explained by [126–130]. The same results can be obtained according to both simple geometrical model and rigorous diffraction theory [128].

Compared to widely used Shack-Hartmann wavefront sensing, pyramid wavefront sensing can achieve higher spatial resolution [131]. However, pyramid wavefront sensing has some limitations. One is that the pyramid prism is difficult to fabricate. In order to solve the problem, Vohnsen group proposed digital pyramid wavefront sensor by adopting digital micro-mirror device (DMD) instead of pyramid prism [132, 133]. Wang group reported sequential operation of pyramid wavefront sensing based on micro-mirror array [134]. Pyramid wavefront sensing relies on dynamic modulation, which often requires a moving component inside the pyramid wavefront sensing system. In order to avoid the modulation demand, tactics relying on extended source [135] or considering non-corrected higher-order or very fast aberrations as a form of modulation [136, 137] offer potential solutions. Moreover, algorithm has also been designed for phase estimation in non-modulated pyramid wavefront sensing system [138]. But these restrictions limit the use of pyramid wavefront sensing in many applications. Additionally, improved algorithms designed for phase reconstruction provide high-performance [139–144]. Besides applications in astronomy, pyramid wavefront sensing has also been used in adaptive optics [145–148], ophthalmics [149–152], optical testing [153], microscopy [154], and so on.

4.4.3   Moiré Deflectometry

Initially developed by Kafri [155], moiré deflectometry is a commonly used method to measure refractive index gradients in flame temperature measurements [156], optical instrument testing [157], and supersonic wind tunnel density measurements [158]. Comparatively to interferometry, moiré deflectometry is able to achieve anti-disturbing measurement in noisy environments with minimal mechanical stability requirements. In contrast to rainbow schlieren [159] and background-oriented schlieren [160], moiré deflectometry appears to perform better in providing quantitative measurements and visualization.

Existence of a non-uniform refractive index gradient causes deflection of light at various angles and the projection data in moiré deflection is an indication of this deviation of light from its original direction, which represents the first order derivative of the refractive index [161]. Several tactics have been reported to extract deflection projections from moiré fringes, such as image intensity-based detection approach [162], Fourier Transform Profilometry (FTP) [163], and the phase-shift method [164]. Fresnel diffraction theory has been utilized to analyze the moiré effect while moiré fringes are defined as the geometric multiplication of gratings intensities [165]. Meanwhile, diffraction phenomena have been reported to further influence the moiré image significantly, as a result the distance between gratings must be set as the Talbot distance [166–168]. Furthermore, the formation of moiré fringes has been explained in another perspective as [161] in scalar diffraction theory and the intensity distribution of zeroth-order as well as first-order moiré patterns are proved to be of a strict cosinusoidal intensity distribution. Herein, the theoretical analysis of moiré deflectometry is briefly explained.

A typical optical schematic diagram of moiré deflectometry is demonstrated in Fig. 4.26a. O refers to the refracting object, G1 and G2 represent two identical Ronchi gratings illuminated with collimated monochromatic coherent light. Δ is the distance between two gratings which oriented at angles $+\partial/2$ and $-\partial/2$, respectively, relative to the y axis. f refers to the focal length of lens L1 and L2. Filter regards as a pinhole filter. OP indicates the plane of observation. Consequently, a 4-f system is generated by L1, L2, and Filter. The field before G1 is expressed as Eq. (4.89), where $\varphi(x, y)$

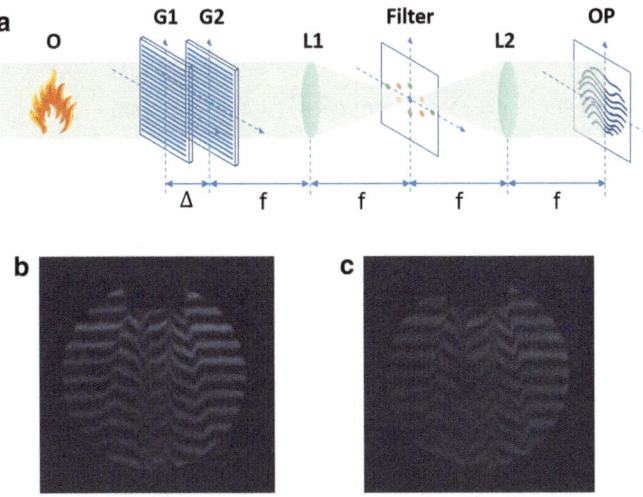

Fig. 4.26 Moiré deflectometry. **a** Optical system; **b** zeroth-order filtering; **c** first-order filtering

is the phase projection and k is defined as $2\pi/\lambda$ according to wavelength λ.

$$u_1^-(x, y) \propto e^{ik\varphi(x,y)} \tag{4.89}$$

The transmittance of G1 can be depicted by its Fourier expansion

$$g_1(x, y) = \sum_m a_m \exp\left[i\frac{2\pi m}{d}\left(x\cos\frac{\alpha}{2} - y\sin\frac{\alpha}{2}\right)\right] \tag{4.90}$$

where d is the period of grating. As a consequence, the field behind the grating G1 is

$$u_1^+(x, y) = F^{-1}\left[\sum_m a_m U_1^-\left(u - \frac{m}{d}\cos\frac{\alpha}{2}, v + \frac{m}{d}\sin\frac{\alpha}{2}\right)\right] \tag{4.91}$$

U_1^- is the angular spectrum of $u_1^-(x, y)$ and (u, v) is the Fourier spatial frequency components. The angular spectrum of $u_1^+(x, y)$ is

$$U_1^+(u, v) = \sum_m a_m U_1^-\left(u - \frac{m}{d}\cos\frac{\alpha}{2}, v + \frac{m}{d}\sin\frac{\alpha}{2}\right) \tag{4.92}$$

Angular spectrum for the plane before G2 is

$$U_2^-(u, v) = \exp\left(ik\Delta\sqrt{1 - \lambda^2(u^2 + v^2)}\right)\sum_m a_m U_1^-\left(u - \frac{m}{d}\cos\frac{\alpha}{2}, v + \frac{m}{d}\sin\frac{\alpha}{2}\right) \tag{4.93}$$

The transmittance of G2 can be expressed as

$$g_2(x, y) = \sum_n a_n \exp\left[i\frac{2\pi n}{d}\left(x\cos\frac{\alpha}{2} + y\sin\frac{\alpha}{2}\right)\right] \tag{4.94}$$

The concept of angular spectrum propagation as well as inverse Fourier transform are adopted and the field behind G2 is

$$u_2^+(x, y) = \exp(ik\Delta)\sum_m\sum_n a_m a_n$$
$$\times \exp\left\{\frac{i2\pi}{d}\left[(m+n)x\cos\frac{\alpha}{2} - (m-n)y\sin\frac{\alpha}{2}\right]\right\}$$
$$\times \exp\left(-\frac{i\pi\lambda\Delta m^2}{d^2}\right)u_1^-\left(x - \frac{\lambda\Delta m}{d}\cos\frac{\alpha}{2}, y + \frac{\lambda\Delta m}{d}\sin\frac{\alpha}{2}\right) \tag{4.95}$$

Considering the arrangement of 4-f system, the field of OP is as same as the field $u_2^+(x, y)$. For the case of the zeroth-order corresponding to $m + n = 0$, the moiré fringe equations are described by Eq. (4.96), where Q is an integer.

$$y = \begin{cases} \frac{Qd}{2\sin\frac{\alpha}{2}} - \frac{\partial\varphi(x,y)}{\partial x}\frac{Kd^2}{2\lambda}ctg\frac{\alpha}{2} & K \text{ is even} \\ \frac{Qd}{2\sin\frac{\alpha}{2}} + \frac{d}{4\sin\frac{\alpha}{2}} - \frac{\partial\varphi(x,y)}{\partial x}\frac{Kd^2}{2\lambda}ctg\frac{\alpha}{2} & K \text{ is odd} \end{cases} \quad (4.96)$$

As to the case of the first-order, which means $m + n = 1$ or $m + n = -1$, the fringe equation is

$$y = \frac{Qd}{2\sin\frac{\alpha}{2}} - \frac{Kd}{4\sin\frac{\alpha}{2}} - \frac{\partial\varphi(x, y)}{\partial x}\frac{Kd^2}{2\lambda}ctg\frac{\alpha}{2} \quad (4.97)$$

The zeroth-order filtering and first-order filtering imaging of Propane flame are illustrated in Fig. 4.26b and c.

4.4.4 Coded Aperture Phase Imaging

Coded aperture imaging has been used in traditional imaging for various applications ranging from astronomy [169–171] to microscopy [172] using a wider spectrum such as neutron [173], gamma rays [171], X-ray [174, 175], visible [176–178], and infrared light [179] as well as THz wave [180, 181]. Moreover, coded aperture imaging has also been employed in hyperspectral imaging [182, 183], polarization imaging [184, 185], light field imaging [186, 187] as well as phase imaging [188–195]. Figure 4.27a shows a classical design on multiple-shot reference-free coded aperture phase imaging [188–193]. A coded mask (amplitude or phase, binary or digital) is set in front of the image recorder. The captured intensity distribution I can be described according to Eq. (4.98), in which, Ψ is the wavefront under detection, ϕ is the complex amplitude of the coded mask, and H_z is the wavefront propagation operator with a propagation distance z. Repeating the process several times, I_m and ϕ_m are the mth recorded image and coded mask.

$$I_m = |H_z(\phi_m \cdot \Psi)|^2 \quad (4.98)$$

Solving φ from multiple measurements is equivalent to optimization problem described by Eq. (4.99).

$$\hat{\Psi} = \arg\min_{\Psi} \sum_m \sqrt{I_m} - |H_z(\phi_m \cdot \Psi)| \quad (4.99)$$

Unfortunately, this approach requires longer time for image recording since multiple coded images should be captured with a variation in coded masks. In order

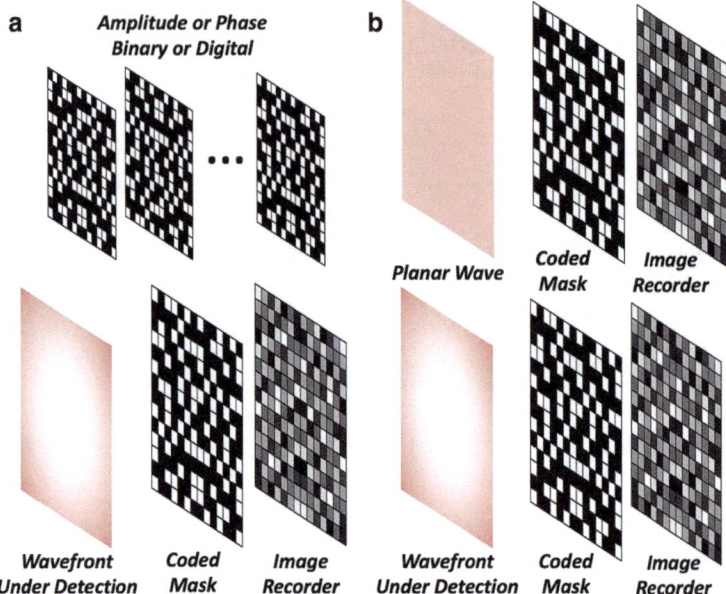

Fig. 4.27 Coded aperture phase imaging **a** Multiple-shot reference-free approach; **b** single-shot reference-based approach

to accelerate the speed of coded aperture phase imaging, Heidrich group provided a single-shot reference-based idea, which can decode the wavefront according to the motion of diffraction pattern [194, 195]. The system is similar to the one shown in Fig. 4.27b. Two intensities should be captured, $I_0(r)$ by a planar wave as a reference, and $I(r)$ by the distorted wave under detection, which can be described according to Eq. (4.100), in which φ is the phase under detection, z is the interval between the coded mask and the image recorder, and k is the wave number as $\frac{2\pi}{\lambda}$, λ is the wavelength, and A is the operator.

$$I(r) = I_0\left(r - \frac{z}{k}\nabla\varphi(r)\right) = AI_0(r) \tag{4.100}$$

Solving φ from multiple measurements is equivalent to the optimization problem described by Eq. (4.101).

$$\hat{\varphi} = \arg\min_{\varphi} I(r) - AI_0(r)^2 \tag{4.101}$$

Though two coded images are to be recorded, the reference one can be calibrated in advance. Therefore, this reference-based coded aperture phase imaging is capable of tracking dynamics.

Fig. 4.28 Phase diversity principle

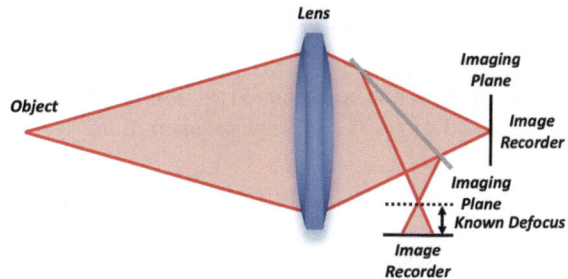

Coded aperture imaging provides a new route to computational optical phase imaging. In principle, such methods share a similar idea with many coherent diffraction imaging techniques such as Gerchberg-Saxton method and ptychography relying on iterative numerical wavefront propagation and sample update. Therefore, coded aperture phase imaging also suffers from heavy iterative phase retrieval computations, limiting its rapid phase reconstruction performance.

4.4.5 Phase Diversity

Phase diversity was first developed by Gonsalves [196], and it has been widely implemented particularly in adaptive optics. Phase diversity is a powerful method for estimating wavefront aberrations from an in-focus image and another out-of-focus image (or even more out of focus images). Unlike the Gerchberg-Saxton method or other coherent diffraction imaging methods that use similar optical systems but require highly coherent light, phase diversity can also be used when the light is partially coherent or even incoherent. It should be noted that the obtained complex amplitude is not the sample complex amplitude, but the OTF of the system. Figure 4.28 shows the phase diversity scheme.

Equation (4.102) describes the imaging principle, in which $i(x, y)$ is the collected imaging, $o(x, y)$ is the object, $PSF(x, y)$ is the point spreading function (PSF), and $\otimes$ represents the convolution.

$$i(x, y) = o(x, y) \otimes PSF(x, y) \tag{4.102}$$

Using Fourier transform, Eq. (4.102) can be transformed to Eq. (4.103), in which, $I(f_x, f_y)$, $O(f_x, f_y)$, and $OTF(f_x, f_y)$ are the Fourier transform of $i(x, y)$, $o(x, y)$, and $PSF(x, y)$, respectively. Moreover, $OTF(f_x, f_y)$ is the optical transfer function (OTF), and it is the autocorrelation of coherent transfer function $CTF(f_x, f_y)$ as described in Eq. (4.104).

$$I(f_x, f_y) = O(f_x, f_y) \cdot OTF(f_x, f_y) \tag{4.103}$$

$$\text{OTF}(f_x, f_y) = \text{CTF}(f_x, f_y) \odot \text{CTF}(f_x, f_y) \tag{4.104}$$

$\text{CTF}(f_x, f_y)$ can be explained according to Eq. (4.105), in which $A(f_x, f_y)$ is the aperture and $\phi(f_x, f_y)$ is the phase aberration.

$$\text{CTF}(f_x, f_y) = A(f_x, f_y) \cdot e^{i\phi(f_x, f_y)} \tag{4.105}$$

However, when the prior known system aberration (such as defocus) is introduced, the corresponding diversity coherent transfer function $\text{CTF}_d(f_x, f_y)$ becomes Eq. (4.106), in which $\Delta\phi(f_x, f_y)$ is the introduced phase aberration due to defocus, and it is also known as the phase diversity term.

$$\text{CTF}_d(f_x, f_y) = A(f_x, f_y) \cdot e^{i[\phi(f_x, f_y) + \Delta\phi(f_x, f_y)]} \tag{4.106}$$

According to the diversity coherent transfer function in Eq. (4.106), the diversity optical transfer function $\text{OTF}_d(f_x, f_y)$ becomes Eq. (4.107).

$$\text{OTF}_d(f_x, f_y) = \text{CTF}_d(f_x, f_y) \odot \text{CTF}_d(f_x, f_y) \tag{4.107}$$

The relation between the object and the image corresponding to the diversity optical transfer function can be described by Eq. (4.108), in which $I_d(f_x, f_y)$ is the Fourier transform of the defocus image $i_d(x, y)$.

$$I_d(f_x, f_y) = O(f_x, f_y) \cdot \text{OTF}_d(f_x, f_y) \tag{4.108}$$

In phase diversity method, $I(f_x, f_y)$ and $I_d(f_x, f_y)$ can be directly obtained from the captured $i(x, y)$ and $i_d(x, y)$, but $O(f_x, f_y)$, $\text{OTF}(f_x, f_y)$, and $\text{OTF}_d(f_x, f_y)$ are unknown. However, additional knowledge between $\text{OTF}(f_x, f_y)$ and $\text{OTF}_d(f_x, f_y)$ can be obtained according to Eqs. (4.105) and (4.106). Therefore, $o(x, y)$ can be obtained from $i(x, y)$ and $i_d(x, y)$ to minimize the error as Eq. (4.109) or (4.110) in either spatial or frequency domain. Therefore, phase retrieval in phase diversity is actually an optimization process.

According to the principle of phase diversity, the key point of phase retrieval is to develop a suitable optimization algorithm to minimize the error metric as the objective function listed in Eqs. (4.109) and (4.110) in spatial and frequency domains, respectively.

$$\varepsilon = \iint |i - o \otimes \text{PSF}|^2 dx dy + \iint |i_d - o \otimes \text{PSF}_d|^2 dx dy \tag{4.109}$$

$$E = \iint |I - O \cdot OTF|^2 df_x df_y + \iint |I_d - O \cdot OTF_d|^2 df_x df_y \tag{4.110}$$

The most classical phase retrieval method for phase diversity is a gradient-search algorithm proposed by Fienup group [197]. Quasi-Newton algorithm as Broyden-Fletcher-Goldfarb-Shanno (BFGS)-based phase retrieval methods have also been proposed [198, 199]. Least-squares [200], maximum-likelihood estimation [201], and marginal estimation [202] have also been adopted to deal with the phase retrieval in phase diversity; and many modern optimization techniques including hybrid particle swarm global optimization [203], particle swarm optimization [204], adaptive cuckoo search optimization [205]. Additionally, simplified phase diversity methods introducing approximations have also been designed [206–208]. However, there are still many more algorithms dealing with phase diversity reconstructions that are not listed here.

Different from commonly used wavefront sensors such as interferometers, shearing interferometers, and Shack-Hartmann sensors, phase diversity method does not rely on extra optical hardware, thus it is a rather simple technique. Compared to phase diversity, curvature wavefront sensing by solving the transport of intensity equation [29, 30] can also retrieve wavefront from several multi-focus images. However, phase diversity and curvature wavefront sensing are completely different methods, which have been clearly illustrated by Fienup [209]. Moreover, as an alternative to conventional focus diversity, Fienup group also reported sub-aperture piston phase diversity [210, 211] and transverse translation phase diversity [212, 213] for wavefront measurements.

References

1. Malacara, D. (ed.): Optical Shop Testing, 3rd edn. Wiley-Interscience, New Jersey (2007)
2. Schmahl, G., Rudolph, D., Niemann, B., Christ, O.: Zone-plate X-ray microscopy. Q. Rev. Biophys. **13**, 297–315 (1980)
3. Hawkes, P.W.: The correction of electron lens aberrations. Ultramicroscopy **156**, A1–A64 (2015)
4. Gerchburg, R., Saxton, W.: A practical algorithm for determination of phase from image and diffraction plane pictures. Optik **35**, 237–246 (1972)
5. Fienup, J.R.: Phase retrieval algorithms: a comparison. Appl. Opt. **21**, 2758–2769 (1982)
6. Faulkner, H.M.L., Rodenburg, J.M.: Movable aperture lensless transmission microscopy: a novel phase retrieval algorithm. Phys. Rev. Lett. **93**, 023903 (2004)
7. Zhang, F., Chen, B., Morrison, G.R., Vila-Comamala, J., Guizar-Sicairos, M., Robinson, I.K.: Phase retrieval by coherent modulation imaging. Nat. Commun. **7**, 13367 (2016)
8. Horstmeyer, R., Yang, C.: A phase space model of Fourier ptychographic microscopy. Opt. Express **22**, 338–358 (2014)
9. Chang, C., Pan, X., Tao, H., Liu, C., Veetil, S.P., Zhu, J.: 3D single-shot ptychography with highly tilted illuminations. Opt. Express **29**, 30878–30891 (2021)
10. Li, P., Batey, D.J., Edo, T.B., Parsons, A.D., Rau, C., Rodenburg, J.M.: Multiple mode x-ray ptychography using a lens and a fixed diffuser optic. J. Opt. **18**, 054008 (2016)
11. Humphry, M.J., Kraus, B., Hurst, A.C., Maiden, A.M., Rodenburg, J.M.: Ptychographic electron microscopy using high-angle dark-field scattering for sub-nanometre resolution imaging. Nat. Commun. **3**, 730 (2012)
12. Truong, N.X., Safaei, R., Cardin, V., Lewis, S.M., Zhong, X.L., Legere, F., Denecke, M.A.: Coherent tabletop EUV ptychography of nanopatterns. Sci. Rep. **8** (2018)

13. Valzania, L., Feurer, T., Zolliker, P., Hack, E.: Terahertz ptychography. Opt. Lett. **43**, 543–546 (2018)
14. Bian, Z., Dong, S., Zheng, G.: Adaptive system correction for robust Fourier ptychographic imaging. Opt. Express **21**, 32400–32410 (2013)
15. Dai, B., Zhu, D., Jaroensri, R., Kulalert, K., Pianetta, P., Pease, R.F.W.: Optical and computed evaluation of keyhole diffractive imaging for lensless x-ray microscopy. J. Vac. Sci. Technol. B **28**, C6Q1 (2010)
16. Bauschke, H.H., Combettes, P.L., Luke, D.R.: Phase retrieval, error reduction algorithm, and Fienup variants: a view from convex optimization. J. Opt. Soc. Am. A **19**, 1334–1345 (2002)
17. Hunt, B.R.: Matrix formulation of the reconstruction of phase values from phase differences. J. Opt. Soc. Am. **69**, 393–399 (1979)
18. Maiden, A.M., Rodenburg, J.M.: An improved ptychographical phase retrieval algorithm for diffractive imaging. Ultramicroscopy **109**, 1256–1262 (2009)
19. Maiden, A.M., Humphry, M.J., Sarahan, M.C., Kraus, B., Rodenburg, J.M.: An annealing algorithm to correct positioning errors in ptychography. Ultramicroscopy **120**, 64–72 (2012)
20. Zhang, F., Peterson, I., Vila-Comamala, J., Berenguer, A.D.F., Bean, R., Chen, B., Menzel, A., Robinson, I.K., Rodenburg, J.M.: Translation position determination in ptychographic coherent diffraction imaging. Opt. Express **21**, 13592–13606 (2013)
21. Claus, D., Rodenburg, J.M.: Multiwavelength Ptychography. Springer, Berlin (2014)
22. Thibault, P., Menzel, A.: Reconstructing state mixtures from diffraction measurements. Nature **494**, 68–71 (2013)
23. Hoppe, W.: Diffraction in inhomogeneous primary wave fields 1. Principle of phase determination from electron diffraction interference. Acta Crystallogr. A **25**, 495–501 (1969)
24. Hoppe, W.: Diffraction in inhomogeneous primary wave fields. 3. Amplitude and phase determination for nonperiodic objects. Acta Crystallogr. A **25**, 508–515 (1969)
25. Hegerl, R., Hoppe, W.: Phase evaluation in generalized diffraction (ptychography). In: Proceeding of the 5th European Congress on Electron Microscopy, pp. 628–629 (1972)
26. Hue, F., Rodenburg, J.M., Maiden, A.M., Midgley, P.A.: Extended ptychography in the transmission electron microscope: Possibilities and limitations. Ultramicroscopy **111**, 1117–1123 (2011)
27. Zhang, F., Rodenburg, J.M.: Phase retrieval based on wave-front relay and modulation. Phys. Rev. B **82**, 121104 (2010)
28. Teague, M.R.: Deterministic phase retrieval: a Green's function solution. J. Opt. Soc. Am. A **73**, 1434–1441 (1983)
29. Roddier, F.: Curvature sensing and compensation: a new concept in adaptive optics. Appl. Opt. **27**, 1223–1225 (1988)
30. Roddier, F.: Wavefront sensing and the irradiance transport equation. Appl. Opt. **29**, 1402–1403 (1990)
31. Gureyev, T.E., Wilkins, S.W.: On X-ray phase retrieval from polychromatic images. Opt. Commun. **147**, 229–232 (1998)
32. Bajt, S., Barty, A., Nugent, K.A., McCartney, M., Wall, M., Paganin, D.: Quantitative phase-sensitive imaging in a transmission electron microscope. Ultramicroscopy **83**, 67–73 (2000)
33. Nugent, K.A., Paganin, D.: Matter-wave phase measurement: A noninterferometric approach. Phys. Rev. A **61**, 063614 (2000)
34. Allman, B.E., McMahon, P.J., Nugent, K.A., Paganin, D., Jacobson, D.L., Arif, M., Werner, S.A.: Imaging-Phase radiography with neutrons. Nature **408**, 158–159 (2000)
35. De Graef, M., Zhu, Y.M.: Quantitative noninterferometric Lorentz microscopy. J. Appl. Phys. **89**, 7177 (2001)
36. Allen, L.J., Oxley, M.P.: Phase retrieval from series of images obtained by defocus variation. Opt. Commun. **199**, 65–75 (2001)
37. Paganin, D., Barty, A., McMahon, P.J., Nugent, K.A.: Quantitative phase-amplitude microscopy. III. The effects of noise. J. Microsc. **214**, 51–61 (2004)
38. Martin, A.V., Chen, F.R., Hsieh, W.K., Kai, J.J., Findlay, S.D., Allen, L.J.: Spatial incoherence in phase retrieval based on focus variation. Ultramicroscopy **106**, 914–924 (2006)

39. Huang, S., Xi, F., Liu, C., Jiang, Z.: Frequency analysis of a wavefront curvature sensor: selection of propagation distance. J. Mod. Opt. **59**, 35–41 (2012)
40. Ishizuka, K., Allman, B.: Phase measurement of atomic resolution image using transport of intensity equation. J. Microsc. **54**, 191–197 (2005)
41. Soto, M., Acosta, E.: Improved phase imaging from intensity measurements in multiple planes. Appl. Opt. **46**, 7978–7981 (2007)
42. Waller, L., Tian, L., Barbastathis, G.: Transport of Intensity phase-amplitude imaging with higher order intensity derivatives. Opt. Express **18**, 12552–12561 (2010)
43. Bie, R., Yuan, X.H., Zhao, M., Zhang, L.: Method for estimating the axial intensity derivative in the TIE with higher order intensity derivatives and noise suppression. Opt. Express **20**, 8186–8191 (2012)
44. Xue, B., Zheng, S., Cui, L., Bai, X., Zhou, F.: Transport of intensity phase imaging from multiple intensities measured in unequally-spaced planes. Opt. Express **19**, 20244–20250 (2011)
45. Zheng, S., Xue, B., Xue, W., Bai, X., Zhou, F.: Transport of intensity phase imaging from multiple noisy intensities measured in unequally-spaced planes. Opt. Express **20**, 972–985 (2012)
46. Zhong, J., Claus, R.A., Dauwels, J., Tian, L., Waller, L.: Transport of Intensity phase imaging by intensity spectrum fitting of exponentially spaced defocus planes. Opt. Express **22**, 10661–10674 (2014)
47. Zuo, C., Chen, Q., Yu, Y., Asundi, A.: Transport-of-intensity phase imaging using Savitzky-Golay differentiation filter-theory and applications. Opt. Express **21**, 5346–5362 (2013)
48. Martinez-Carranza, J., Falaggis, K., Kozacki, T.: Optimum measurement criteria for the axial derivative intensity used in transport of intensity-equation-based solvers. Opt. Lett. **39**, 182–185 (2014)
49. Martinez-Carranza, J., Falaggis, K., Kozacki, T.: Optimum plane selection for transport-of-intensity-equation-based solvers. Appl. Opt. **53**, 7050–7058 (2014)
50. Martinez-Carranza, J., Falaggis, K., Kozacki, T.: Multi-filter transport of intensity equation solver with equalized noise sensitivity. Opt. Express **23**, 23092–23107 (2015)
51. Hu, J., Wei, Q., Kong, Y., Jiang, Z., Xue, L., Liu, F., Kim, D.Y., Liu, C., Wang, S.: Higher order transport of intensity equation methods: comparisons and their hybrid application for noise adaptive phase imaging. IEEE Photon. J. **11**, 4200214 (2019)
52. Takeda, M., Ina, H., Kobayashi, S.: Fourier-transform method of fringe-pattern analysis for computer-based topography and interferometry. J. Opt. Soc. Am. **72**, 156–160 (1982)
53. Ichikawa, K., Lohmann, A.W., Takeda, M.: Phase retrieval based on the irradiance transport equation and the Fourier transform method: experiments. Appl. Opt. **27**, 3433–3436 (1988)
54. Gureyev, T.E., Nugent, K.A.: Rapid quantitative phase imaging using the transport of intensity equation. Opt. Commun. **133**, 339–346 (1997)
55. Paganin, D., Nugent, K.A.: Noninterferometric phase imaging with partially coherent light. Phys. Rev. Lett. **80**, 2586–2589 (1998)
56. Gureyev, T.E., Roberts, A., Nugent, K.A.: Phase retrieval with the transport-of-intensity equation: matrix solution with use of Zernike polynomials. J. Opt. Soc. Am. A **12**, 1932–1941 (1995)
57. Gureyev, T.E., Nugent, K.A.: Phase retrieval with the transport-of-intensity equation. II. Orthogonal series solution for nonuniform illumination. J. Opt. Soc. Am. A **13**, 1670–1682 (1996).
58. Woods, S.C., Greenaway, A.H.: Wave-front sensing by use of a Green's function solution to the intensity transport equation. J. Opt. Soc. Am. A **20**, 508–512 (2003)
59. Zuo, C., Chen, Q., Asundi, A.: Boundary-artifact-free phase retrieval with the transport of intensity equation: fast solution with use of discrete cosine transform. Opt. Express **22**, 9220–9244 (2014)
60. Zuo, C., Chen, Q., Li, H., Qu, W., Asundi, A.: Boundary-artifact-free phase retrieval with the transport of intensity equation II: applications to microlens characterization. Opt. Express **22**, 18310–18324 (2014)

61. Volkov, V.V., Zhu, Y., Graef, M.D.: A new symmetrized solution for phase retrieval using the transport of intensity equation. Micron **33**, 411–416 (2002)
62. Parvizi, A., Muller, J., Funken, S.A., Koch, C.T.: A practical way to resolve ambiguities in wavefront reconstructions by the transport of intensity equation. Ultramicroscopy **154**, 1–6 (2015)
63. Schmalz, J.A., Gureyev, T.E., Paganin, D.M., Pavlov, K.M.: Phase retrieval using radiation and matter-wave fields: Validity of Teague's method for solution of the transport-of-intensity equation. Phys. Rev. A **84**, 023808 (2011)
64. Zuo, C., Chen, Q., Huang, L., Asundi, A.: Phase discrepancy analysis and compensation for fast Fourier transform based solution of the transport of intensity equation. Opt. Express **22**, 17172–17186 (2014)
65. Huang, L., Zuo, C., Idir, M., Qu, W., Asundi, A.: Phase retrieval with the transport-of-intensity equation in an arbitrarily shaped aperture by iterative discrete cosine transforms. Opt. Lett. **40**, 1976–1979 (2015)
66. Streibl, N.: Phase imaging by the transport equation of intensity. Opt. Commun. **49**, 6–10 (1984)
67. Gureyev, T.E., Roberts, A., Nugent, K.A.: Partially coherent fields, the transport-of-intensity equation, and phase uniqueness. J. Opt. Soc. Am. A **12**, 1942–1946 (1995)
68. Gureyev, T.E., Paganin, D.M., Stevenson, A.W., Mayo, S.C., Wilkins, S.W.: Generalized eikonal of partially coherent beams and its use in quantitative imaging. Phys. Rev. Lett. **93**, 068103 (2004)
69. Gureyev, T.E., Pogany, A., Paganin, D.M., Wilkins, S.W.: Linear algorithms for phase retrieval in the Fresnel region. Opt. Commun. **231**, 53–70 (2004)
70. Gureyev, T.E., Nesterets, Y.I., Paganin, D.M., Pogany, A., Wilkins, S.W.: Linear algorithms for phase retrieval in the Fresnel region. 2. Partially coherent illumination. Opt. Commun. **259**, 569–580 (2004)
71. Zysk, A.M., Schoonover, R.W., Carney, P.S., Anastasio, M.A.: Transport of intensity and spectrum for partially coherent fields. Opt. Lett. **35**, 2239–2241 (2010)
72. Petruccelli, J.C., Tian, L., Barbastathis, G.: The transport of intensity equation for optical path length recovery using partially coherent illumination. Opt. Express **21**, 14430–14441 (2013)
73. Zuo, C., Chen, Q., Tian, L., Waller, L., Asundi, A.: Transport of intensity phase retrieval and computational imaging for partially coherent fields: the phase space perspective. Opt. Lasers Eng. **71**, 20–32 (2015)
74. Zuo, C., Chen, Q., Qu, W., Asundi, A.: High-speed transport-of-intensity phase microscopy with an electrically tunable lens. Opt. Express **21**, 24060–24075 (2013)
75. Waller, L., Kou, S.S., Sheppard, C.J.R., Barbastathis, G.: Phase from chromatic aberrations. Opt. Express **18**, 22817–22825 (2010)
76. Blanchard, P.M., Fisher, D.J., Woods, S.C., Greenaway, A.H.: Phase-diversity wave-front sensing with a distorted diffraction grating. Appl. Opt. **39**, 6649–6655 (2000)
77. Waller, L., Luo, Y., Yang, S.Y., Barbastathis, G.: Transport of intensity phase imaging in a volume holographic microscope. Opt. Lett. **35**, 2961–2963 (2010)
78. Yu, W., Tian, X., He, X., Song, X., Xue, L., Liu, C., Wang, S.: Real time quantitative phase microscopy based on single-shot transport of intensity equation (ssTIE) method. Appl. Phys. Lett. **109**, 071112 (2016)
79. Zuo, C., Chen, Q., Qu, W., Asundi, A.: Noninterferometric single-shot quantitative phase microscopy. Opt. Lett. **38**, 3538–3541 (2013)
80. Li, Y., Di, J., Ma, C., Zhang, J., Zhong, J., Wang, K., Xi, T., Zhao, J.: Quantitative phase microscopy for cellular dynamics based on transport of intensity equation. Opt. Express **26**, 586–593 (2018)
81. Li, Y., Di, J., Wu, W., Shang, P., Zhao, J.: Quantitative investigation on morphology and intracellular transport dynamics of migrating cells. Appl. Opt. **58**, G162–G168 (2019)
82. Gupta, A.K., Mahendra, R., Nishchal, N.K.: Single-shot phase imaging based on transport of intensity equation. Opt. Commun. **477**, 126347 (2020)

83. Ma, C., Lin, X., Suo, J., Dai, Q., Wetzstein, G.: Transparent object reconstruction via coded transport of intensity. Paper presented at the 2014 IEEE Conference on Computer Vision and Pattern Recognition, Columbus, 23–28 June 2014

84. Tian, X., Yu, W., Meng, X., Sun, A., Xue, L., Liu, C., Wang, S.: Real-time quantitative phase imaging based on transport of intensity equation with dual simultaneously recorded field of view. Opt. Lett. **41**, 1427–1430 (2016)

85. Gong, Q., Wei, Q., Xu, J., Kong, Y., Jiang, Z., Qian, W., Zhu, Y., Xue, L., Liu, F., Liu, C., Wang, S.: Digital field of view correction combined dual-view transport of intensity equation method for real time quantitative imaging. Opt. Eng. **57**, 063102 (2018)

86. Shan, Y., Gong, Q., Wang, J., Xu, J., Wei, Q., Liu, C., Xue, L., Wang, S., Liu, F.: Measurements on ATP induced cellular fluctuations using real-time dual view transport of intensity phase microscopy. Biomed. Opt. Express **10**, 2337–2354 (2019)

87. Chen, C., Lu, Y.N., Huang, H., Yan, K., Jiang, Z., He, X., Kong, Y., Liu, C., Liu, F., Xue, L., Wang, S.: PhaseRMiC: phase real-time microscope camera for live cell imaging. Biomed. Opt. Express **12**, 5261–5271 (2021)

88. Xing, X., Zhu, L., Chen, C., Sun, N., Yang, C., Yan, K., Xue, L., Wang, S.: Transformer oil quality evaluation using quantitative phase microscopy. Appl. Opt. **61**, 422–428 (2022)

89. Barone-Nugent, E.D., Barty, A., Nugent, K.A.: Quantitative phase-amplitude microscopy I: optical microscopy. J. Microsc. **206**, 194–203 (2002)

90. McMahon, P.J., Barone-Nugent, E.D., Allman, B.E., Nugent, K.A.: Quantitative phase-amplitude microscopy II: differential interference contrast imaging for biological TEM. J. Microsc. **206**, 204–208 (2002)

91. Bellair, C.J., Curl, C.L., Allman, B.E., Harris, P.J., Roberts, A., Delbridge, L.M.D., Nugent, K.A.: Quantitative phase amplitude microscopy IV: imaging thick specimens. J. Microsc. **214**, 62–69 (2004)

92. Zuo, C., Li, J., Sun, J., Fan, Y., Zhang, J., Lu, L., Zhang, R., Wang, B., Huang, L., Chen, Q.: Transport of intensity equation: a tutorial. Opt. Lasers Eng. **135**, 106187 (2020)

93. Hartmann, J.: Bemerkungen uber den Bau und die Justirung von Spektrographen. Zeitschrift fuer Instrumentenkunde **20**, 47–58 (1900)

94. Shack, R.V., Platt, B.C.: Production and use of a lenticular Hartmann screen. J. Opt. Soc. Am. **61**, 656 (1971)

95. Platt, B.C., Shack, R.V.: History and principles of Shack-Hartmann wavefront sensing. J. Refractive Surg. **17**, S573–S577 (2001)

96. Malacara-Hernández, D., Malacara-Doblado, D.: What is a Hartmann test? Appl. Opt. **54**, 2296–2301 (2015)

97. Levine, B.M.: Hartmann sensors for optical testing. Proc. SPIE, 3134 (1997)

98. Flöter, B., Juranic, P., Kapitzki, S., Keitel, B., Mann, K., Plönjes, E., Schäfer, B., Tiedtke, K.: EUV Hartmann sensor for wavefront measurements at the Free-electron LASer in Hamburg. New J. Phys. **12**, 083015 (2010)

99. Davies, R., Kasper, M.: Adaptive optics for astronomy. Annu. Rev. Astron. Astrophys. **50**, 305–351 (2012)

100. Booth, M.J.: Adaptive optical microscopy: the ongoing quest for a perfect image. Light Sci. Appl. **3**, e165 (2014)

101. Hamilton, D.K., Sheppard, C.J.R.: Differential phase contrast in scanning optical microscopy. J. Microsc. **133**, 27–39 (1984)

102. Hamilton, D.K., Sheppard, C.J.R., Wilson, T.: Improved imaging of phase gradients in scanning optical microscopy. J. Microsc. **135**, 275–286 (1984)

103. Hamilton, D.K., Wilson, T.: Two-dimensional phase imaging in the scanning optical microscope. Appl. Opt. **23**, 348–352 (1984)

104. Atkinson, M.R., Dixon, A.E.: Single-pinhole confocal differential phase contrast microscopy. Appl. Opt. **33**, 641–653 (1994)

105. Amos, W.B., Reichelt, S., Cattermole, D.M., Laufer, J.: Re-evaluation of differential phase contrast (DPC) in a scanning laser microscope using a split detector as an alternative to differential interference contrast (DIC) optics. J. Microsc. **210**, 166–175 (2003)

106. Stewart, W.C.: On differential phase contrast with an extended illumination source. J. Opt. Soc. Am. **66**, 813–818 (1976)

107. Tian, L., Wang, J., Waller, L.: 3D differential phase-contrast microscopy with computational illumination using an LED array. Opt. Lett. **39**, 1326–1329 (2014)

108. Kheireddine, S., Smith, Z.J., Nicolau, D.V., Wachsmann-Hogiu, S.: Simple adaptive mobile phone screen illumination for dual phone differential phase contrast (DPDPC) microscopy. Biomed. Opt. Express **10**, 4369–4380 (2019)

109. Mehta, S.B., Sheppard, C.J.R.: Quantitative phase-gradient imaging at high resolution with asymmetric illumination-based differential phase contrast. Opt. Lett. **34**, 1924–1926 (2009)

110. Tian, L., Waller, L.: Quantitative differential phase contrast imaging in an LED array microscope. Opt. Express **23**, 11394–11403 (2015)

111. Chen, M., Tian, L., Waller, L.: 3D differential phase contrast microscopy. Biomed. Opt. Express **7**, 3940–3950 (2016)

112. Lu, H., Chung, J., Ou, X., Yang, C.: Quantitative phase imaging and complex field reconstruction by pupil modulation differential phase contrast. Opt. Express **24**, 25345–25361 (2016)

113. Chen, M., Phillips, Z.F., Waller, L.: Quantitative differential phase contrast (DPC) microscopy with computational aberration correction. Opt. Express **26**, 32888–32899 (2018)

114. Kellman, M., Chen, M., Phillips, Z.F., Lustig, M., Waller, L.: Motion-resolved quantitative phase imaging. Biomed. Opt. Express **9**, 5456–5466 (2018)

115. Lee, D., Ryu, S., Kim, U., Jung, D., Joo, C.: Color-coded LED microscopy for multi-contrast and quantitative phase-gradient imaging. Biomed. Opt. Express **6**, 4912–4922 (2015)

116. Lin, Y.-Z., Huang, K.-Y., Luo, Y.: Quantitative differential phase contrast imaging at high resolution with radially asymmetric illumination. Opt. Lett. **43**, 2973–2976 (2018)

117. Chuang, Y.-H., Lin, Y.-Z., Vyas, S., Huang, Y.-Y., Yeh, J.A., Luo, Y.: Multi-wavelength quantitative differential phase contrast imaging by radially asymmetric illumination. Opt. Lett. **44**, 4542–4545 (2019)

118. Bonati, C., Laforest, T., Kunzi, M., Moser, C.: Phase sensitivity in differential phase contrast microscopy: limits and strategies to improve it. Opt. Express **28**, 33767–33783 (2020)

119. Streibl, N.: Three-dimensional imaging by a microscope. J. Opt. Soc. Am. A **2**, 121–127 (1985)

120. https://www.sci-microscopy.com

121. Ragazzoni, R.: Pupil plane wavefront sensing with an oscillating prism. J. Mod. Optic. **43**, 289–293 (1996)

122. Ragazzoni, R., Farinato, J.: Sensitivity of a pyramidic wave front sensor in closed loop adaptive optics. Astron. Astrophys. **350**, L23–L26 (1999)

123. Esposito, S., Riccardi, A.: Pyramid wavefront sensor behaviour in partial correction adaptive optics system. Astron. Astrophys. **369**, L9–L12 (2001)

124. Welford, W.T.: A note on the theory of the Foucault knife-edge test. Opt. Commun. **1**, 443–445 (1970)

125. Riccardi, A., Bindi, N., Ragazzoni, R., Esposito, S., Stefanini, P.: Laboratory characterization of a Foucault-like wavefront sensor for adaptive optics. Proc. SPIE **3353**, 941–951 (1998)

126. Esposito, S., Feeney, O., Riccardi, A.: Laboratory test of a pyramid wavefront sensor. Proc. SPIE **4007**, 416–422 (2000)

127. Verinaud, C.: On the nature of the measurements provided by a pyramid wavefront sensor. Opt. Commun. **233**, 27–38 (2004)

128. Burvall, A., Daly, E., Chamot, S.R., Dainty, C.: Linearity of the pyramid wavefront sensor. Opt. Express **14**, 11925–11934 (2006)

129. Korkiakoski, V., Vérinaud, C., Le Louarn, M., Conan, R.: Comparison between a model-based and a conventional pyramid sensor reconstructor. Appl. Opt. **46**, 6176–6184 (2007)

130. Wang, J., Bai, F., Ning, Y., Huang, L., Wang, S.: Comparison between non-modulation four-sided and two-sided pyramid wavefront sensor. Opt. Express **18**, 27534–27549 (2010)

131. Plantet, C., Meimon, S., Conan, J.-M., Fusco, T.: Revisiting the comparison between the Shack-Hartmann and the pyramid wavefront sensors via the Fisher information matrix. Opt. Express **23**, 28619–28633 (2015)

132. Akondi, V., Castillo, S., Vohnsen, B.: Digital pyramid wavefront sensor with tunable modulation. Opt. Express **21**, 18261–18272 (2013)
133. Akondi, V., Vohnsen, B., Marcos, S.: Virtual pyramid wavefront sensor for phase unwrapping. Appl. Opt. **55**, 8363–8367 (2016)
134. Yao, K., Wang, J., Liu, X., Li, H., Wang, M., Cui, B., Yu, S.: Pyramid wavefront sensor using a sequential operation method. Appl. Opt. **54**, 3894–3901 (2015)
135. Ragazzoni, R., Diolaiti, A., Vernet, E.: A pyramid wavefront sensor with no dynamic modulation. Opt. Commun. **208**, 51–60 (2002)
136. Costa, J.B., Ragazzoni, R., Ghedina, A., Carbillet, C., Verinaud, M., Feldt, M., Esposito, S., Puga, E., Farinato, J.: Is there need of any modulation in the pyramid wavefront sensor? Proc. SPIE **4839**, 288–298 (2003)
137. Costa, J.B.: Modulation effect of the atmosphere in a pyramid wavefront sensor. Appl. Opt. **44**, 60–66 (2005)
138. Frazin, R.A.: Efficient, nonlinear phase estimation with the nonmodulated pyramid wavefront sensor. J. Opt. Soc. Am. A **35**, 594–607 (2018)
139. Korkiakoski, V., Vérinaud, C., Le Louarn, M.: Improving the performance of a pyramid wavefront sensor with modal sensitivity compensation. Appl. Opt. **47**, 79–87 (2008)
140. Shatokhina, I., Obereder, A., Rosensteiner, M., Ramlau, R.: Preprocessed cumulative reconstructor with domain decomposition: a fast wavefront reconstruction method for pyramid wavefront sensor. Appl. Opt. **52**, 2640–2652 (2013)
141. Shatokhina, I., Ramlau, R.: Convolution- and Fourier-transform-based reconstructors for pyramid wavefront sensor. Appl. Opt. **56**, 6381–6390 (2017)
142. Hutterer, V., Ramlau, R.: Nonlinear wavefront reconstruction methods for pyramid sensors using Landweber and Landweber-Kaczmarz iterations. Appl. Optics **57**, 8790–8804 (2018)
143. Wang, S., Wei, K., Zheng, W.: Modulation-nonmodulation pyramid wavefront sensor with direct gradient reconstruction algorithm on the closed-loop adaptive optics system. Opt. Express **26**, 20952–20964 (2018)
144. Landman, R., Haffert, S.Y.: Nonlinear wavefront reconstruction with convolutional neural networks for Fourier-based wavefront sensors. Opt. Express **28**, 16644–16657 (2020)
145. LeDue, J., Jolissaint, L., Véran, J.-P., Bradley, C.: Calibration and testing with real turbulence of a pyramid sensor employing static modulation. Opt. Express **17**, 7186–7195 (2009)
146. Carbillet, M., Riccardi, A.: Low-light-level charge-coupled devices for pyramid wavefront sensing on 8 m class telescopes: what actual gain? Appl. Opt. **49**, G167–G173 (2010)
147. Wang, S., Rao, C., Xian, H., Zhang, J., Wang, J., Liu, Z.: Laboratory demonstrations on a pyramid wavefront sensor without modulation for closed-loop adaptive optics system. Opt. Express **19**, 8135–8150 (2011)
148. Liu, Y., Mu, Q., Cao, Z., Hu, L., Yang, C., Xuan, L.: Precise calibration of pupil images in pyramid wavefront sensor. Appl. Opt. **56**, 3281–3286 (2017)
149. Iglesias, I., Ragazzoni, R., Julien, Y., Artal, P.: Extended source pyramid wave-front sensor for the human eye. Opt. Express **10**, 419–428 (2002)
150. Chamot, S.R., Dainty, C., Esposito, S.: Adaptive optics for ophthalmic applications using a pyramid wavefront sensor. Opt. Express **14**, 518–526 (2006)
151. Daly, E.M., Dainty, C.: Ophthalmic wavefront measurements using a versatile pyramid sensor. Appl. Opt. **49**, G67–G77 (2010)
152. Brunner, E., Shatokhina, J., Shirazi, M.F., Drexler, W., Leitgeb, R., Pollreisz, A., Hitzenberger, C.K., Ramlau, R., Pircher, M.: Retinal adaptive optics imaging with a pyramid wavefront sensor. Biomed. Opt. Express **12**, 5969–5990 (2021)
153. Esposito, S., Pinna, E., Puglisi, A., Tozzi, A., Stefanini, P.: Pyramid sensor for segmented mirror alignment. Opt. Lett. **30**, 2572–2574 (2005)
154. Iglesias, I.: Pyramid phase microscopy. Opt. Lett. **36**, 3636–3638 (2011)
155. Kafri, O.: Noncoherent methods for mapping phase objects. Opt. Lett. **5**, 555–557 (1980)
156. Keren, E., Bar-Ziv, E., Glatt, I., Kafri, O.: Measurements of temperature distribution of flames by moiré deflectometry. Appl. Opt. **20**, 4263–4266 (1981)

157. Kafri, O., Glatt, I.: Moiré deflectometry: a ray deflection approach to optical testing. Opt. Eng. **24**, 944–960 (1985)
158. Song, Y., Zhang, B., He, A.: Algebraic iterative algorithm for deflection tomography and its application to density flow fields in a hypersonic wind tunnel. Appl. Opt. **45**, 8092–8101 (2006)
159. Xiao, X., Puri, I.K., Agrawal, A.K.: Temperature measurements in steady axisymmetric partially premixed flames by use of rainbow schlieren deflectometry. Appl. Opt. **41**, 1922–1928 (2002)
160. Goldhahn, E., Seume, J.: The background oriented schlieren technique: sensitivity, accuracy, resolution and application to a three-dimensional density field. Exp. Fluids **43**, 241–249 (2007)
161. Song, Y., Chen, Y.Y., He, A., Zhao, Z.: Theoretical analysis for moiré deflectometry from diffraction theory. J. Opt. Soc. Am. A **26**, 882–889 (2009)
162. Canabal, H., Quiroga, J.A., Bernabeu, E.: Automatic processing in moiré deflectometry by local fringe direction calculation. Appl. Opt. **37**, 5894–5901 (1998)
163. Wang, M.: Fourier transform moiré tomography for highsensitivity mapping asymmetric 3-D temperature field. Opt. Laser Technol. **34**, 679–685 (2002)
164. Ranjbar, S., Khalesifard, H.R., Rasouli, S.: Nondestructive measurement of refractive index profile of optical fiber preforms using moiré technique and phase shift method. Proc. SPIE **6025**, 602520 (2006)
165. Keren, E., Kafri, O.: Diffraction effects in moiré deflectometry. J. Opt. Soc. Am. A **2**, 111–120 (1985)
166. Bar-Ziv, E.: Effect of diffraction on the moiré image for temperature mapping in flames. Appl. Opt. **23**, 4040–4044 (1984)
167. Bar-Ziv, E.: Effect of diffraction on the moiré image I. Theory. J. Opt. Soc. Am. A **2**, 371–379 (1985)
168. Bar-Ziv, E., Sgulim, S., Manor, D.: Effect of diffraction on the moiré image. II. Experiment. J. Opt. Soc. Am. A **2**, 380–385 (1985)
169. Dicke, R.H.: Scatter-hole cameras for x-rays and gamma rays. Astrophys. J. **153**, L101–L106 (1968)
170. Caroli, E., Stephen, J.B., Cocco, G.D., Natalucci, L., Spizzichino, A.: Coded aperture imaging in X- and gamma-ray astronomy. Space Sci. Rev. **45**, 349–403 (1987)
171. Stephen, J.B.: Techniques of coded aperture imaging for gamma-ray astronomy. Adv. Space Res. **11**, 407–418 (1991)
172. Adams, J.K., Boominathan, V., Avants, B.W., Vercosa, D.G., Ye, F., Baraniuk, R.G., Robinson, J.T., Veeraraghavan, A.: Single-frame 3D fluorescence microscopy with ultraminiature lensless FlatScope. Sci. Adv. **3**, e1701548 (2017)
173. Cieślak, M.J., Gamage, K.A.A., Glover, R.: Coded-aperture imaging systems: past, present and future development—a review. Radiat. Meas. **92**, 59–71 (2016)
174. Brady, D.J., Marks, D.L., MacCabe, K.P., O'Sullivan, J.A.: Coded apertures for x-ray scatter imaging. Appl. Opt. **52**, 7745–7754 (2013)
175. Haboub, A., MacDowell, A.A., Marchesini, S., Parkinson, D.Y.: Coded aperture imaging for fluorescent x-rays. Rev. Sci. Instrum. **85**, 063704 (2014)
176. Asif, M.S., Ayremlou, A., Sankaranarayanan, A., Veeraraghavan, A., Baraniuk, R.G.: FlatCam: thin, lensless cameras using coded aperture and computation. IEEE T. Comput. Imag. **3**, 384–397 (2017)
177. Jiang, Z., Kong, Y., Qian, W., Wang, S., Liu, C.: Resolution and signal-to-noise ratio enhancement for synthetic coded aperture imaging via varying pinhole array. Appl. Opt. **58**, 6157–6164 (2019)
178. Jiang, Z., Yang, S., Huang, H., He, X., Kong, Y., Gao, A., Liu, C., Yan, K., Wang, S.: Programmable liquid crystal display based noise reduced dynamic synthetic coded aperture imaging camera (NoRDS-CAIC). Opt. Express **28**, 5221–5238 (2020)
179. Slinger, C., Gordon, N., Lewis, K., McDonald, G., McNie, M., Payne, D., Ridley, K., Strens, M., De Villiers, G., Wilson, R.: Coded aperture systems as non-conventional lensless imagers for the visible and infrared. Proc. SPIE **6737**, 67370D (2007)

180. Furxhi, O., Jacobs, E.L., Preza, C.: Image plane coded aperture for terahertz imaging. Opt. Eng. **51**, 091612 (2012)
181. Shrekenhamer, D., Watts, C.M., Padilla, W.J.: Terahertz single pixel imaging with an optically controlled dynamic spatial light modulator. Opt. Express **21**, 12507–12518 (2013)
182. Cao, X., Yue, T., Lin, X., Lin, S., Yuan, X., Dai, Q., Carin, L., Brady, D.J.: Computational snapshot multispectral cameras: toward dynamic capture of the spectral world. IEEE Signal Proc. Mag. **33**, 95–108 (2016)
183. Yuan, X., Brady, D.J., Katsaggelos, A.K.: Snapshot compressive imaging: theory, algorithms, and applications. IEEE Signal Proc. Mag. **38**, 65–88 (2021)
184. Tsai, T.H., Brady, D.J.: Coded aperture snapshot spectral polarization imaging. Appl. Opt. **52**, 2153–2161 (2013)
185. Ren, W., Fu, C., Wu, D., Xie, Y., Arce, G.R.: Channeled compressive imaging spectropolarimeter. Opt. Express **27**, 2197–2211 (2019)
186. Liu, J., Zaouter, C., Liu, X., Patten, S.A., Liang, J.: Coded-aperture broadband light field imaging using digital micromirror devices. Optica **8**, 139–142 (2021)
187. Rosen, J., Vijayakumar, A., Kumar, M., Rai, M.R., Kelner, R., Kashter, Y., Bulbul, A., Mukherjee, S.: Recent advances in self-interference incoherent digital holography. Adv. Opt. Photonics **11**, 1–66 (2019)
188. Horisaki, R., Ogura, Y., Aino, M., Tanida, J.: Single-shot phase imaging with a coded aperture. Opt. Lett. **39**, 6466–6469 (2014)
189. Horisaki, R., Tanida, J.: Multidimensional object acquisition by single-shot phase imaging with a coded aperture. Opt. Express **23**, 9696–9704 (2015)
190. Horisaki, R., Egami, R., Tanida, J.: Experimental demonstration of single-shot phase imaging with a coded aperture. Opt. Express **23**, 28691–28697 (2015)
191. Egami, R., Horisaki, R., Tian, L., Tanida, J.: Relaxation of mask design for single-shot phase imaging with a coded aperture. Appl. Opt. **55**, 1830–1837 (2016)
192. Wang, B.Y., Han, L., Yang, Y., Yue, Q.Y., Guo, C.S.: Wavefront sensing based on a spatial light modulator and incremental binary random sampling. Opt. Lett. **42**, 603–606 (2017)
193. Wu, Y., Sharma, M.K., Veeraraghavan, A.: Wish: wavefront imaging sensor with high resolution. Light-Sci. Appl. **8**, 44 (2019)
194. Wang, C., Dun, X., Fu, Q., Heidrich, W.: Ultra-high resolution coded wavefront sensor. Opt. Express **25**, 13736–13746 (2017)
195. Wang, C., Fu, Q., Dun, X., Heidrich, W.: Modeling classical wavefront sensors. Opt. Express **28**, 5273–5287 (2020)
196. Gonsalves, R.A.: Phase retrieval and diversity in adaptive optics. Opt. Eng. **21**, 215829 (1982).
197. Paxman, R.G., Fienup, J.R.: Optical misalignment sensing and image reconstruction using phase diversity. J. Opt. Soc. Am. A **5**, 914–923 (1988)
198. Johnson, P.M., Goda, M.E., Gamiz, V.L.: Multiframe phase-diversity algorithm for active imaging. J. Opt. Soc. Am. A **24**, 1894–1900 (2007)
199. Yue, D., Xu, S., Nie, H.: Co-phasing of the segmented mirror and image retrieval based on phase diversity using a modified algorithm. Appl. Opt. **54**, 7917–7924 (2015)
200. Lee, D.J., Roggemann, M.C., Welsh, B.M., Crosby, E.R.: Evaluation of least-squares phasediversity technique for space telescope wave-front sensing. Appl. Opt. **36**, 9186–9197 (1997)
201. Paxman, R.G., Schulz, T.J., Fienup, J.R.: Joint estimation of object and aberrations by using phase diversity. J. Opt. Soc. Am. A **9**, 1072–1085 (1992)
202. Blanc, A., Mugnier, L.M., Idier, J.: Marginal estimation of aberrations and image restoration by use of phase diversity. J. Opt. Soc. Am. A **20**, 1035–1045 (2003)
203. Zhang, P.G., Yang, C.L., Xu, Z.H., Cao, Z.L., Mu, Q.Q., Xuan, L.: Hybrid particle swarm global optimization algorithm for phase diversity phase retrieval. Opt. Express **24**, 25704–25717 (2016)
204. Qi, X., Ju, G., Xu, S.: Efficient solution to the stagnation problem of the particle swarm optimization algorithm for phase diversity. Appl. Opt. **57**, 2747–2757 (2018)
205. Li, D., Xu, S., Qi, X., Wang, D., Cao, X.: Variable step size adaptive cuckoo search optimization algorithm for phase diversity. Appl. Opt. **57**, 8212–8219 (2018)

206. Mocœur, I., Mugnier, L.M., Cassaing, F.: Analytical solution to the phase-diversity problem for real-time wavefront sensing. Opt. Lett. **34**, 3487–3489 (2009)
207. Smith, C.S., Marinică, R., den Dekker, A.J., Verhaegen, M., Korkiakoski, V., Keller, C.U., Doelman, N.: Iterative linear focal-plane wavefront correction. J. Opt. Soc. Am. A **30**, 2002–2011 (2013)
208. Zhang, D., Xu, S., Liu, N., Wang, X.: Detecting wavefront amplitude and phase using linear phase diversity. Appl. Opt. **56**, 6293–6299 (2017)
209. Fienup, J.R., Thelen, B.J., Paxman, R.G., Carrara, D.A.: Comparison of phase diversity and curvature wavefront sensing. Proc. SPIE **3353**, 930–940
210. Bolcar, M.R., Fienup, J.R.: Sub-aperture piston phase diversity for segmented and multi-aperture systems. Appl. Opt. **48**, A5–A12 (2009)
211. Moore, D.B., Fienup, J.R.: Subaperture translation estimation accuracy in transverse translation diversity phase retrieval. Appl. Opt. **55**, 2526–2536 (2016)
212. Guizar-Sicairos, M., Fienup, J.R.: Phase retrieval with transverse translation diversity: a nonlinear optimization approach. Opt. Express **16**, 7264–7278 (2008)
213. Brady, G.R., Guizar-Sicairos, M., Fienup, J.R.: Optical wavefront measurement using phase retrieval with transverse translation diversity. Opt. Express **17**, 624–639 (2009)

Chapter 5
Typical Applications of Computational Phase Imaging

Computational phase imaging is a new method of imaging that utilizes phase measurements in combination with sophisticated image reconstruction algorithms to optimize the entire imaging process, from acquisition to reconstruction. As a computation-intensive inverse problem, the reconstruction problem eliminates bulky and expensive optical components usually used in traditional optical imaging. Computing phase imaging has several advantages and provides a lot of flexibility in imaging, which makes it highly interdisciplinary in nature and a preferred tool in diverse fields, from pathology to manufacturing and materials science. Recent developments have allowed the development of several unique applications that are not possible with traditional optical imaging. For example, in experimental mechanics, optical imaging techniques such as photo-elasticity [1–4], digital speckle pattern interferometry [5] and moiré [6] were used to measure deformations and inner stresses. In all these methods, the deformation or stress is detected by measuring the induced phase shift $\Delta\varphi(x, y)$ in the transmitted or reflected light. In many cases the measurement results are fringes in the form of $|\cos[\Delta\varphi(x, y)]|^2$, and the phase shifting technique is typically used to obtain quantitative values, making the optical structure of measurement systems quite complex. A computational phase imaging technique can be used to make these measurements more convenient and robust. Computer phase imaging has also been widely used in biological research, where cultured cells or tissues are highly transparent due to their low and uniform absorbing properties, and therefore cannot be observed directly with common wide field optical microscopes. Traditionally, these biological samples were dyed first, and then observed with a wide field microscope, confocal laser scanning microscope, and fluorescence microscope. Despite the fact that dying material enhances the contrast of generated images, it also impacts the function and activity of cells and tissues, thereby making some important biological processes such as cell division and fat metabolism difficult to observe. With computational phase microscopy, many of these physical limitations of conventional imaging can be overcome by using indirect image reconstruction, providing a higher contrast image of biological

© The Author(s), under exclusive license to Springer Nature Singapore Pte Ltd. 2022
C. Liu et al., *Computational Optical Phase Imaging*, Progress in Optical Science and Photonics 21, https://doi.org/10.1007/978-981-19-1641-0_5

samples without causing them any damage, making it a very useful tool for biological research.

A number of applications, from microscopic to macroscopic, that use computational optical phase imaging are discussed in this chapter, including stress and deformation measurement, optical testing, biomedical imaging, adaptive optics, refocusing and tracking, 3D imaging and sensing. The capabilities of computational phase imaging make it ideal for meeting the unique needs of each application in terms of spatial and temporal resolution, field of view, and accuracy, which makes it very attractive in imaging applications.

5.1   Measurement of Inner Stress and Deformation

Inner stress of mechanical components and buildings can reduce their structural strength and service life [1] and stress inside crystals and glasses can degrade their optical quality by causing optical birefringence [2]. Thus, inner stress measurement is quite important in the field of mechanical engineering, experimental mechanics, and optical engineering. Nanoindentation technique was widely used for residual stress detection because of its simplicity in setup and operation [3], however, it can only analyze the residual stress in superficial layers. Hydrofluoric acid etching approach provides 3D information of residual stress by causing an irreversible destruction to the sample detected [4]. Non-destructive techniques such as X-ray diffraction [5], Raman spectroscopy [6], and photo-elasticity methods [7, 8] also have several limitations. X-ray diffraction method is not suitable for measurement of amorphous quartz glass, and Raman spectroscopy cannot provide 3D information in most of the cases. Photoelastic method only provides an average value of residual stress along optical axis. On the other hand, since the strain and stress are related to the first and second order derivatives of the displacement, stress measurement can also be realized by measuring the spatial derivatives of displacement as well as the displacement itself. Though the derivatives can be obtained numerically from the displacement data, the procedure is prone to errors. Shearography [9] is a conventional method for the measurement of the spatial derivatives of displacement, where a point in image plane receives contributions from two points of the object and the interference between images of undeformed and deformed objects generate a fringe pattern related to the slope of displacement. The main drawback of conventional shearography is that only the slope in the direction of shearing is available; furthermore, the measurement of the second order derivatives of displacement, such as curvature and twist, is always very difficult or impossible. Some other methods [10, 11] proposed to overcome the limitations of traditional shearography are seldom applied due to their complicated setup or poor image quality. Computational phase imaging is an alternative to the existing challenges in this field and the following section discusses its application in inner stress and deformation measurement with several illustrated examples.

5.1.1 Measurement of 3D Stress Around Laser Induced Damage

High power lasers have become highly indispensable for researches that require power outputs that are higher than 10^{13} W [12], however, any minor defect in the used optical elements can cause serious undesirable modulation in laser beam because of strong optical non-linear effects—leading to small-scale foci of ultra-high intensities that can damage optical elements [13]. Furthermore, laser induced damages are capable to generate more subsequent damages on downstream optical elements [14]. Such damages can rapidly degrade the performance of laser driver. Laser damage repair is commonly realized by evaporating tiny fragments or melting cracks with focused laser beam to restore the transparency of the damaged region [15, 16], however, the repairing can induce additional stress around repaired spots and can cause new cracks, if the induced stress is too strong. Challenges in dealing with a strong residual stress that has evolved from laser repair techniques limits its wider application in the field of high power lasers. Any technique that can timely measure and evaluate the extent of residual stress around any damaged spot before and after its repair can definitely enhance the quality of the repair and can guide us toward further technological improvement.

In photo-elastic methods with double principle stress model [17], the light beam illuminating the optical element is divided into two components along directions of two orthogonal principle stresses $\sigma_1(x, y)$ and $\sigma_2(x, y)$. These two components have a phase difference of $\varphi(x, y) = dC[\sigma_1(x, y) - \sigma_2(x, y)]$ after passing through the sample under inspection, where d is the average thickness of sample and C is a constant for a given material. According to photo-elasticity theory, $\sigma_1(x, y) - \sigma_2(x, y)$ is the maximal inner shearing stress $\tau(x, y)$. For uniform birefringence plate with a known thickness d, its inner average $\tau(x, y)$ can be obtained as $\tau(x, y) = \varphi(x, y)/dC$ [18–20]. However, for sample with non-uniform birefringence, the shearing stress $\tau(x, y, z)$ is a fast varying function of coordinates (x, y, z), while what is obtained in common photo-elasticity device is the integration of $\tau(x, y, z)$ in z direction $\int \tau(x, y, z) dz$. Hence, such a measurement of laser repaired damage with common photo-elasticity device is of little help to understand the stress generation mechanism and consequent technological improvement. By using computational phase imaging and traditional photo-elastic technique in combination, an accurate measurement of $\tau(x, y, z)$ is possible while achieving a z-axial resolution of 10 μm experimentally.

Figure 5.1a shows the bird view of a laser damage spot with radial principal-stress $\sigma_\parallel(x, y, z)$ and circumferential principal-stress $\sigma_\perp(x, y, z)$ [21, 22]. By numerically slicing damaged spot into many thin layers, each layer is regarded as a 2D sample. Inside the kth layer, $\sigma_\parallel(x, y, z_k)$ and $\sigma_\perp(x, y, z_k)$ generates birefringence around the damage spot and the light polarized in radial and circular directions have refractive indices of $n_k^\parallel(x, y)$ and $n_k^\perp(x, y)$, respectively. A light beam illuminates the material in Fig. 5.1b from the top, and since the change in refractive index is always much less than a few thousands of original refraction index n_0 [23, 24], the intensity of the laser beam illuminating each layer is roughly the same as that is incident on the

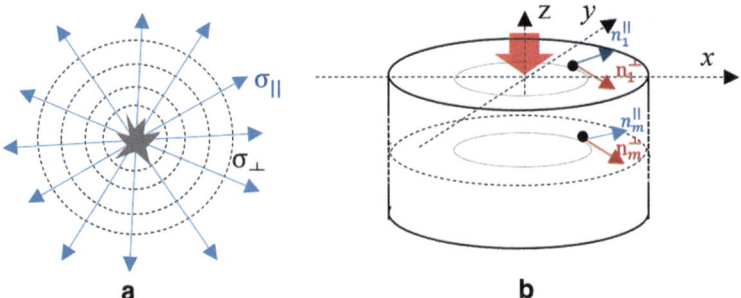

Fig. 5.1 Principle of 3D inner stress measurement with partially coherent interferometry. **a** Two principal-stress axes around laser induced damage crack; **b** birefringence index distribution of numerical slices along z-direction

first layer. Assuming an intensity of E_0 and a linear polarization angle of θ with respect to x-axis for the illumination laser beam, the radially and circularly polarized components reflected from the interface of k and $k + 1$ layers can be approximated by Eq. (5.1).

$$
E_k^{\parallel}(x, y) = E_0 \cos\theta \frac{n_k^{\parallel}(x, y) - n_{k+1}^{\parallel}(x, y)}{n_k^{\parallel}(x, y) + n_{k+1}^{\parallel}(x, y)} e^{\left[2i\frac{2\pi}{\lambda}\int_0^{z_k} n_k^{\parallel}(x,y,z)dz\right]}
$$

$$
E_k^{\perp}(x, y) = E_0 \sin\theta \frac{n_k^{\perp}(x, y) - n_{k+1}^{\perp}(x, y)}{n_k^{\perp}(x, y) + n_{k+1}^{\perp}(x, y)} e^{\left[2i\frac{2\pi}{\lambda}\int_0^{z_k} n_k^{\perp}(x,y,z)dz\right]} \qquad (5.1)
$$

Using approximations of weakly scattering, the intensity of reflected light after passing through a polarizer at an angle of $\theta + \frac{\pi}{2}$ relative to x-axis is equal to

$$
I_k(x, y) = \frac{C^2}{16n_0^2} E_0^2 \sin^2 2\theta \left[\tau_k(x, y) - \tau_{k+1}(x, y)\right]^2 = \frac{C^2}{16n_0^2} E_0^2 \sin^2 2\theta [\Delta\tau_k(x, y)]^2
$$
$$
(5.2)
$$

If the original illumination is circularly polarized [25], the reflected intensity can be rewritten as $I_k(x, y) = \frac{C^2}{16n_0^2} E_0^2 [\Delta\tau_k(x, y)]^2$, and then $\Delta\tau_k(x, y) = \frac{4n_0}{C}\sqrt{\frac{I_k(x,y)}{I_0}}$. Since $\Delta\tau_k(x, y)$ is the difference between shearing stresses of k and $k + 1$ slices, we can obtain stress $\tau_k(x, y)$ of the kth slice in Fig. 5.1b with $\tau_k(x, y) = \sum_{i=1}^{k-1} \Delta\tau_i(x, y)$ after measuring the reflected light intensity $I_k(x, y)|_{k=1,2\ldots}$ from each interface and then calculating $\Delta\tau_k(x, y)$. Lights reflected from all slices mix together when leaving the top surface in Fig. 5.1b, and white light interferometry can be used to measure their intensities separately. Since the coherent length is limited, only the light reflected from a given depth has an optical length equal to that of the reference beam and can form interference fringes. Scanning the reference mirror in axial direction and using traditional band pass filtering method to process the recorded interferogram, the intensity of light reflected from different layers can be measured separately [26].

The experimental setup is shown in Fig. 5.2, where a linearly polarized beam from a super continuum laser source was split into two beams after expansion, one illuminates the damaged sample from the opposite side of damaged site, and the other one illuminates the reference object. Light reflected from both arms are combined by a beam splitter and then collected by a lens. A detector is placed at the common imaging plane of both sample and reference object. The polarizer and analyzer are orthogonal to each other. Since the central wavelength of the laser source is 850 nm and full width of its spectrum at half maximum is 100 nm, the coherence length of the illumination is about 7.2 μm. Thus only the light reflected from the kth interface, whose optical path length is equal to that of reference object (or optical path length difference smaller than 7.2 μm) can interference with reference light to form regular interference fringes. Thus, by scanning the reference object along the optical axis step by step, intensities of light reflected from all interfaces in Fig. 5.1b can be measured separately, followed by $\Delta\tau_k(x, y)$ at all depths. A refractive index matching oil is used on the back surface of the optical element to eliminate the reflection from glass-air interface so that it does not spoil the signal light reflected from the birefringence volume of the sample. The quarter wavelength plate in Fig. 5.2 adjusts the polarization incident on reference grating and the intensity of the reference light reaching the detector can be adjusted by rotating the quarter wavelength plate. The sample is a piece of fused-silica glass with a laser induced damage on its surface. The image of the top right insert is the photo of the laser induced damage used in this experiment.

Figure 5.3a shows a set of recorded interferograms where we can see both damage pit and regular interference fringes, and the red broken lines show the edges of damage caves. Since only reflected light with optical path length equal to that of reference beam can generate interference fringes, the fringe visibility is quite low. Figure 5.3b shows the measured intensities $I_k(x, y)$ which are reflected from interfaces

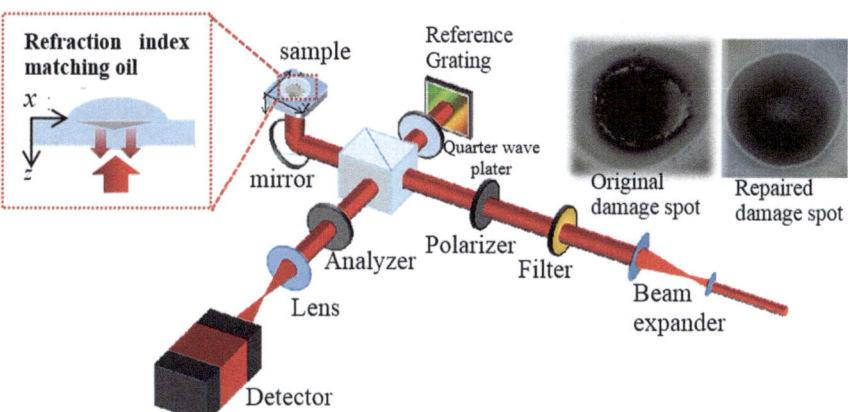

Fig. 5.2 Experimental setup for residual stress measurement on laser induced damage site

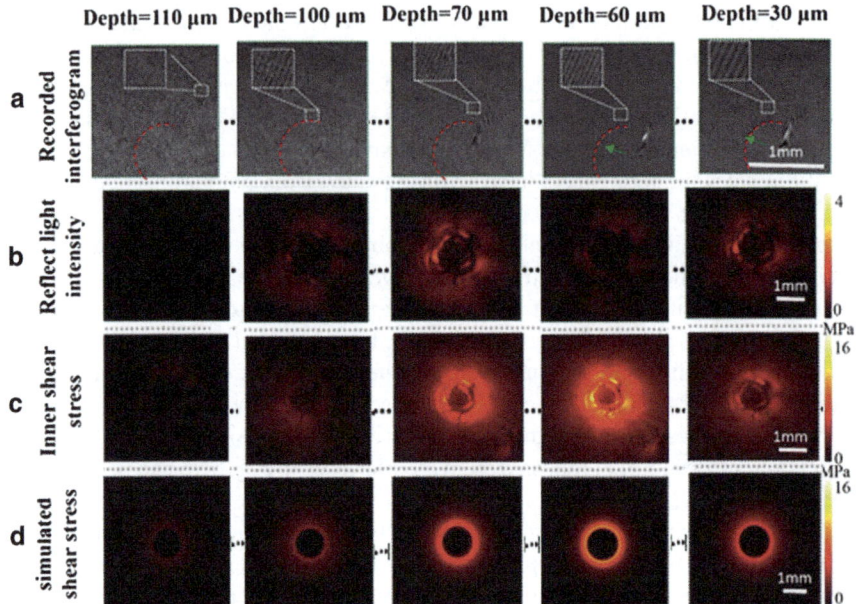

Fig. 5.3 Stress measurements of different layers. **a** Recorded interferograms; **b** reflected light intensity; **c** measured inner shear stress; **d** simulated stress corresponding to various depths. White bars in (**a**), (**b**), (**c**), and (**d**) represent 1 mm

at varying depth. The residual shearing stress $\tau_k(x, y)$ of kth layer can be obtained by calculating $\tau(x, y) = \sum_{i=1}^{k-1} \Delta\tau_k(x, y)$ by using $\Delta\tau_i(x, y)|_{i=1, 2, 3...}$ computed using above mentioned equations. Some of the obtained $\tau_k(x, y)$ were shown in Fig. 5.3c, where we can find that the residual shearing stress was almost rotationally symmetric at each depth, and became stronger as we go from the surface to deeper into the optical element and has reached the maximum at a depth of about 60 μm. It then became weaker with increasing depth and approaches to zero at a depth of about 110 μm.

The inner stress around the damage cave produced by long laser pulse can be well predicted theoretically and numerically for rotationally symmetric laser beam and uniform materials. Finite Element Method (FEM) was adopted to calculate the inner stress around laser induced damage cave with parameters same as that used in experiment for Fig. 5.3c. The material was assumed as fused silica, and the shearing stresses are calculated at the same depths and are shown in Fig. 5.3d. A comparison of both results shows that the general trend of change in stress with increasing depth matches with the experimental results obtained in Fig. 5.3c.

For further comparison and analysis, residual stress is calculated for 12 slices and are plotted three-dimensionally in Fig. 5.4a. The average stress along the radial and depth directions is plotted in Fig. 5.4b and c, respectively. Solid red line represents the average shear stress at an unrepaired site and it peaks at a depth of 60 μm as shown in Fig. 5.4c. The average radial distribution at the depth 60 μm is plotted

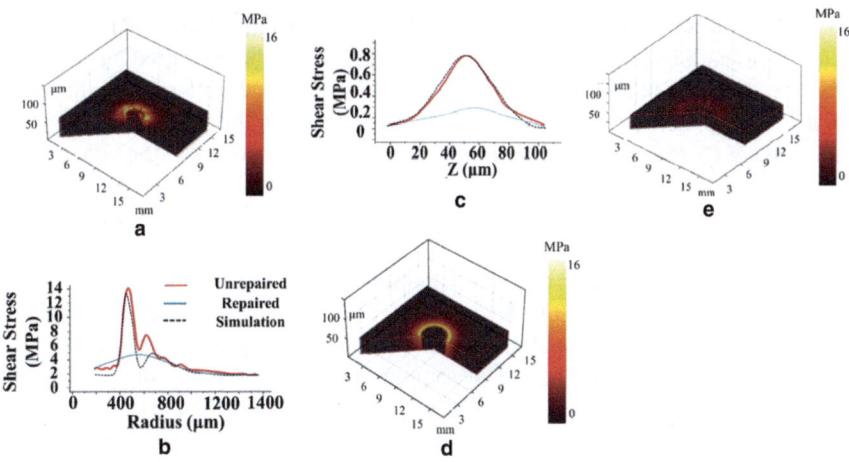

Fig. 5.4 3D inner stress measurements. **a** 3D plot of the measured residual stress around laser induced damage; **b** average stress distribution along radical direction; **c** average stress distribution along depth; **d** 3D plot of the simulated residual stress around laser induced damage; **e** 3D plot of the measured residual stress around laser induced damage after laser repair

with solid red line in Fig. 5.4b, where the maximum stress of 14 MPa appears at a radial distance of about 380 μm, and another smaller peak at 600 μm. Further, the residual stress has decreased exponentially. The 3D simulated shearing stress is shown in Fig. 5.4d, which is quite similar to Fig. 5.4a. Average shear stress obtained in simulations (dotted line in Fig. 5.4b and c) follows the same trend as that of the experimental results represented by solid red line. Both curve peaks approximately at the same position. This verifies the validity of the suggested method as it is in good agreement with the simulation results.

The laser induced damage was measured again with the same setup after it was repaired with CO_2 laser. The photo of the repaired damage site is shown in Fig. 5.2, where we can find that the damage site has become quite smooth. Three dimensionally measured residual stress is shown in Fig. 5.4e, where the measured residual stress is much weaker than that was shown in Fig. 5.4a. The average stress that is changing with depth is shown with a solid green line in Figs. 5.4b and c and its maximum still appears at a depth of 60 μm. The residual stress of repaired damage is remarkably weaker than that of original laser damage, nevertheless it is still quite strong.

5.1.2 Measurement of Deformation with Digital Holography

Deformation and stress measurements are quite important for studying the property of mechanical components and buildings etc., and this kind of measurement can be realized with various optical techniques including photo elasticity [20], digital speckle pattern interference (DSPI) [27], and moiré [28]. Deformation or inner stress will change the refractive index and therefore the optical length of reflected 1 or transmitted light, and thus refractive index or deformation can be measured by detecting the phase change in reflected 1 or transmitted light with an accuracy of 0.1 light wavelength or more [27]. Stress analysts are often interested in the spatial derivatives of displacement as well as the displacement itself, because strain and stress are related to the first and second order derivatives of displacement. Digital holography can measure displacement and its derivatives simultaneously [29] and such a measurement setup is shown in Fig. 5.5a. Parallel light beam from He–Ne laser is split into two beams by a cubic splitter, one beam illuminates the sample under detection, and the other directly illuminates the detector as a reference beam. The reflected and the reference light forms traditional off-axis hologram $I(x, y)$ on the chip of CCD. One hologram $I_1(x, y)$ is recorded before the sample was deformed, and another hologram $I_2(x, y)$ was recorded after the deformation took place. The complex amplitude of reflected light on the $\xi-\eta$ plane of sample surface can be computed as follows [29].

$$\psi(m, n) = \frac{ia}{\lambda d} e^{\frac{i\pi}{\lambda d}(m^2 \Delta \xi^2 + n^2 \Delta \eta^2)} F\left[I_H(k, l) e^{\frac{i\pi}{\lambda d}(k^2 \Delta x^2 + l^2 \Delta y^2)}\right] \qquad (5.3)$$

Complex amplitudes $\psi_1(\xi, \eta)$ and $\psi_2(\xi, \eta)$ correspond to un-deformed and deformed states of object that can be reconstructed from $I_1(x, y)$ and $I_2(x, y)$, respectively. Because the deformation is at the scale of several wavelengths in most cases, there is only a phase difference between $\psi_1(\xi, \eta)$ and $\psi_2(\xi, \eta)$, that is, $\psi_2(\xi, \eta) = \psi_1(\xi, \eta)e^{i\varphi(\xi, \eta)}$. By assuming the axial deformation of sample as $h(\xi, \eta)$ for the situation in Fig. 5.5, we can write, $\varphi(\xi, \eta) = \frac{2\pi}{\lambda}h(\xi, \eta)$. Since both $\psi_1(\xi, \eta)$ and $\psi_2(\xi, \eta)$ can be computed separately using Eq. (5.3), we can get another complex amplitude $\psi'(\xi, \eta)$ using Eq. (5.4) by multiplying $\psi_2(\xi, \eta)$ with the conjugate of $\psi_1(\xi, \eta)$. The axial deformation $h(\xi, \eta)$ can be measured by computing the phase of $\psi'(\xi, \eta)$.

$$\psi'(\xi, \eta) = \psi_2(\xi, \eta)\psi_1^*(\xi, \eta) = |\psi_1(\xi, \eta)|^2 e^{i\varphi(\xi, \eta)} \qquad (5.4)$$

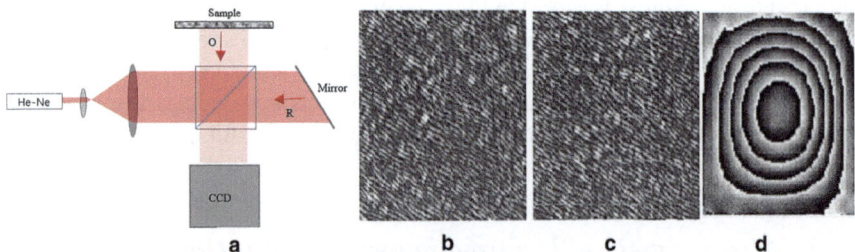

a   b   c   d

Fig. 5.5 Digital holography for displacement measurements. **a** Experimental setup for out of plane deformation measurement; **b** Hologram recorded before deformation; **c** Hologram recorded after deformation; **d** measured deformation

Shifting $\psi'(\xi, \eta)$ laterally to obtain $\psi'(\xi + \delta_\xi, \eta + \Delta\eta)$ and multiplying it with the conjugate of $\psi'(\xi, \eta)$, we have

$$\Omega(\xi, \eta) = \left|\psi_1(\xi + \delta_\xi, \eta + \delta_\eta)\right|^2 |\psi_1(\xi, \eta)|^2 e^{i[\varphi(\xi + \delta_\xi, \eta + \delta_\eta) - \varphi(\xi, \eta)]}$$

$$= A(\xi, \eta) e^{i\Delta\varphi(\xi, \eta)} = A(\xi, \eta) e^{i\frac{2\pi}{\lambda}[h(\xi + \delta_\xi, \eta + \delta_\eta) - h(\xi, \eta)]}$$

$$\approx A(\xi, \eta) e^{i\frac{2\pi}{\lambda}\left[\frac{\partial h(\xi, \eta)}{\partial \xi}\delta_\xi + \frac{\partial h(\xi, \eta)}{\partial \eta}\Delta\eta\right]} \tag{5.5}$$

With appropriate δ_ξ and δ_η, we can obtain the slope of $h(\xi, \eta)$ in any direction. The curvature of the displacement $h(\xi, \eta)$ can be obtained if we shift $\Omega(\xi, \eta)$ laterally in the ξ direction to get $\Omega(\xi + \delta_\xi, \eta)$ and multiply it with the conjugate of $\Omega(\xi, \eta)$.

$$\sigma(\xi, \eta) = A(\xi + \delta_\xi, \eta) A(\xi, \eta) e^{i[\Delta\varphi(\xi + \delta_\xi, \eta) - \Delta\varphi(\xi, \eta)]}$$

$$= A(\xi + \delta_\xi, \eta) A(\xi, \eta) e^{i\frac{2\pi}{\lambda}[\Delta\varphi(\xi + \delta_\xi, \eta) - \Delta\varphi(\xi, \eta)]}$$

$$\approx A(\xi + \delta_\xi, \eta) A(\xi, \eta) e^{i\frac{2\pi}{\lambda}\left[\frac{\partial h^2(\xi, \eta)}{\partial \xi^2}\delta_\xi^2\right]} \tag{5.6}$$

So the curvature of the displacement in ξ direction can be obtained by calculating the phase of $\sigma(\xi, \eta)$. To obtain the normal twist of $h(\xi, \eta)$, we can shift $\Omega(\xi, \eta)$ by distances of δ_ξ and δ_η in ξ and η directions to get $\Omega(\xi + \delta_\xi, \eta)$ and $\Omega(\xi, \eta + \delta_\eta)$, respectively, and the normal twist of the displacement $h(\xi, \eta)$ can be obtained in the same way.

$$\Lambda(\xi, \eta) = A(\xi + \delta_\xi, \eta) A(\xi, \eta + \delta_\eta) e^{i[\Delta\varphi(\xi + \delta_\xi, \eta) - \Delta\varphi(\xi, \eta + \delta_\eta)]}$$

$$\approx A(\xi + \delta_\xi, \eta) A(\xi, \eta + \delta_\eta) e^{i\frac{2\pi}{\lambda}\left[\frac{\partial h^2(\xi, \eta)}{\partial \delta_\xi \partial \delta_\eta}\delta_\xi \delta_\eta\right]} \tag{5.7}$$

An experiment is carried out to show the performance of above discussed method using a setup shown in Fig. 5.5a. A 10mW He–Ne laser is used as the light source, and the specimen is a sanded aluminum plate of 1 mm thickness, with an area of 537 cm^2, and this plate is clamped along its boundary. The distance between the recording plane and the object plane is 2.0 m. The CCD target used in this experiment contains 768×576 pixels and has an area of 7×5 mm^2. After a digitized hologram $I_1(k, l)$ was recorded at the undeformed state of the specimen, the aluminum plate was loaded with a screw in the middle, and then the second hologram $I_2(k,l)$ was captured. Since the sample was illuminated vertically with a collimated laser beam, the experimental setup is sensitive to out-of-plane deformation. Holograms $I_1(k, l)$ and $I_2(k, l)$ are shown in Fig. 5.5b and c, respectively, and interference fringes that are characteristics of the off-axis geometry are observable in them. By substituting the two digitized holograms $I_1(k, l)$ and $I_2(k, l)$ into Eq. (5.3), we can obtain two reconstructed wavefronts $\psi_1(\xi, \eta)$ and $\psi_2(\xi, \eta)$, and the phase map $\varphi(\xi, \eta)$ corresponding to the out-of-plane displacement of sample was obtained by substituting $\psi_1(\xi, \eta)$ and $\psi_2(\xi, \eta)$ into Eq. (5.5). The computed phase map $\varphi(\xi, \eta)$ was shown in Fig. 5.5d.

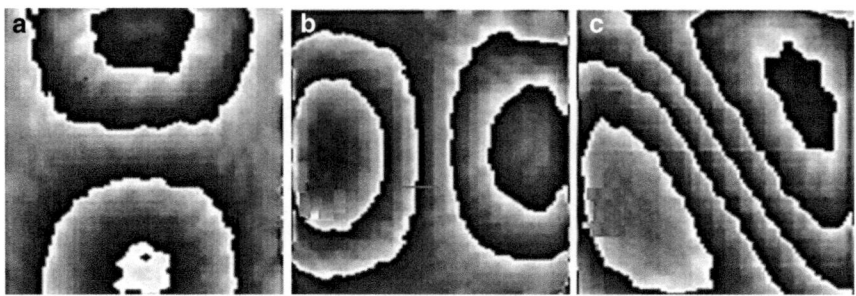

Fig. 5.6 Displacement measurement. Measured slopes in **a** vertical and **b** horizontal directions and **c** normal twist of out plane deformation

With $\psi_1(\xi, \eta)$ and $\psi_2(\xi, \eta)$, the slope of the displacement in any direction can be computed by substituting them into Eq. (5.6) with appropriate δ_ξ and δ_η. Figure 5.6 shows the obtained phase maps of curvatures and the normal twist of displacement. Figure 5.6a and b are computed curvatures in η and ξ directions, respectively, and Fig. 5.6c is the normal twist of the out-of-plane deformation $h(\xi, \eta)$. By doing phase-unwrapping [30–32] on these phase maps, the deformation $h(\xi, \eta)$ and its slopes, curvatures and normal twist can be obtained directly.

5.2 Applications of CDI in Optical Engineering

Many optical measurement methods work by detecting the departure of the wave-front of reflected or transmitted light from an ideally regular spherical or planar wavefront, and the measured departure indicates the properties of the illumination beam or the quality of the optical element. Traditional methods are mainly based on Shark-Hartmann sensing and interferometry. However, coherent diffraction imaging (CDI) has shown remarkable capability in measuring seriously distorted wave-fronts with very simple optical alignment, although there is still no direct correspondence between recorded data and the reconstructed complex amplitude. The following section will introduce CDI methods for measuring the quality of laser beams and optical elements, using two typical examples.

5.2.1 Diagnosing the High Power Laser Beam Online with Coherent Modulation Imaging (CMI)

High power laser facility employed in inertial confinement fusion (ICF) is shown in Fig. 5.7 schematically. The beam generated by seed laser source passes through hundreds of large diameter optical components during its transmission over several

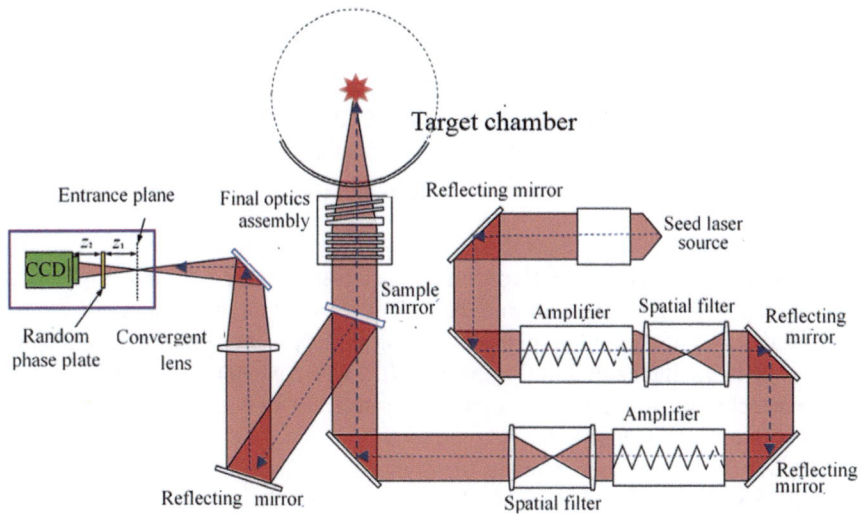

Fig. 5.7 Experimental setup for online diagnosis of high power laser beam with CMI

hundreds of meters before reaching the target [33]. In this process, various factors, such as inaccuracies in material uniformity and surface profiles of large diameter optical elements or distortion introduced into laser slabs when illuminated by flash lamps, severely affect the transmission of the beam and ultimately cause actual experiments to fail [34–36]. So, a strict control on the quality of wave-front is required in ICF in order to focus a sufficiently high amount energy in the region of interest [37], and hence the monitoring of the phase of large diameter laser beams becomes a task of great importance in the field of ICF [38]. Hartmann wave-front sensors can be used to monitor the phase of laser beams online [39–42], however, in order to shrink the diameter of the laser beam to the size of its sensor chip, another complex and large device is required, in addition its resolution limited by its subaperture and the number of micro-lenses used is insufficient. A large aperture interferometer used in conventional interferometry is having enough accuracy and resolution [43] but it is practically not viable as it is challenging to provide a large enough space and stable enough environment for its element-alignment and operation within the high power facility. In the absence of proper phase measurement devices, the phase changes in large diameter laser beam cannot be monitored online, and thus, the performance of the entire facility cannot be evaluated strictly. Coherent Modulation Imaging (CMI) method, which was proposed in 2010 by Fucai Zhang [44], can be used as an alternative technique to monitor the laser beam quality by measuring its phase change online [45, 46].

Coherent modulation imaging has very compact optical structure and has the capability to measure focused laser beam. Thus it is an ideal method for the online diagnosis of high power laser beam. Experiments were carried out on National Laser Facility of Israel with the optical alignment shown in Fig. 5.7, where the laser beam with diameter of 120 mm and pulse duration of 10 ns is firstly sampled by a reflecting mirror and then focused by an aspheric condenser lens with a focal length of 2750 mm. Because of the limited space available, the focused beam was reflected again to illuminate a binary random phase plate and a CCD camera in the downstream of random phase plate records the diffraction patterns formed. The distances of random phase plate to focal spot and CCD are 80.3 mm and 72.3 mm, respectively. The wavelength is 1053 nm.

Figure 5.8a and b show the measured amplitude and phase distribution of the random phase plate with ePIE algorithm, respectively. Figure 5.8c is the recorded diffraction pattern, Fig. 5.8d and e are reconstructed modulus and phase of focused laser beam incident on random phase plate. Figure 5.8f is the computed focal spot, and the maximum intensity is of the order of 10^{23}, which cannot be recorded directly by any detector. Figure 5.8g and h are the reconstructed modulus and phase of laser beams before the condenser lens.

As CMI is a single shot imaging method, each high power laser pulse can be measured for its phase and modulus. Figure 5.9 shows the phase and modulus of six laser pulses, where we can find that the intensity of each laser pulse is roughly unchanged, but the phases are quite different from each other. Results in Figs. 5.8 and 5.9 are the first application of CDI in the field of high power laser facility.

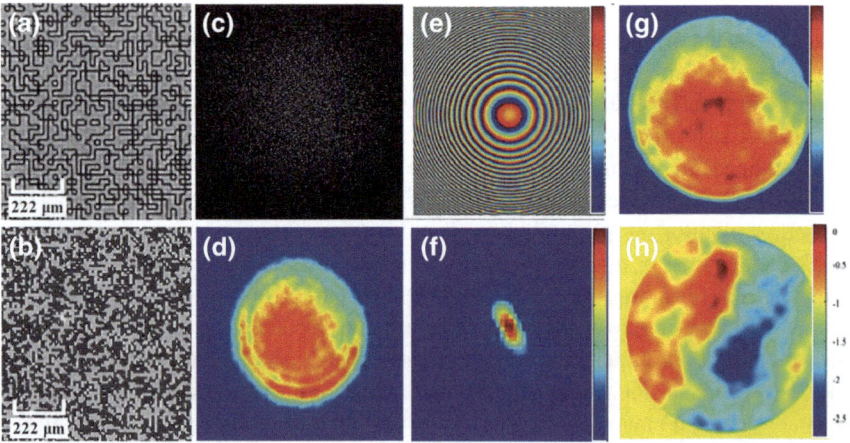

Fig. 5.8 Measurement of high power laser with CMI. **a** Measured random phase plate amplitude; **b** measured random phase plate phase; **c** recorded diffraction pattern; **d** reconstructed modulus of focused laser beam incident on random phase plate; **e** reconstructed phase of focused laser beam incident on random phase plate; **f** computed focal spot; **g** reconstructed modulus of focused laser beam before the condenser lens; **h** reconstructed phase of focused laser beam before the condenser lens

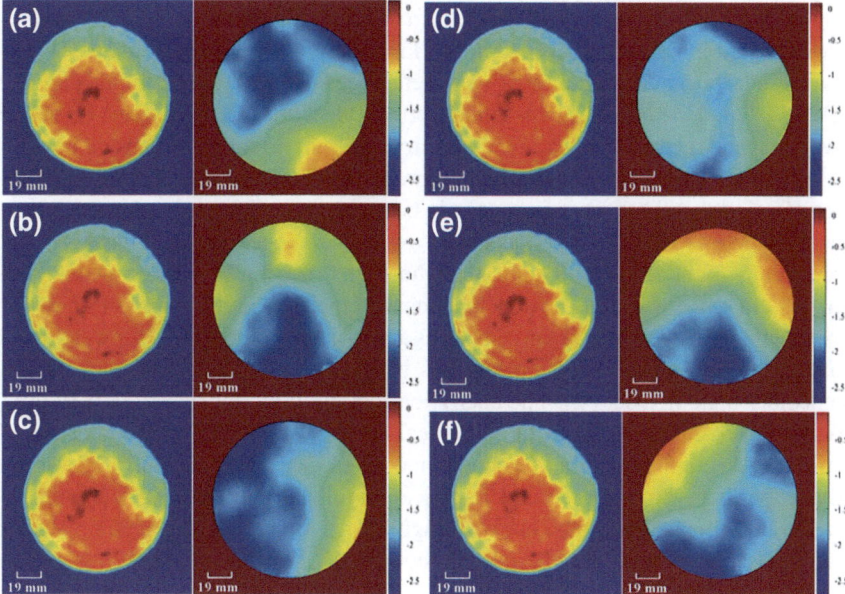

Fig. 5.9 Measurement of six high power laser pulses. **a–f** Reconstructed modulus and phase of six laser pulses

5.2.2 Inspection on the Quality of Optical Element with PIE

The aberration of optical lens arising from non-uniformity of materials or manufacturing error can obviously distort the wave-front of imaging light beam which can result in a resolution much lower than the designed value. A measurement of the aberrations of optical imaging elements is quite crucial for both quality evaluations and further aberration compensations for optimal resolution in astronomical and biomedical applications using adaptive techniques [47–50]. With computational optical imaging, it is now possible to numerically correct the blurred image from an imperfect lens with known aberrations and obtain the theoretical resolution determined by its numerical aperture, and hence quantitative measurement of the aberration of optical elements is highly desirable. Different methods have been proposed for aberration recovery [51]. Among these, star testing [52] and knife/wire testing [53] have been adopted widely for imaging quality evaluations and lens alignment. Depending on the star or shadow, different types of aberrations can be distinguished. Although both methods provide high sensitivity in aberration detection by utilizing simple optical systems and convenient operations, they can only provide qualitative or semi-quantitative aberration information, thus limiting their applicability to quantitative aberration recovery. As the most widely used technique for quantitative aberration detections, interferometry can extract the phase difference between the imaging and reference beams precisely from the fringes patterns and provide quantitative aberration measurements on various kind of optical elements [54–56]. Unfortunately, interferometry relies on an extra reference beam, not only complicating the optical system, but also easily introducing errors due to the often non-ideal reference wave-front and the inevitable system vibration. The lateral shear interferometry with common-path optical system is designed to avoid the disadvantages of extra reference beam [57–59], but the wave-front can only be reconstructed from the shearing phase gradients obtained in two orthogonal shearing directions, and noise or measurement errors at any point can spread out to the whole aperture in **integration reconstruction**, resulting in remarkable error in final measurements. Apart from interferometry-based methods, Shack-Hartmann sensing as a non-interferometry tool for wave-front sensing and aberration detection that has been widely used in astronomical and biomedical imaging [60–62]. The aberration measured by this system is, however, too low in resolution to meet the numerical correction method's requirements. In addition, both commercial interferometry and Shack-Hartmann sensing methods are rather expensive, limiting their wider application for aberration recovery. In order to develop inexpensive methods, but with high aberration recovery precision, Waller et al. devised a method for aberration recovery by imaging a weak diffuser with Fourier ptychographic microscopy [63–67]. The method worked well with common optical objectives and lenses with a short working distance. However, for lenses with a long working distance, the sample must be placed far away from the lens or an additional optical element will be required, making the aberration measurement quite inconvenient. A different technique for aberration retrieval is phase diversity [68–70], but its low convergence and slow processing speed prevent its widespread application.

Quantitative measurement of optical element aberrations can be realized at a high precision and resolution by using extended ptychographic iterative engine (ePIE). By capturing two sets of diffraction arrays, one with and one without the optical element under detection, the wavefronts of the laser beam illuminating the lens and leaving the lens can be retrieved using the classical ePIE method [71–74], with the phase difference between them representing the transmission phase of the optical element. This subtraction of the measured transmission phase from that of an ideal lens corresponds to the phase distortion of the optical element being measured, allowing different aberrations to be accurately determined. The ePIE-based aberration recovery approach only needs compact common-path configuration without any extra reference beam, thus not only simplifies the system and the operations, but also avoids the error induced by the non-ideal reference beam and system vibration. Using Wegener-filter-like updating formula in ePIE, the reconstructed images are always in high quality and free of speckle noise [75], and the reconstruction is fast and accurate due to the large number of recorded diffraction patterns and the fact that there is high data redundancy [76]. Furthermore, the ePIE-based method can extract the aberration of optical elements in high precision using an affordable and compact system [77].

Figure 5.10 schematically represents the optical setup for measuring the phase aberration of an imaging lens with ePIE algorithm. Figure 5.10a illustrates the expansion of a thin laser beam into a wide parallel beam that passes through an aperture and hits a thin transverse object $O(x, y)$ to form diffraction patterns $I(x, y)$ on the target CCD. The aperture size was adjusted to match the aperture of the lens to be detected in advance. With recorded diffraction intensities $I_n(x, y)|_{n=1:N}$, the complex amplitude of illuminating beam $P_0(x, y)$ incident on the surface of scanning object can be iteratively reconstructed with ePIE algorithm, and by numerically back propagating $P_0(x, y)$ to the aperture plane, the complex amplitude of light just leaving the aperture can be obtained as $P_0'(x, y)$. By placing the lens under detection exactly behind the aperture as shown in Fig. 5.10b and repeating the above data acquisition and iterative computation, the complex amplitude leaving the lens under detection can be reconstructed as $P_1'(x, y)$. By computing the phase $\varphi(x, y)$ of $P_1'(x, y)P_0'^*(x, y)$ we can finally get the transmission function of the lens under detection as $e^{i\varphi(x,y)}$. By subtracting a standard spherical wave-front $\varphi_0(x, y) = \frac{2\pi}{\lambda}\sqrt{x^2 + y^2 + f^2}$, where f is the designed focal length of lens, the phase aberration of this detected lens is $\Delta\varphi(x, y) = \varphi(x, y) - \varphi_0(x, y)$, and then by doing Zernike decomposition the aberration coefficients of various geometric aberrations can be computed. The data flow chart of this method is shown in Fig. 5.11.

A proof-of-principle-experiment was carried out using Fig. 5.11 to demonstrate the feasibility of above illustrated method. A He–Ne laser beam (Shangpu Optics, China) with a wavelength of 632.8 nm was used as a probe light, and the diameter of the aperture on the screen in Fig. 5.10 is 4.0 mm. The scanning object was a stem slice of corn stalk (Jialian Biotech, China) fixed on a translation stage (Thorlabs, US). A CCD camera (Pike 421B, AVT, Germany) with 2048 × 2048 pixels was used for diffraction pattern recording, and the size of each pixel size was 7.4 × 7.4 μm^2. The sample was scanned to 8 × 8 positions with an overlapping ratio ~80% to record 64 diffraction patterns, before and after a concave lens was placed into optical length. Sample to aperture distance and that to CCD camera are set at 189 mm and 218 mm, respectively. The focal length of the lens under detection is

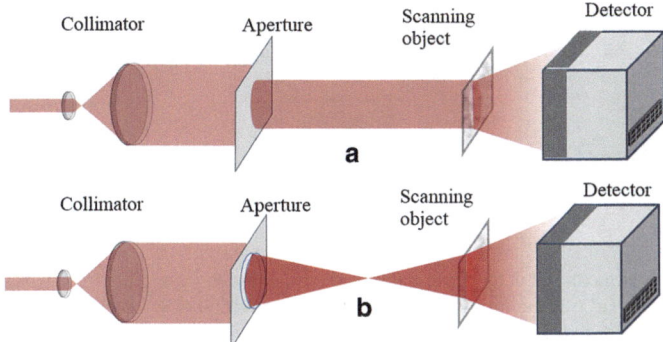

Fig. 5.10 Schematic diagram of experimental setup

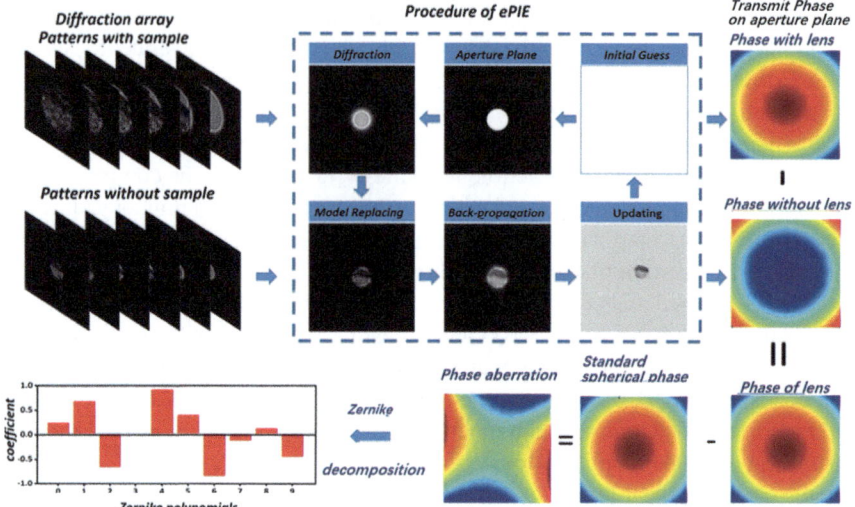

Fig. 5.11 Flow-chart of the ePIE based aberration recovery approach

$f = 100$ mm (Daheng Optics, China). Figure 5.12 shows the experimental result. Figure 5.12a1 shows some diffraction patterns obtained without placing the lens to be detected in the optical length and Fig. 5.12a2 shows the reconstructed modulus

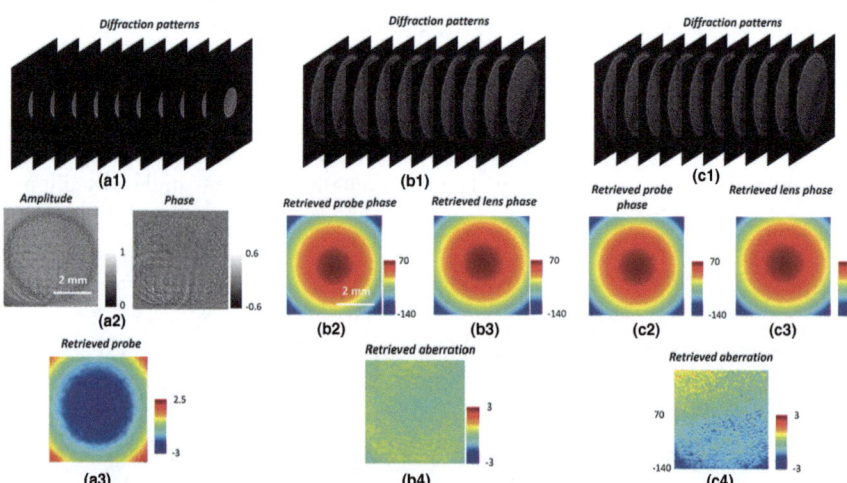

Fig. 5.12 Experiments for lens phase measurements. **a** Probe light reconstruction without lens under detection: (**a1**) diffraction patterns, (**a2**) reconstructed sample amplitude and phase, (**a3**) reconstructed probe light phase. **b** and **c** Probe light reconstruction with lens under detection placed at on and off-axial positions. (**b1**) and (**c1**) diffraction patterns, (**b2**) and (**c2**) reconstructed probe light phase, (**b3**) and (**c3**) reconstructed lens phases with aberrations, (**b4**) and (**c4**) reconstructed aberrations

and phase of the scanning object. Figure 5.12a3 shows the phase of light beam leaving the aperture plane. Figure 5.12b1 shows some recorded diffraction patterns obtained when a lens is placed on the optical path. Figure 5.12b2 is the reconstructed wave-front of the light leaving the lens under detection. Figure 5.12b3 is the phase difference between Figs. 5.12b2 and a3, which is essentially the transmission function of the lens under detection. By subtracting a standard spherical wave-front with radius of 100 mm, the obtained phase aberration of lens under detection is shown in Fig. 5.12b4. Figure 5.12c1–c4 show another set of experimental results when the lens under detection was transversely shifted to a distance of 500 μm in y direction. We can find that the measured phase aberration shows an obvious phase ramp, which coincides with the prediction of common geometric optics.

To quantitatively check the accuracy of above experiments, a cylindrical lens was placed exactly behind the lens to introduce a known phase aberration. According to the result in Fig. 5.12b4, the phase aberration of the lens under detection was rather small and the extracted aberration is mostly generated by the cylindrical lens. By comparing the measured phase aberration and the theoretical phase vale of cylindrical lens, the accuracy of the measurement could be quantitatively certificated. Figure 5.13 shows two sets of experimental results corresponding to cylindrical lenses placed in two different orientations. Figure 5.13a1–a7 are experimental results when the cylinder lens was placed in the direction of y-axis, and we can find from the comparison between the measured phase values and the theoretical values of cylinder lens in Fig. 5.13a6 that the measurement error was much smaller than 1.0%. Figure 5.13b1–b7 are experimental results corresponding to cylindrical lens placed at 45° relative to y-axis. We can also find that the measurement errors in these two sets experiments are also much smaller than 1.0%.

5.3 Computational Optical Phase Imaging in Biomedical Imaging

Microscopy is an essential tool in biomedical applications since it can provide direct observations on the specimens in different perspectives. Therefore, aiming at offering specimen details in various imaging modes, many kinds of microscopy have been designed. In the authors' view, they can roughly be divided into two categories as "linear" and "non-linear" microscopy. For the "linear" microscopy, the collected light wavelengths are the same as the illumination light wavelengths; while for the "non-linear" microscopy, the emission light wavelengths are no longer the same as the excitation light wavelengths.

For the "linear" microscopy focusing on the intensity signals, bright- and dark-field microscopy are two representative techniques, and they are often used in biomedical observations due to their simple systems and low costs. However, both techniques have rather poor imaging contrast, especially for label-free live cells which are almost transparent specimens. Though labeling can significantly improve the

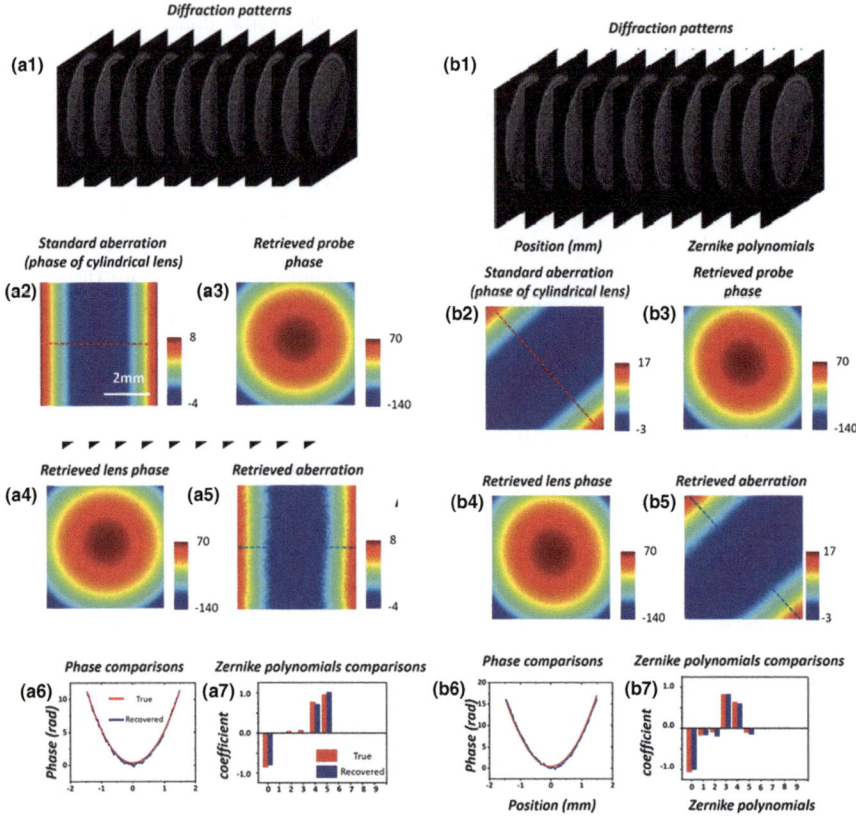

Fig. 5.13 Experiments for lens aberration measurements introduced by cylindrical lens. **a–b** measurements with two different cylindrical lenses: (1) diffraction patterns, (2) theoretical cylindrical lens phases as setting aberrations, (3) reconstructed probe light phases, (4) reconstructed lens phases with aberrations, (5) reconstructed aberrations, (6) sectional phase comparison, (7) Zernike polynomials of the setting and reconstructed aberrations in (2) and (5)

imaging contrast, the introduced dye not only complicates the specimen pre-treatment operations, but also influences the specimen activity. Phase contrast microscopy and differential interference contrast microscopy are alternative techniques for label-free imaging and belong to the "linear" microscopy, focusing on the phase signals. The latter one has become an often used tool in live cell imaging not only because of the high imaging contrast, but also it avoids the halo often generated in the phase contrast microscopy. Unfortunately, phase contrast microscopy and differential interference contrast microscopy are qualitative but not quantitative. The specimen phases are coded in the phase contrast intensities and there exists only a non-linear relation between the phase contrast intensity and the phase that can be hardly extracted quantitatively. In order to solve the problem, many quantitative phase imaging techniques such as digital holographic/interferometric microscopy, ptychography, and transport of intensity phase microscopy have been designed. Different from the phase contrast and differential interference contrast microscopy, these quantitative phase imaging techniques not only maintain the high imaging contrast, but also retrieve the quantitative specimen phases. In other words, quantitative phase imaging introduces metrology into microscopy, thus upgrades the classical qualitative observations into the quantitative measurements. Therefore, these quantitative optical phase imaging techniques are becoming popular tools in biomedical applications, since they can provide more information on the targets compared to traditional qualitative optical phase imaging techniques.

Besides "linear" microscopy, various kinds of "non-linear" microscopy are also designed and applied, such as classical fluorescence microscopy, second/third harmonic generation microscopy, multi-photon microscopy, Raman microscopy (coherent anti-Stokes Raman scattering microscopy, stimulated Raman scattering microscopy) and so on. These techniques can specifically determine the targets via labeling or non-linear optical effect, therefore they are the mainstream tools especially in biomedical applications. Nevertheless, non-linear microscopy can only provide local-field images since dyes can only be labeled to specific targets, and non-linear optical effects can only happen where satisfying phase matching is achieved. By contrast, linear microscopy can provide whole-field images. Both "linear" and "non-linear" microscopy is useful in different biomedical applications. For example, in cell and tissue imaging for morphology detection and structure recognition, "linear microscopy" is a better choice due to their relatively simple systems and whole-field imaging capability; while in order to distinguish the cell and tissue architectures, "non-linear microscopy" is preferred due to it imaging specificity. In some applications, both techniques can be used in conjunction. Thus, compared to bright-field and dark-field microscopy, as well as traditional phase contrast microscopy and differential interference contrast microscopy, quantitative optical phase imaging provides high contrast and high accuracy for observations on label-free specimens as well as for measurements on various aspects. Additionally, quantitative optical phase imaging provides quantitative whole-field images as a background of the local-field images generated by non-linear microscopy. Computational optical phase imaging is becoming an increasingly important tool in biomedical imaging.

This section revisits studies on static optical phase imaging, dynamic optical phase imaging, and hybrid optical phase imaging for direct specimen observation in biomedical imaging, as illustrated in Fig. 5.14. These studies serve as a guide to selecting appropriate computational optical phase imaging techniques for various applications.

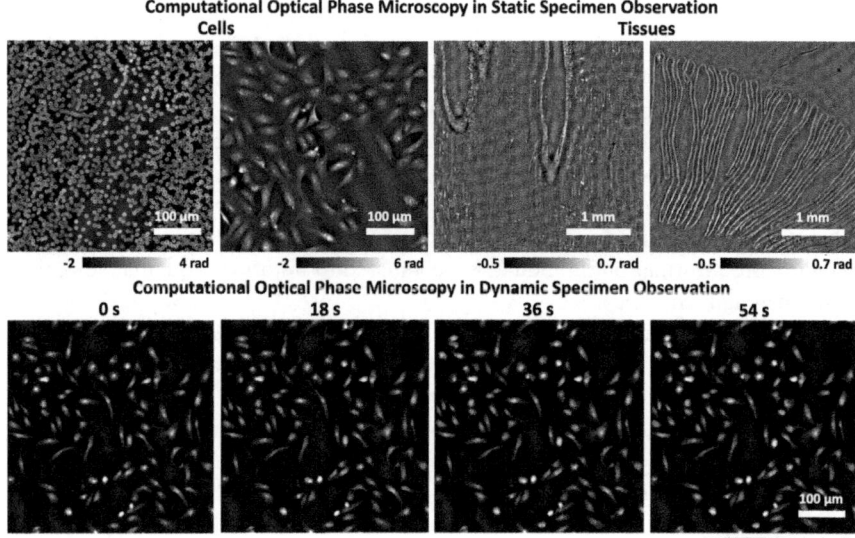

Fig. 5.14 Computational optical phase microscopy in static and dynamic specimen observation

5.3.1   Computational Optical Phase Microscopy in Static Specimen Observation

The most outstanding advantage of computational optical phase microscopy is that it can provide images with much higher contrast than commonly used bright- and dark-field microscopy. Therefore, computational optical phase microscopy is a suitable tool for cell imaging, especially for those label-free cells which are almost transparent objects.

Many computational optical phase imaging works have been reported focusing on red blood cell observations. Because of the unique refractive index distribution, the morphology of the red blood cells can be directly measured from the extracted phase distributions using computational optical phase imaging. Moon group adopted off-axis digital holographic microscopy to retrieve the quantitative phase distributions of red blood cells [78–80] and provided many quantitative parameters of cellular morphology, such as projected surface area, average phase value, circularity, mean phase of center part, sphericity, elongation, and pallor (phase value difference between cell edge and center). The quantitative information is further used to calculate the mean corpuscular hemoglobin, mean corpuscular hemoglobin surface density [81] and dry mass [82]. Based on these quantitative parameters, red blood cells can be accurately classified into stomatocyte, discocyte, and echinocyte types [80], thus providing references for disease diagnostics. Moreover, the same group also used the off-axis digital holographic microscopy to analyze the morphology and the mean corpuscular hemoglobin changes in human red blood cells stored in different periods, and they found that the red blood cell morphology changed with the storage time, but the hemoglobin content did not change substantially [83, 84].

In addition to being used in automatic classification and quantification, the red blood cell morphology is also an important indicator for disease diagnosis. Malaria caused by a protozoan parasite induces morphological changes in the red blood cells, therefore, optical imaging can be effectively used to diagnose malaria [85]. Among various optical imaging techniques, computational optical phase imaging has a high imaging contrast while extracting quantitative phase distributions of red blood cells which leads to malaria diagnostics. Lee [86] and Zhu group [87] adopted the quantitative interferometric microscopy to observe the malaria parasite infected red blood cells, and proved that computational optical phase microscopy is a preferred tool in malaria diagnostics. Besides malaria, it has also been used to diagnose sickle cell anemia. Wax [88] and Park group [89] adopted the off-axis quantitative interferometric microscopy to measure red blood cells and recognized different types of sickle red blood cells. Moreover, Javidi group designed a three-dimensionally printed shearing digital holographic microscopy for red blood cell phase imaging and sickle cell anemia diagnostics [90]. In addition, Doblas et al. provided a proof-of-concept study that computational optical phase microscopy can distinguish the red blood cells of diabetic and healthy human, since they found that the red blood cell phases are closely correlated with the glycated haemoglobin and the blood glucose values [91].

The classical two-dimensional computational optical phase imaging can only provide the phase integral along the imaging axis rather than the three-dimensional refractive index distribution, which limits its applications to the biomedical field. In order to extract the inner structure of cells, three-dimensional computational optical phase imaging techniques have been designed and successfully applied. Considering the fact that red blood cells have a unique refractive index distribution, Mesquita group successfully reconstructed the three-dimensional refractive index distribution of the red blood cell from its multi-focal images [92]. However, it is still not a general three-dimensional computational optical phase imaging, since most of the cells do not have unique refractive index distributions. Choi et al. [93] designed the tomographic phase microscopy, which collects the wavefront passing through the specimen in multiple directions and then reconstructs the three-dimensional refractive index distribution through the inverse Radon transform similar to computed tomography. Such method can be widely used in different kinds of specimens, even their refractive index distributions are not unique. Moreover, Park group designed a commercialized optical diffraction tomography to reconstruct the three-dimensional refractive index distributions of red blood cells, and provided morphological parameters as surface area, volume, sphericity and hemoglobin content [94]. It should be noted that the parameters such as three-dimensional refractive index distribution and surface area can hardly be obtained only using classical two-dimensional computational optical phase imaging. Moreover, the same group has also measured the three-dimensional refractive index distributions of human red blood cells from cord blood of newborn infants and maternal blood, and analyzed their differences on shape, hemoglobin content, volume, and surface area [95]. Since the parasite infected red blood cells often have non-unique refractive index distributions, classical two-dimensional computational optical phase imaging can only reflect partial information. In order to solve the problem, Park group also adopted the optical diffraction tomography to reconstruct the three-dimensional refractive index distributions of the red blood cells parasitized by *Plasmodium falciparum* [96] and *Babesia microti* [97], and the same group also employed the optical diffraction tomography as the hematologic screening for diseases and syndromes such as iron-deficiency anemia, reticulocytosis, hereditary spherocytosis, and diabetes mellitus [98]. It is generally known that three-dimensional computational optical phase imaging is more suitable for more complex cells such as white blood cells [99], HeLa cell [92], human myelocytic leukemia cells [100], leukemic T-cells [100], breast and lung cancer cells [101], since this technique can extract inner structures for cellular analysis at sub-cellular level. Moreover, using three-dimensional computational optical phase imaging, it also proved that the refractive index of the nuclear is not always higher than that of the cytoplasm [92, 100, 101], which is different from the previous knowledge.

In biomedical applications, it is rather important to obtain phase distributions on a large number of specimens for statistical analysis. However, when imaging on micro-scale cells, often micro-objectives are equipped in computational optical phase imaging systems to significantly magnify the specimens, however, the field of view is remarkably limited. In order to balance the amplification and the field of view, Fourier ptychographic microscopy is a suitable solution. It often uses a lower magnification micro-objective to ensure a rather large field of view, and through synthetic aperture scanning, the frequency information can be expanded to obtain satisfactory image resolution. Yang group used Fourier ptychographic microscopy for blood smear imaging and successfully distinguished red and white blood cells in an extremely large field of view of 120 mm^2 [102]. To further expand the field of view, Ozcan group proposed a lensfree digital holographic microscopy for dense blood cell imaging with a density of 0.4×10^6 cells/μL [103].

Besides red blood cells, computational optical phase microscopy has also been used in other cells, and among them, one successful application is sperm morphology analysis. Sperm analysis is a diagnostic tool for assessing male reproductive health in humans and animals, and in sperm analysis, sperm morphology is one key parameter. Bright- and dark-field microscopy can hardly provide high-quality sperm imaging unless labeled, which requires extra specimen pre-treatment operations and suffers from staining damages. Though phase contrast and differential interference contrast microscopy can enhance the imaging contrast significantly, they can hardly reflect the actual sperm morphology quantitatively. While computational optical phase microscopy provides a suitable tool since it can provide high contrast and quantitative phase imaging on label-free sperms. As Shaked group proved that the computational optical phase imaging could still obtain the high-contrast images to accurately reflect the sperm morphology without staining [104, 105] compared to the staining method recommended by WHO [106]. Ferraro group [107–109] and John group [110] have successfully analyzed the sperm morphology according to the sperm phase distributions retrieved using the digital holographic microscopy and transport of intensity microscopy, respectively. Additionally, Ozcan group [111] and Demirci group

[112] adopted the lensfree digital holographic microscopy in order to observe more sperms in the fixed field of view. Shaked group [113] also designed holographic virtual staining which can provide virtually stained cells from phase imaging using deep learning, and it was also successfully used in sperm imaging, thus providing a simple label-free imaging tool for sperm analysis.

Besides cells, computational optical phase microscopy has also been widely used in tissue observations. In classical tissue imaging, various staining techniques such as hematoxylin and eosin (H&E) staining are adopted to significantly improve the imaging contrast and widely used in many applications such as cancer diagnostics. However, these methods require complicated staining process, therefore, label-free tissue imaging is preferred in the biomedical observations. Second/third-harmonic generation microscopy and Raman microscopy can provide high contrast and even chemical-specific images in the label-free mode. Unfortunately, they require extremely complicated and expensive equipment, limiting their wider applications in biomedical fields. As a simple and cost-effective solution, computational optical phase imaging provides another tool for high-contrast imaging, and several works have been reported. Popescu group adopted the spatial light interference microscopy which is a quantitative phase contrast microscopy to observe the tissue biopsies [114–120], and they reported that computational optical phase imaging can be used to diagnose or even to predict breast cancer [114–116], prostate cancer [117, 118] and pancreatic ductal adenocarcinoma [119]. Moreover, Popescu group and Jung group also adopted the common-path and off-axis quantitative interferometric microscopy to observe the histological sections of the mouse kidney to predict the stage of acute kidney disease [120]. Moreover, they also studied the variability of computational optical phase imaging with the staining dye, to characterize the influence of H&E staining on the phase delay and the optical properties [121]. All these results prove that computational optical phase imaging is an effective tool not only for high-contrast tissue imaging but also for disease diagnostics. While both interference-based spatial light interference microscopy and quantitative interferometric microscopy used in these studies require micro-objectives to achieve higher imaging resolution, the micro-objectives sacrifice the field of view since normal tissue biopsies are much larger than micro-objective fields of view.

As a result of the trade-off between resolution and field of view, Fourier ptychographic microscopy becomes a promising technique for tissue imaging [122]. It uses low magnified micro-objectives and, therefore, often has a large field of view. In order to further enlarge the field of view, Ozcan group adopted the lensfree digital holography for tissue imaging to pursue both intensity and phase distributions with a field of view of ~20 mm^2 [123–132]. As the tissue is directly set on the CMOS chip, the field of view of the tissue can reach the size of the CMOS chip, while the magnification is near to 1, thus the imaging resolution is often poor when using classical phase recovery methods directly. In order to improve the imaging resolution, sub-pixel shifting [125], multi-height [126, 127], and synthetic aperture [128]-based phase recovery methods have been adopted; moreover, even deep learning-based phase recovery method has also been designed [129]. Such works have been successfully used for tissue imaging. Additionally, Ozcan group has also extended the lensfree digital holography into color imaging [130, 131] and CLARITY imaging [132]. Though the Fourier ptychographic microscopy and the lensfree digital holography can obtain high imaging resolution and large field of view, their fields of view are still fixed and can hardly be further extended. Ptychographic iterative engine is a preferred tool to further enlarge the field of view. Though it still requires scanning, it can obtain specimen intensity and phase images with extremely high resolution, theoretically reaching the diffraction limit. It has the large field of view that depends only on the scanning range. Considering their advantages, ptychographic iterative engine has been successfully used in tissue imaging [133] and even three-dimensional applications [134, 135]. It still should be noted that though computational optical phase imaging can provide high-quality tissue imaging, H&E staining still has its specific advantages, such as physicians are familiar with the imaging mode. Ozcan group adopted deep learning to obtain virtually stained image from phase images, named as PhaseStain [136, 137]. With such method, high-quality stained image can be obtained but avoiding the complicated staining operations.

Here, two examples on computational optical phase microscopy in static specimen observation are provided. The first example is based on quantitative interferometric microscopy [138]. The Fig. 5.15 illustrates the quantitative interferometric microscopic system based on a Mach–Zehnder interferometer in off-axis mode. Using a He–Ne laser (Thorlabs, US) with a wavelength of 632.8 nm as a coherent source, a micro-objective (Daheng Optics, China) for specimen magnification, and a CCD camera (AVT Pike 421B, Germany) with a pixel size of 7.4 μm for interferogram recording, the optical system was constructed. Other optical elements, such as mirrors, beam splitters, and pinholes, are also provided by Daheng Optics; the sample was a blood smear.

Figure 5.16 shows the results of quantitative interferometric microscopy. Figure 5.16a is the captured off-axis interferogram; and Fig. 5.16b is the spectrum of Fig. 5.16a by implementing the fast Fourier transform. The red box in Fig. 5.16b indicates the first order information selection in the frequency domain. The wrapped phase was first extracted as Fig. 5.16c using fast Fourier transform-based phase extraction method. Using phase unwrapping method, Fig. 5.16d exhibits the unwrapped phase, clearly shows the biconcave structures of red blood cells.

Such work is an example of computational optical phase microscopy in static specimen observation. However, this suffers from the trade-off between the field of view and spatial resolution. In order to expand the field of view, ptychographical

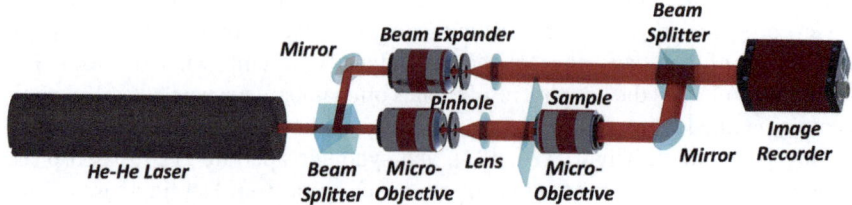

Fig. 5.15 Quantitative interferometric microscopic system

Fig. 5.16 Quantitative interferometric microscopy results. **a** Captured off-axis interferogram; **b** spectrum and the first order information selection in the frequency domain; **c** extracted wrapped phase; **d** retrieved unwrapped phase where the color bar indicates the phase and the length of while line is 10 μm

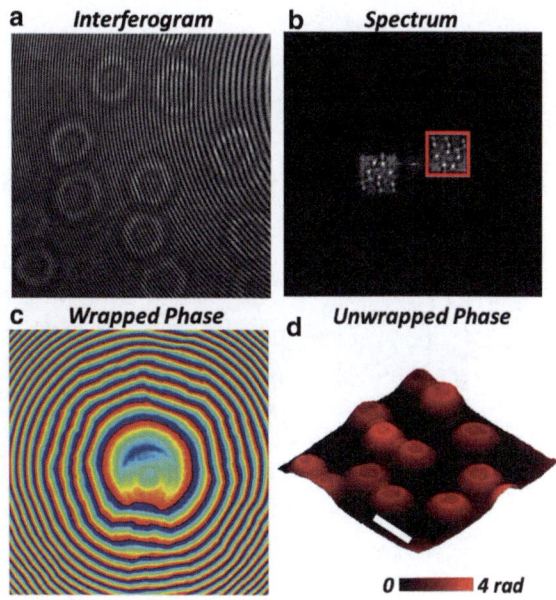

iterative engine is a suitable method as reported in another work [139]. Figure 5.17a exhibits the scheme of ptychographical iterative engine. A probe light illuminates on the object and generates diffraction patterns that are captured by imaging recorder. Object under detection should be scanned in two dimensions, often with extremely large field of view. Both object amplitude and phase distributions can be retrieved iteratively as in the example shown in Fig. 5.17b, which demonstrates that such technique can provide quantitative phase imaging in extremely large field of view.

Unfortunately, classical ptychographical iterative engine relies on 2-D mechanical scanning, which is not only slow in scanning, but also inaccurate in scanning precision. In order to solve such problem, improved digital micro-mirror device (DMD)-based ptychographical iterative engine [140] was reported as shown in Fig. 5.18. Figure 5.18a demonstrates its optical system, where the laser is expanded into a planar laser beam and illuminates the surface of DMD at an incident angle of 12°. Micro-mirrors within a proper region are switched to the "on" state to reflect the laser beam to the sample, forming an illuminating probe. When the position of this circular region is programmed to rapidly move on the DMD sensor, the illuminating probe will quickly scan the sample and form diffraction patterns that are captured by the image recorder. Unlike the classical mechanical scanning-based ptychographical iterative engine, where the sample fixed on a translation stage is scanned to many positions relative to the static illuminating beam and static detector, in Fig. 5.18a, the sample, the detector, and the light beam incident on the DMD are all stationary during the whole data acquisition. The data acquisition is controlled by switching on/off the micro-mirrors within a moving region to form the scanning illumination on the sample and record the diffractions. This can be automatically and rapidly

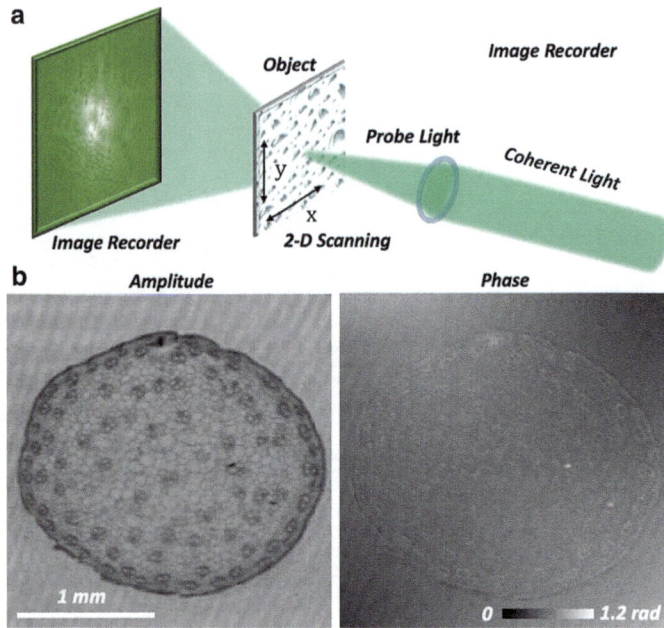

Fig. 5.17 Classical mechanical scanning-based ptychographical iterative engine. **a** Optical system; **b** reconstructed results

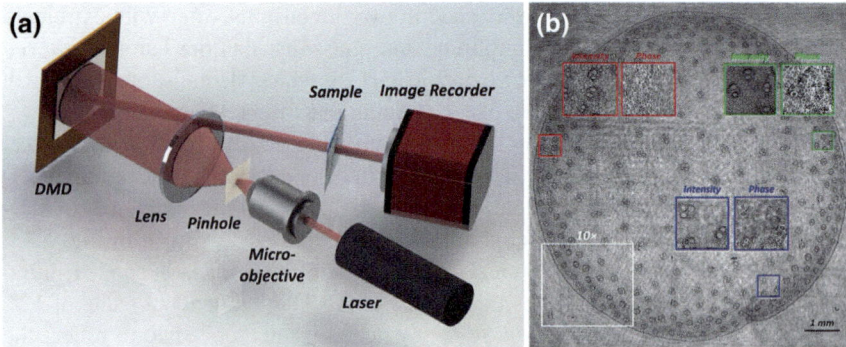

Fig. 5.18 Improved DMD-based ptychographical iterative engine. **a** Optical system; **b** reconstructed results

realized with a computer program. The maximum field of view reachable is limited by the size of the DMD sensor and the size of the imaging chip. The sample used is a paraffin section of a stem cut of monocot plant. The size of the observed field of view in the proposed ptychographical iterative engine is ~45 time larger than the field of view of commercial 10 × objective. Figure 5.18b shows the reconstructed amplitude of the whole sample. The fine details of the sample can be clearly observed in the zoomed-in modulus and phase images that are also inserted in Fig. 5.18b. Moreover, large FoV of 147.08 mm^2 is reached within 20 s, which is about 45 times faster than the classical mechanical scanning-based ptychographical iterative engine.

Static optical phase imaging is mostly used for fixed specimens such as cells and tissues. Moreover, among various computational optical phase imaging techniques such as digital holographic microscopy, transport of intensity phase microscopy, Fourier ptychographic microscopy, ptychographic iterative engine, and so on, ptychography is the most suitable one for static optical phase imaging: in static cell imaging, Fourier ptychographic microscopy can both obtain large field of view and higher resolution. In static tissue imaging with rather large sizes, ptychographic iterative engine not only can reach extremely high resolution, theoretically reaching the diffraction limit, but also has extremely large field of view only determined by specimen scanning; moreover, it also has the capability for three-dimensional tomographic imaging. While for live cell imaging, in order to catch their dynamics, dynamic optical phase imaging techniques are required, which is discussed in the following section.

5.3.2 Computational Optical Phase Imaging in Dynamic Specimen Observation

In addition to static observations, computational optical phase imaging can also be used to measure dynamic specimens such as live cells. Compared with static measurements, dynamic measurements can reveal more about the specimen. For instance, the dynamics of live red blood cells can provide insight into human disease etiology, the dynamic cardiomyocyte change can measure how well the heart beats, and the cellular dynamics can also reveal cell growth and death. As a result, a variety of computational optical phase imaging techniques have been reported to extract more information from specimens.

As a simple but important cell, the mechanics of the red blood cell were studied in detail in a dynamic perspective using dynamic optical phase imaging. Based on the quantitative interferometric microscopy such as the Hilbert phase microscopy and the diffraction phase microscopy, Popescu et al. measured the dynamics of red blood cells [141–144]. Specifically, they could quantify the fluctuations of the red blood cell membranes in nanoscale accuracy with high temporal and spatial resolution. They not only measured the mechanical changes of the red blood cells during their morphological changes [145], but also studied the non-equilibrium membrane fluctuations of the red blood cells powered by adenosine 5′-triphosphate (ATP) [146]. Additionally, Magistretti group and Marquet group also adopted the digital holographic microscopy to measure the red blood cell membrane fluctuations [147], and also analyzed the spatially-resolved eigenmode decomposition of the red blood cell membrane fluctuations [148]. Moreover, Wang group used the quantitative interferometric microscopy to monitor the red blood cell shape changes due to the introduced lithium ion and lead ion [149]. Popescu group and Moon group also evaluated the stored red blood cell fluctuations in order to monitor red blood cell functionality during storage time [150, 151]. The two-dimensional computations of optical phase imaging on red blood cells have also been extended to three-dimensional images using fast scanning, high-speed CMOS cameras, and parallel computing reconstruction [152], which requires a much faster capturing rate compared to static three-dimensional imaging.

Besides red blood cells, dynamics of the cardiomyocytes were also analyzed using computational optical phase imaging. Cardiomyocytes have a beat cycle composed of contraction and relaxation, and during this beat cycle, cardiomyocytes change the shape rapidly, thus the shape changes can be monitored using dynamic optical phase imaging. Popescu group used the spatial light interference microscopy [153] and Moon group used the digital holographic microscopy [154–156] to measure the dynamic phase distributions to reflect both the contraction and relaxation processes during the beat cycle. These works prove that dynamic optical phase imaging not only can used for beating pattern characterization, but also can be applied for preclinical drug assessment in future for cardiac liability.

Besides direct cell observations, dynamic optical imaging was also adopted for cell cycle observations and measurement, including cell growth, division, and death. Popescu group adopted the spatial light interference microscopy to observe the cell growth [157, 158]. Shaked group monitored the cell phase during the whole cell cycle relying on the interferometric phase microscopy [159]. Allier group used the lenless digital holographic microscopy to observe the cell division [160], and further unveiled the cell functions including cell-substrate adhesion, cell spreading, division, orientation, and death [161]. Magistretti group and Marquet group used the digital holographic microscopy to study the fission yeast cell cycle [162]. Moreover, they also designed early cell death detection criterion based on the computational optical phase imaging [163]. Chen group also measured the cell phase distributions to monitor the cell volume changes during apoptosis still using digital holographic microscopy [164].

Other applications were also reported using dynamic optical phase imaging. Popescu group studied the active intracellular transport in metastatic cells using the spatial light interference microscopy [165] and further developed phase correlation imaging to quantitatively describe the cell fluctuations [166]. Shaked group recognized cancer cells from healthy cells according to the fluctuation profiles measured by interferometric phase microscopy [167, 168]. Magistretti group and Marquet group provided the dynamic cell morphometry using digital holographic microscopy [169, 170], and further analyzed the transmembrane water fluxes in neurons [171], screened the chloride channel [172], and studied the cytoarchitectural alterations during simulated microgravity [173, 174]. Ferraro group analyzed the cytotoxicity of light [175] and chemicals [176] using digital holographic microscopy. Monneret group adopted the quadriwave lateral shearing interferometry as a self-referenced interferometric technique to monitor the dry mass of cells [82].

Besides the above dynamic optical phase imaging methods relying on interferometry including digital holographic microscopy, quantitative interferometric microscopy and spatial light interference microscopy, other tactics were also used to measure the cell dynamics. Mesquita group provided the defocus microscopy to monitor the cell fluctuations from multi-focal images through scanning [177–180], however, they did not provide effective solutions for simultaneous multi-focal imaging thus limiting its real-time applications. In order to improve the temporal resolution of such method, Zhao group inserted the flipping imaging module into the microscope to record the multi-focal images in single-shot [181, 182]; and Wang group also designed point spreading function engineering method relying on a spatial light modulator to obtain the single-shot multi-focal images as well [183], both to obtain real-time phase imaging. However, their fields of view are significantly limited. To solve the problem, Wang group also proposed the dual-view transport of intensity phase microscopy in which two identical image recorders are set on different imaging planes to simultaneously capture two multi-focal images thus to retrieve the real-time phase images [184, 185], and such method was successfully used to analyze the ATP induced cellular fluctuations [186].

Besides the above mentioned applications in dynamic specimen observations that are mainly focusing on specimen dynamics, a few works have been focusing on scanning the sample, especially in microfluidic chips, to increase the detection region. Ferraro group adopted the digital holographic microscopy for label-free analysis on human blood cells in microfluidic flow [187]. Schonbrun group proposed fluidics-based focus-stack collecting microscope, in which multi-focal images of red blood cells flowed through the tilted microfluidic channel, and the phase can be retrieved from these multi-focal images through solving the Poisson equation, the same way as transport of phase microscopy [188]. Wang group adopted off-axis quantitative interferometric microscopy to image the gravity-driven microfluidic flow of red blood cells to pursue a rather fast scanning speed [189, 190]. Moreover, optoflu-idic time-stretch quantitative phase imaging relying on extremely fast speed single pixel detector is also designed which can significantly accelerate the imaging speed [191–198].

Computational optical phase microscopy for dynamic specimen observation is briefly discussed here using two examples. The first one is a PhaseRMiC [199]—a Phase Real-time Microscope Camera, for live cell phase imaging. PhaseRMiC can be directly connected to a commercial microscope for simultaneous multi-focus image recording as shown in Fig. 5.19a. Figure 5.19b describes the PhaseRMiC configuration: the wavefront from the microscope is first divided using a prism beam splitter and then collected with a board-level camera integrated with two CMOS imaging chips (1280 × 1024, 4.8 μm, Daheng Imaging, China). The distances between two CMOS imaging chips and the beam splitter are designed as 2 mm for simultaneously recording two under- and over-focus images with defocus distance of 1 mm; and the in-focus image can be approximated as the average of the under- and over-focus images. All the elements are integrated using 3D printed structures to construct a camera-like device with a compact size of 136 × 91 × 60 mm^3 and is also cost-effective comparable to many commercial microscope cameras as exhibited in Fig. 5.19c.

After recording both under- and over-focus images, via solving the transport of intensity equation, the phase can be extracted. It is worth noting that PhaseRMiC

Fig. 5.19 PhaseRMiC. **a**
PhaseRMiC prototype; **b**
PhaseRMiC optical and
mechanical configurations; **c**
PhaseRMiC in applications

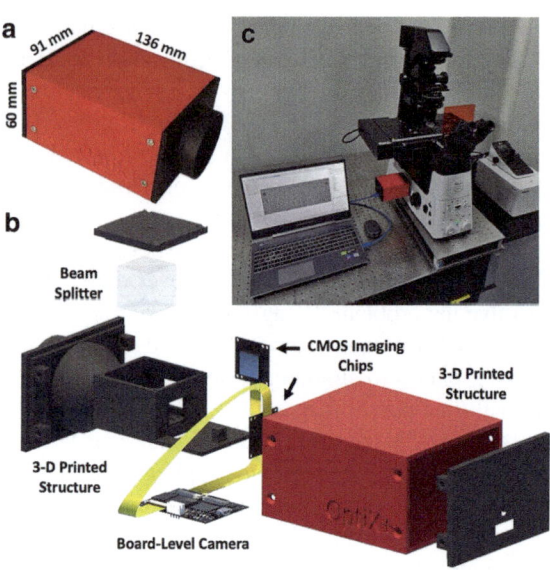

not only can be well used in static phase imaging, but also especially suits for
dynamic imaging. For example, the dynamic phase fluctuations of live Vero cells
were measured. Vero cells were cultured with 10 mL Dulbecco's modified eagle
medium with 0.05% fetal bovine serum for 12–14 h till these cells were completely
adherent and extended. After removing the culture medium, 2 mL phosphate buffer
saline was introduced for cell cleaning for three times and finally sucked out. Additionally, to promote rapid cell change, trypsin solution was added, and the cell shapes
changed from rhombic to round in 3 min. Dynamics of live Vero cells were recorded
using PhaseRMiC with the frame rate of 20 fps. Figure 5.20 lists the retrieved phases
at different time which clearly demonstrate that Vero cell changed from rhombic to
round due to the introduced trypsin solution. The application proves that PhaseRMiC
can be well used in live cell phase imaging.

Another example is gravity-driven high-throughput phase cytometer based on
quantitative interferometric microscopy [189, 190], which has the same optical
system as demonstrated in Fig. 5.21. The sample was no longer a static slice, but
a simply fabricated microfluidic channel with diluted red blood cell solutions. As
shown in Fig. 5.21, the microfluidic channel is vertically located, and by manually
pumping red blood cells (from rabbit) diluted with blood preservation solution, these
red blood cells quickly flowed through the microfluidic channel driven by gravity.
Using a high speed camera, red blood cell generated interferograms could be recorded
for cellular phase retrieval. Such design aims at scanning more samples but with much
faster speed compared to mechanical sample scanning.

Figure 5.22 reveals the measured results of such gravity-driven high-throughput
phase cytometer. After extracting the red blood cell phases using the classical fast
Fourier transform, more red blood cell details, such as area, volume, roundness and

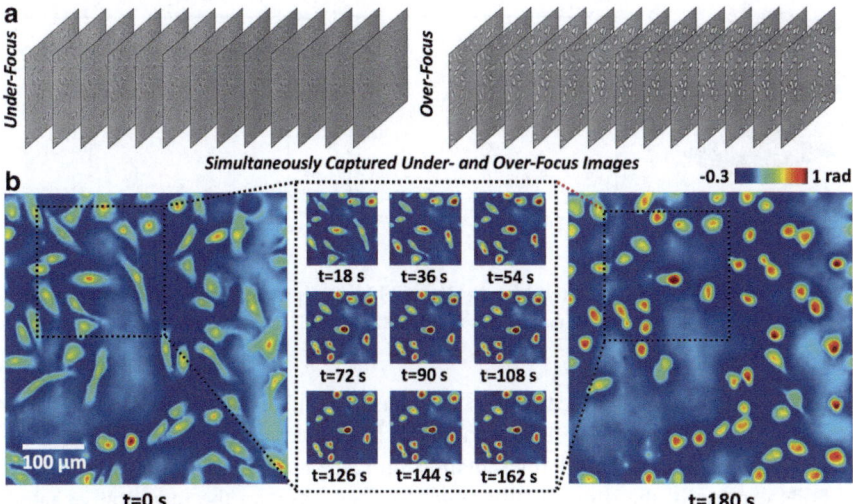

Fig. 5.20 Phase imaging on live Vero cells using PhaseRMiC. **a** Simultaneously captured under- and over-focus images; **b** reconstructed phase distributions at different time. The white bar in **b** indicates 100 μm

Fig. 5.21 Simply fabricated microfluidic channel for gravity-driven high-throughput phase cytometer

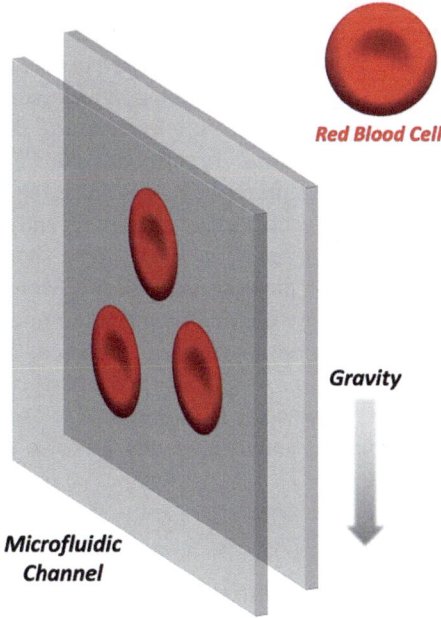

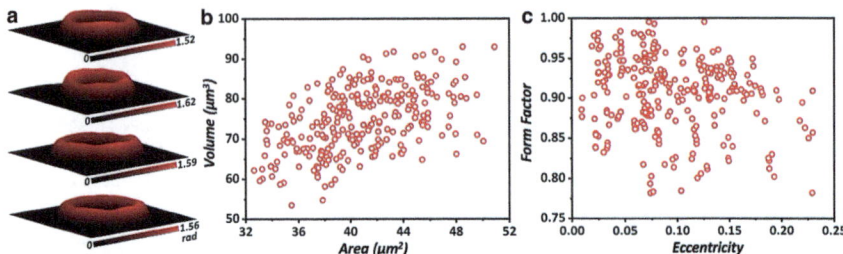

Fig. 5.22 Statistical analysis on red blood cells. **a** Measured phase distributions of several red blood cells; **b** cellular areas and volumes; **c** cellular eccentricities and form factors

so on can be quantitatively measured. Such phase cytometer as well as its derivatives provides a potential route for high-throughput label free cellular detection.

Among these tactics used for dynamic optical phase imaging, interferometry methods especially the digital holographic microscopy and the quantitative interferometric microscopy are widely used. It is because of the fact that these methods can often retrieve the phase distributions from single-shot interferograms or holograms, thus supporting the dynamic specimen observations. Moreover, real-time transport of intensity phase microscopy has also been designed aiming at recording multi-focal images in single shot to obtain real-time optical phase imaging. Unfortunately, ptychography including ptychographic iterative engine and Fourier ptychographic microscopy is less reported in dynamic optical phase imaging, though it often provides high imaging resolution and accuracy with extremely large field of view. It not only relies on multiple recording through scanning, but also requires time-consuming iterations in phase retrieval. Therefore, ptychography is not a suitable solution in dynamic optical phase imaging. However, it should be noted that though single-shot ptychography has been recently proposed [200–207], it suffers from the limited field of view. Therefore, single-shot ptychography is still not preferred in the biological and medical applications. Moreover, in this section, designs and applications of many interferometry-based microscopy and some transport of intensity phase microscopy have been discussed, and it is proved that these methods can support real-time image only in the perspective of the hardware. Besides, fast phase retrieval processed by the software is still required to obtain rapid or real-time optical phase imaging and display, which is also a challenge in the dynamic optical phase imaging.

5.3.3 Computational Optical Phase Imaging in Hybrid Imaging

Though computational optical phase imaging has been successfully used for cell and tissue imaging in label-free mode, it can hardly replace the widely used fluorescence imaging or non-linear optical imaging techniques since it loses the specificity in target observations. It is generally known that fluorescence dyes can be specifically labeled on the organelles and the non-linear optical imaging techniques can locate special structures of tissues. In computational optical phase imaging, though according to the refractive index distributions, nucleus or other organelles can be recognized in the cell, it is still difficult to distinguish different organelles. While using the "non-linear" microscopy, they can be easily distinguished according to fluorescence signals or non-linear optical signals. Therefore, "non-linear" microscopy is still the mainstream technique in biomedical imaging due to its specificity. However, "non-linear" microscopy can only provide local-field images, which could not reflect the overall information in the whole field of view.

In order to maintain the imaging specificity but also to collect the overall specimen information, the combination of the computational optical phase imaging and the fluorescence imaging is a preferred solution. Actually, fluorescence imaging has already been combined with bright-field/phase contrast/differential interference contrast imaging for many applications not only to reflect the specimen outlines, but also to provide the specific specimen details. Compared to those classical "linear" microscopy methods as bright-field, phase contrast, and differential interference contrast imaging, computational optical phase imaging not only reveals specimen images in high contrast, but also provides quantitative specimen phase distributions. However, the combination of the computational optical phase imaging and the fluorescence imaging is still less reported. Feld group designed a combining setup equipped with both the diffraction phase and fluorescence microscopy [208]. Wax group studied the cells undergoing apoptosis from the apoptotic enzyme activity observed using the fluorescence resonance energy transfer and the whole cell morphological changes measured via the diffraction phase microscopy [209]. Magistretti group and Marquet group proposed the multi-modal microscopy combining the epifluorescence microscopy and the digital holography to monitor the cell morphology and the intracellular ionic homeostasis simultaneously [210]. Also combining with the digital holographic microscopy and the fluorescence microscopy, Shaked group designed a system which providing both quantitative phase distribution and molecular specificity using a single camera [211]. Matoba group designed a common-path multi-modal fluorescence and phase imaging system [212], which could reconstruct high-quality images distributed in the axial direction through wavefront propagation. Moreover, Blandin group proposed a common path setup coupled fluorescence imaging with in-line holographic imaging [213], and additionally, Ozcan group also updated their lensless digital holography system with both phase and fluorescence imaging capability [214], both to pursue extremely large fields of view in observations. Zheng group designed the multi-channel microscopy for whole-slide multiplane, multispectral, and phase imaging [215], and Nienhaus group proposed a dual-mode phase and fluorescence imaging system [216], both extracting specimen phases via transport of intensity phase microscopy. Moreover, Feld group also combined Raman microscopy and computational optical phase imaging for chemical sensing and morphology measurements, respectively [217].

Besides two-dimensional imaging, computational optical phase imaging and fluorescence imaging are also combined in the three-dimensional imaging applications. Park group designed the three-dimensional fluorescence and refractive index tomography to bridge the gap between the molecular specificity and the quantitative imaging [218]. Moreover, the same group also combined the optical diffraction tomography and the structured illumination microscopy to obtain super-resolved three-dimensional fluorescence and phase imaging of live cells [219]. Chen group and Shi group also combined them to build the super-resolution fluorescence-assisted diffraction computational tomography system and revealed the three-dimensional landscape of the cellular organelle interactome [220]. More than the combination of the fluorescence imaging and the optical phase imaging, computational optical phase microscopy has also been combined with atomic force microscopy reported by Zhong group [221] and Wax group [222].

These works have proved that computational optical phase imaging can be combined with fluorescence imaging/non-linear optical imaging to pursue both molecular specificity and overall specimen information. Many fluorescence imaging techniques require specimen labeling in pre-treatment, while computational optical phase imaging often works in label-free mode, therefore, it seems contradictory to combine them. However, it should be noted that they work in different wavelengths: computational optical phase imaging provides the quantitative phase distributions in its corresponding wavelength; while fluorescence imaging provides the fluorescence information in an emission wavelength but with the light excitation in another excitation wavelength, and both the excitation and emission wavelengths should be different from that used in computational optical phase imaging. Therefore, optical phase imaging and fluorescence imaging/non-linear optical imaging are compatible, and can be used in many biomedical applications as the combination of specific local-field imaging and the overall whole-field imaging.

5.3.4 Computational Optical Phase Imaging in Extended Applications

Besides the above applications in direct specimen observation, computational optical phase imaging is used for other applications in biomedical fields, such as scattering, spectroscopy, photo-thermal sensing and tracking, etc.

Static light scattering is known to provide the spatial refractive index distribution of biological specimens non-invasively, and dynamic light scattering is known to provide specimen mechanics. However, in many scattering studies, the scattering signals collected from multiple specimens often provide the information averaged from groups, thus losing the scattering details of individual specimens. In order to analyze the light scattering from the individual specimens, Popescu group designed the Fourier transform light scattering method which actually first uses the interferometry-based quantitative phase microscopy to retrieve the complex specimen distribution, and then computes its corresponding far-field scattering field from the retrieved complex distribution through Fourier transform [223, 224]. From both the static and dynamic phase distributions, the Fourier transform light scattering not only can provide the static and dynamic light scattering information, but also has rather high spatial resolution, thus can be further used in various applications. Popescu group adopted the Fourier transform light scattering to study the scattering properties of the rat organ tissues such as the scattering mean-free path and the anisotropy factor [225]; and the same group also employed it to study the actin cytoskeleton on the spatiotemporal fluctuations [226]. Besides, Park group used the Fourier transform light scattering to measure the refractive indexes and diameters of particles and even the structures of the clusters [227, 228]; and they are also applied in determining the sickle red blood cells [229] and identify the bacteria [230, 231].

Since biological specimens often have different phase responses to different wavelengths, the dispersion effect can be applied such as in hemoglobin concentration measurements. Park group proposed both the static and dynamic spectroscopic phase microscopy in three different wavelengths to obtain the wavelength-dependent phase distributions relying on multiple lasers and filters, respectively [232]. Moreover, Popescu group adopted the spatial light modulator to construct the spectroscopic diffraction phase microscopy [233], in which the wavelengths can be precisely selected via changing the transmission aperture of the spatial light modulator, and the spectral resolution is much higher than those based on multiple laser sources/filters [232]. Moreover, computational optical phase imaging is also applied in photo-thermal sensing. Due to the thermal-induced refractive index gradients in the heated region, thermal-induced optical distortion will be attached to the wavefront passing through the sample. This wavefront distortion can be measured to reconstruct the actual temperature distribution even with the accuracy of less than 1 K. Based on it, Shaked group [234, 235], Baffou group [236–238], and Park group [239] adopted interferometry-based quantitative phase microscopy to measure and analyze the photo-thermal effect from the nanoparticles and structures on the cells.

5.3.5 Summary

The purpose of this section is to discuss computational optical phase imaging in biomedical applications, including the applications on static specimen imaging, dynamic specimen imaging, hybrid imaging, and further extended applications. A variety of computational optical phase imaging techniques are suitable for various applications. For instance, holography and interferometry are suitable for dynamic imaging due to their single-shot phase retrieval capabilities; ptychography, on the other hand, is excellent for static cell imaging due to its well-balanced field of view and resolution, although it requires multiple image recordings. Transport of intensity phase microscopy and quantitative differential phase contrast microscopy can be simply integrated with commercial microscopes and they have been successfully extended to dynamic phase imaging. According to the authors, although optical phase imaging can provide sample details in the phase perspective, they cannot be used independently to provide rich information on the samples since the label-free tactic simplifies preparation but eliminates specificity. Computational optical phase imaging is therefore an auxiliary technique, such as providing background information for fluorescence imaging. A recent study shows that, through deep learning, the retrieved phase can provide specific sample information such as virtue phase stain, which opens up the possibility of computational optical phase imaging in specific biomedical applications. There are also other references on this topic but in the context of various applications [240–243], which can be useful as a complement to this section.

5.4 Computational Optical Phase Imaging for Adaptive Optics

In addition to direct imaging on biomedical specimens, computational optical phase imaging has been used in many other areas, including adaptive optics. Using computational optical phase imaging, the wavefront distortions caused by the optical system and the specimen can be measured and compensated to improve the imaging quality. In addition, both transmission and reflection matrices can be measured via computational optical phase imaging to reconstruct high-quality images of targets in complex scattering media. In this section, we briefly review these works on computational optical phase imaging in adaptive optics.

5.4.1 Computational Optical Phase Imaging in Optical Aberration Detection

As specimens can be directly imaged via optical microscopy, the imaging quality is inevitably influenced by the optical system. The error in system design, assembly, and adjustment often introduces regular optical aberrations such as spherical aberration, astigmatism, field curvature, distortion, and chromatic aberration. However, even using perfect optical system, the imaging quality is still influenced by the specimens, especially the thick ones. The specimen refractive index variations inevitably introduce often irregular optical aberrations to the light passing through the specimen. Both the optical aberrations from the optical systems and the specimens are coded in the wavefront, thus decreasing the imaging resolution and depth, as well as deteriorating the imaging quality.

It is generally known that the introduced optical aberrations are exactly distorted phases, therefore, in order to improve the imaging quality, the optical aberration should be first determined using computational optical phase imaging, and then compensated using devices such as deformable mirrors and spatial light modulators. While for computational optical phase sensing, the most classical approach is interferometry, which can precisely determine the optical aberrations of optical elements and systems [51]. Besides such method, Shack-Hartmann sensing is also widely used as a non-interferometric method [244, 245]. Though it suffers from low spatial resolution, since optical aberrations are present in low frequency components, it can still measure the optical aberrations in high accuracy but using much simpler system compared to interferometry. Besides these methods, digital holography [246], phase diversity [69], Fourier ptychography [247], and ptychographic iterative engine [77] have also been adopted for optical aberration detection. Figure 5.23 reveals the scheme of computational optical phase sensing in adaptive optical microscopy.

However, in many of above works, the optical aberrations were measured offline. In other words, the optical aberration sensing and the specimen imaging cannot be implemented simultaneously. These offline methods still work well when the systems under detection are further used in thin specimen detection, and the aberration induced by the system measured in prior can still be adopted for distortion compensation and imaging improvement. While for those optical aberrations generated from the thick specimens, the offline methods are not preferred, since the measured aberration often does not include the specimen induced distortion. In order to solve the problem, adaptive optical microscopy inspired by astronomical imaging has been proposed, which first measures the optical aberrations induced by both optical systems and specimens, and then uses deformable mirrors or spatial light modulators to compensate the optical aberrations even in real time [248]. Methods in two categories have been proposed and employed in computational optical phase sensing in adaptive optical microscopy. One estimates the wavefront distortion based on image analysis, thus they can be defined as sensorless approaches, such as modal wavefront sensing [249–251] and pupil segmentation [49, 252–254]. They do not rely on any wavefront sensors, therefore, they often have simplified optical systems thus

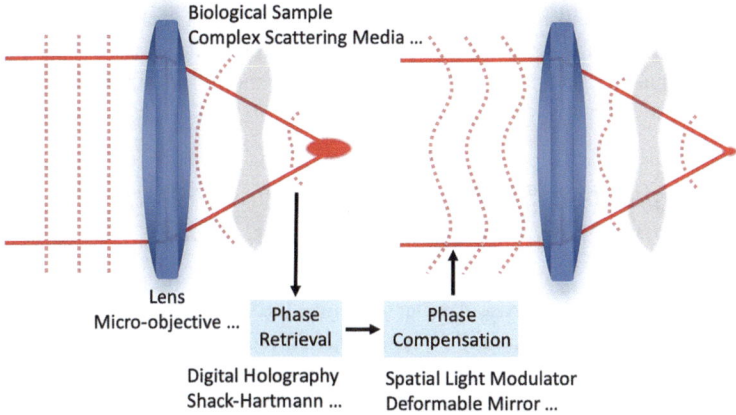

Fig. 5.23 Computational optical phase sensing in adaptive optical microscopy

cutting down the equipment cost. Unfortunately, they often require a large number of images for optical aberration estimation, thus not suitable for rapid imaging in dynamic conditions; moreover, they sometimes suffer from poor accuracy in wavefront estimation. Therefore, many computational optical phase imaging techniques have been used in adaptive optical microscopy. The most widely used computational optical phase imaging technique is Shack-Hartmann sensing, in such methods, the wavefront from an isolated point source of light should be measured [248]. Unfortunately, light from the whole specimens as many super-positioned point sources often leads to ambiguous measurements. Therefore, in computational optical phase imaging used in adaptive optical microscopy, the measured wavefront should be generated from single isolated point source. Many methods have been designed to only collect such light. The most classical one is based on the confocal microscopy with suitable pinhole [255]. Moreover, for multi-photon microscopy, as another kind of confocal microscopy without the pinhole, its focus can provide an ideal point source for computational optical phase sensing because of the confinement of fluorescence emission to the focal region [256, 257]. In addition, guide-stars as nanoparticles [258, 259] and labeled centrosomes [260] can act as the isolated point sources. Moreover, in order to avoid the effects of out-of-focus light, coherence-gated wavefront sensing, which is the combination of the low coherence phase shifting interferometry and the Shack-Hartmann sensor, can only measure the wavefront corresponding to light from the focal region [261–264]. Therefore, it is especially used in the conditions of thick and strongly scattering specimens.

As a widely used technique especially in thick specimen imaging, adaptive optical microscopy can significantly improve the imaging quality with higher spatial resolution and larger axial depth. Though many methods including interferometry, phase diversity, Fourier ptychography, and ptychographic iterative engine can measure the optical aberrations, they are often offline methods. Shack-Hartmann sensor is a preferred tool for online wavefont sensing and widely used in the adaptive optical microscopy due to its advantages as non-interference, simple configuration, and single-shot wavefront extraction capability. Though it suffers from low spatial resolution which does not fit for direct phase imaging for tissues and cells where higher imaging resolutions are required, it is definitely suitable for wavefront aberration reconstruction as the optical aberrations are present in low frequency components.

5.4.2 Computational Optical Phase Imaging in Deep Imaging Within Complex Scattering Media

Though adaptive optical microscopy can minimize the specimen induced aberrations, thus to improve the imaging quality, it is still difficult to achieve high imaging quality in deep tissue. Many adaptive optical microscopes consider the wavefront directly passing through the specimen, thus only suppresses the low-order perturbation, while becomes ineffective in multiple light scattering. Therefore, in order to obtain high-quality imaging even in deep tissue, multiple light scattering should be measured and analyzed [265–267].

Many methods have been proposed for deep imaging based on point optimization [268, 269], optical memory effect [270], correlation [271, 272], nanoparticle guide-stars [273, 274], ultrasound guidance [275–277], and so on. Among them, those based on the transmission/reflection matrix are more general, since when the transmission/reflection matrix can be measured, the image can be simply obtained from the inverse of the transmission/reflection matrix. For transmission matrix, it was firstly demonstrated and measured by Gigan group based on a spatial phase modulator together with a full-field interferometric measurement on a camera [278, 279]. In addition, Gigan group also designed reference-less method to retrieve the transmission matrix [280]; and Park group proposed wavefront shaping-based transmission matrix reconstruction approach [281]. Using these approaches, many applications have been proposed, and two of such successful applications using transmission matrix are deep focusing and fiber imaging. For deep focusing, focal points can be generated in the complex scattering media by compensating the influence on the wavefront induced by the complex scattering media with the previously measured transmission matrix. Based on this principle, Park group obtained deep focusing [282–284] and even subwavelength focusing [285, 286] in the complex scattering media. For fiber imaging, a fiber first guides light to illuminate the specimen, then collects the signal reflected by the specimen and finally guides it to the sensor, therefore, the specimen can be reconstructed through retrieving the specimen reflected light using the fiber transmission matrix. Based on such principle, fluorescence imaging [287–294], confocal imaging [295], and two-photon fluorescence imaging [296] can all be obtained using different kinds of fibers. Additionally, the fiber deformation can also be analyzed using its transmission matrix [297]. Moreover, Park group also designed the reference-free holographic image sensor [298], the scattering optical elements [299], and the dynamic 3D holographic display [300] based on transmission matrix. Besides those methods for transmission matrix reconstruction, interferometry-based optical phase sensing can also be used to measure the transmission matrix. Choi group extended the 3D phase tomography as a computational optical phase sensing technique to measure the complex distribution, therefore, the transmission matrix can be accurately measured [301, 302]. The same group has proposed multi-mode optical fiber-based endoscopy based on fiber transmission matrix [303–307], which has been successfully used for wide-field imaging. Besides transmission matrix, reflection matrix is also used for deep tissue imaging. Choi group designed closed-loop accumulation of single scattering to identify the single scattering wave that incident on and reflected from an object for deep tissue imaging [308–310]. In these methods, reflection matrix is also measured based on interferometry-based optical phase sensing. Moreover, the eigenchannels in the transmission/reflection matrix can also be obtained to analyze the interference of multiple scattering wave, thus have the potential to provide high quality imaging [311–314].

Therefore, beyond the adaptive optical microscopy only considering low-order perturbation, the computational optical phase sensing can also deal with multiple scattering, which can be potentially used for deep imaging. In transmission and reflection matrix measurements, interferometry-based optical phase sensing has been successfully used. Compared to Shack-Hartmann sensors often used in adaptive optical microscopy, interferometry has much higher spatial resolution, which can retrieve the details in the transmission/reflection matrix, thus providing higher quality in deep focusing and imaging.

5.4.3 Summary

Unlike classical intensity-based imaging techniques, computational optical phase imaging can measure the "transparent" specimens such as non-labeled cells or even optical aberrations. Therefore, computational optical phase imaging is mainly used in two perspectives. One is to pursue the high-contrast image of label-free specimens in both static and dynamic detection modes, or even combining with other techniques such as fluorescence imaging (Sect. 5.3). Another is to retrieve the wavefront distortion induced by the optical systems and specimens and then compensate them for improving the imaging quality (Sect. 5.4). Different computational optical phase imaging techniques have been reported in various applications, mainly including holography and interferometry, transport of intensity phase imaging, Fourier ptychography, ptychographic iterative engine, and Shack-Hartmann sensing. Table 5.1 lists their features and suitable application scopes.

Among these computational optical phase imaging techniques, holography and interferometry-based methods have been widely used in live cell imaging since many of these methods can support single-shot phase retrieval. Moreover, these methods are also extended to three-dimensional imaging, therefore, they are not only used for three-dimensional cell and tissue refractive index reconstruction, but also applied for transmission and reflection matrix retrieval for deep imaging. However, holography and interferometry can hardly be coupled with commercial microscopes. Transport of intensity phase imaging can be easily combined with commercial microscopes, and it can be updated to support single-shot phase retrieval. Therefore, it can be successfully used in real-time live cell imaging. Though it can reconstruct the phase faster, the phase retrieval accuracy is still limited. The key advantage of Fourier ptychography is that it can achieve both large field of view and high spatial resolution simultaneously. Therefore, it can be used for imaging cells and tissues in large size but still with high spatial resolution. Unfortunately, Fourier ptychography requires multi-angular illumination scanning as scanning is done in the frequency domain, and additionally, the iteration is required in phase retrieval, making it unsuitable for live cell imaging. Similarly, ptychographic iterative engine is also preferred in tissue imaging in extremely large field of view only limited by the scanning, while not by the imaging system. Different from Fourier ptychography, ptychographic iterative engine relies on the field of view scanning as scanning is in the spatial domain. But it still needs numerical iteration to retrieve the phase and hence not suitable for live cell imaging. With Shack-Hartmann sensing, optical aberrations can

Table 5.1 Comparison of main computational optical phase imaging techniques

	Holography and interferometry	Transport of intensity phase imaging	Fourier ptychography	Ptychographic iterative engine	Shack–Hartmann sensor
Real-time phase retrieval	Yes	Yes	No	No	Yes
Phase retrieval speed	Normal	Fast	Slow	Slow	Normal
Phase retrieval accuracy	High	Normal	High	Very high	Normal
Spatial resolution	Normal	Normal	Very high	High	Low
Field of view	Normal	Normal	Large	Very large	Normal
Complicity	Normal	Simple	Normal	Complicated	Simple
Compatible with commercial microscopes	No	Yes	Yes	No	Yes
Suitable application scopes	Real-time live cell imaging and transmission/reflection matrix retrieval for deep imaging	Real-time live cell imaging	Fixed cell and tissue imaging with high resolution and large field of view	Fixed cell and tissue imaging with extremely large field of view	Adaptive optical microscopy for optical aberration retrieval

be detected in real time, and it can be easily coupled with commercial microscopes, but its spatial resolution is low. Shack-Hartmann sensing is therefore mostly used in adaptive optical microscopy for optical aberration retrieval involving low frequency components, rather than cell and tissue phase imaging, which relies on much higher frequency information.

5.5 Refocusing and Tracking

5.5.1 Background

Images that are in focus tend to have sharper edges and more detail than those that are out of focus. This is because defocus aberration leads to blurring, which erodes the quality of the image. Focusing is required for both macro photography and microscopy in order to achieve excellent image quality. In most photography and microscopy, the focal plane and the in-focus image are determined by subjective judgment. Although manual focusing is quite simple, it is inaccurate and slow. In comparison with manual focusing, autofocusing determines the focal plane more accurately and with greater speed; additionally, rather than relying on subjective judgment, it relies on objective judgment for the determination of the focal plane and in-focus image, which can result in more robust autofocusing results.

Different autofocusing methods have been developed and successfully implemented in cameras with both high focusing precision and fast processing speeds. There have also been many methods developed for microscopy, including laser reflection and image analysis ones [315, 316]. With laser reflectance autofocus, the sample slice is located to a fixed plane in an extremely fast time by tracking the angle of laser reflectance over the surface. However, if the thickness of the cover-slip varies from the standard value, autofocus becomes invalid. This problem can be overcome by image-analysis-based autofocusing, so it has become more popular. In such methods, first a series of multi-focal images should be captured by scanning the sample stage or the micro-objective; then the corresponding characteristic variables of these captured multi-focal images according to in-focus criteria (such as contrast, resolution, frequency components, entropy, and so on) should be extracted; and finally, the in-focus image can be determined according to the extremum estimated by the in-focus criteria. The image-analysis-based autofocus methods avoid the error caused by sample slice variations because they can determine the focal position directly from the captured intensities, and have therefore been widely adopted in many commercial microscopes due to their high accuracy and robustness. Further, some other autofocusing methods have also been developed, such as pinhole assisted [317, 318] and deep learning-based [319–321]. Although a little different from classical image-analysis-based autofocusing, the focal plane is still determined by image features; therefore, they can still be regarded as image-analysis-based autofocusing.

Most autofocus methods are meant for bright-/dark-field microscopy, phase contrast, and differential interference microscopy, and fluorescence microscopy, which all reflect sample details in terms of intensity. In optical phase imaging, autofocus is still required to reconstruct the in-focus phase distribution, thus avoiding defocus blurs. For example, it is crucial for non-imaging digital holography, since the in-focus wavefront can only be reconstructed in high quality when the defocus distance is precisely known. Furthermore, even in imaging phase microscopy, such as transport of intensity phase microscopy, the error in focal plane determination may cause defocus blurs in phase retrieval. As with classical intensity-based imaging, autofocusing in optical phase imaging is still based on what is known as the "overall focusing", which still analyzes the overall information from the entire field of view to determine the focal plane and reconstructed in-focus phase. By contrast, since the wavefront can be retrieved in phase imaging, sample scanning or micro-objective scanning that is required by autofocusing can be replaced by numerical wavefront propagation. Therefore, the term "refocusing" is used in optical phase imaging instead of "autofocusing" to emphasize the numerical wavefront propagation-based focusing. Since mechanical focusing can be replaced by a numerical focusing, optical phase images have some advantages, such as a simple system. While there are many in-focus criteria designed for intensity-based imaging, it is also necessary to develop in-focus criteria that consider both intensity and phase to suit optical phase imaging.

The use of refocusing can also result in a deeper depth of field in addition to phase imaging with quality in-focus images. As a result of the limited depth of view of an imaging system for a sample distributed in a large depth of view (such as a tilted plane), only part of the sample can be imaged in focus, and the other parts come out of focus. In principle, by scanning an imaging system or a sample multiple times, it is possible to create images with an extended depth of view, however, the process is complex and time-consuming. Using phase imaging, various regions of interest located in different depths of view can be refocused using numerical wavefront propagation, generating images with extended depth of view even with single-shot recording. Thus, different from "overall focusing", which aims to achieve in-focus images with high quality, refocusing in optical phase imaging can also be employed to reconstruct high-quality images with extended depth of field. This refocusing can be referred to as "local focusing" in optical phase imaging.

Three-dimensional particle tracking can be viewed similarly to "overall focusing" and "local focusing" in that it can determine the focal plane for these specific particles. This is referred to as "specific focusing" here. There are several applications for three-dimensional particle tracking, such as virus/protein/mRNA tracking, and it is also the prototype for three-dimensional single molecule localization microscopy [322]. There are many methods for tracking three-dimensional particles, such as three-dimensional imaging, multi-focal imaging, point spread function engineering, etc. The majority of these methods require complicated and expensive equipment such as confocal microscopes and spatial light modulators, and they all rely on fluorescence labeling. The optical phase imaging technique provides a label-free solution to three-dimensional particle tracking. Wavefront retrieval using optical phase imaging allows precise three-dimensional determination of particle position, since the phase information quantitatively reflects the particle's axial position is encoded in the wavefront. Three-dimensional particle tracking can be achieved by obtaining sequential particle positions in three dimensions.

Figure 5.24 compares the "overall focusing", "local focusing", and "specific focusing". This section introduces the principles and applications. We will introduce and discuss refocusing in phase imaging for in-focus image reconstruction, as in digital holography, ptychography, and transport of intensity phase microscopy. In addition, refocusing for depth of view extension is also discussed with an imaging sample located in a tilted plane. Additionally, we discuss three applications of phase imaging-aided three-dimensional particle tracking, including particle image velocimetry, sperm tracking, and sub-diffraction target tracking.

Fig. 5.24 Comparison on "overall focusing", "local focusing", and "specific focusing". **a** "Overall focusing" for in-focus imaging; **b** "Local focusing" for depth of view extension; **c** "Specific focusing" for three-dimensional particle tracking

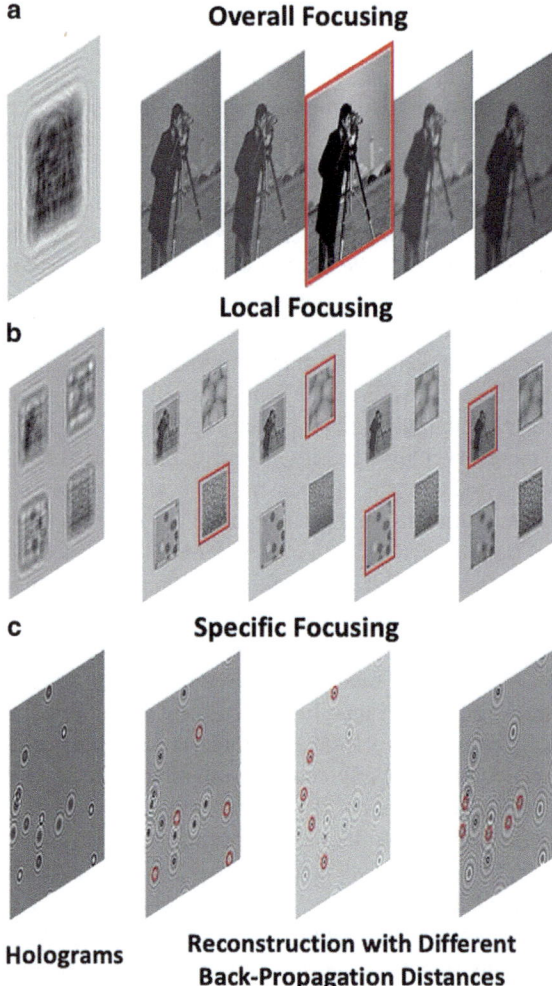

5.5.2 Refocusing in Optical Phase Imaging for In-Focus Image Reconstruction

The most fundamental function for refocusing in optical phase imaging is to reconstruct the in-focus wavefront. In this part, refocusing in digital holography, ptychography, and transport of intensity phase microscopy is introduced and discussed in detail.

When a hologram recording plane is conjugated to a sample plane, it is considered imaging digital holography. By using proper phase retrieval algorithms, the in-focus phase images can be directly obtained from the hologram without requiring any numerical wavefront propagation (most quantitative interferometric microscopy

[323] is associated with imaging digital holography). However, there are many configurations where the sample planes are not conjugated to the recording planes, especially with lens-free digital holography, and these are defined as non-imaging digital holography. For reconstructing the in-focus sample information, back-propagation is often necessary. In most experiments, however, the defocus distance can rarely be precisely known or measured, therefore, in refocusing in digital holography, multi-focal images should still be obtained through numerical wavefront propagation, which is similar to classical image-analysis-based autofocusing by recording multi-focal images through sample or micro-objective scanning. The in-focus reconstruction and the defocus distance can both be determined using different evaluation functions.

A scheme shown in Fig. 5.25a exhibits the procedures of refocusing in digital holography. Setting amplitude and phase distributions are shown in Fig. 5.25b, and the phase-shifting on-axis holograms is shown in Fig. 5.25c. After using phase-shifting phase retrieval algorithm, the wavefront distribution can be reconstructed. Then using numerical wavefront propagation, a series of reconstructed intensity distributions corresponding to different propagation distances are shown in Fig. 5.25d. When the propagation distance is closer to the actual defocus distance, the reconstructed intensity often has sharper edges and better details. In order to determine the in-focus reconstruction as well as to obtain the precise defocus distance, here a simple evaluation function on reconstructed intensity gradient is used, and the in-focus plane can be determined as shown in Fig. 5.25d.

In order to pursue precise focal plane determination, numerous in-focus criteria have been demonstrated, and some representative works are listed in the following. Gillespie and King used both self-entropy of the magnitude and phase as the in-focus criterion especially in the presence of noise [324]. Liebling and Unser designed the Fresnelet-sparsity criterion for autofocus aiming at digital Fresnel holograms [325]. Dubois group adopted integrated amplitude modulus as the in-focus criterion for both pure amplitude and pure phase objects but with opposite extremums [326]. They also updated the integrated amplitude modulus into high-pass filtered complex amplitude modulus to obtain both minimum corresponding to in-focus imaging for both amplitude and phase objects [327]. Moreover, the same group also proposed in-focus criterion suitable for color digital holographic microscopy based on the phase in the Fourier domain [328]. Li et al. [329] detected focus from digital in-line holograms based on spectral 11 norms. Langehanenberg et al. [330] compared criteria based on weighted spectral analysis, variance of gray value distribution, cumulated edge detection by gradient calculation, and cumulated edge detection by Laplace filtering. Memmolo et al. [331] proposed the Tamura coefficient as the in-focus evaluation function only requiring image gray-level standard deviation and mean. The same group employed Gini's index as a sparsity measurement coefficient as an effective autofocus criterion [332]. Lyu et al. [333] adopted the first longitudinal difference as the autofocus criterion for both amplitude- and phase-contrast objects. Zhang et al. [334] designed the sparsity of the gradient as a robust and accurate autofocus criterion. Lam group proposed the in-focus criteria using the structure tensor and its eigenvalues [335] as well as connected domain [336]. Gao et al. [337] also designed

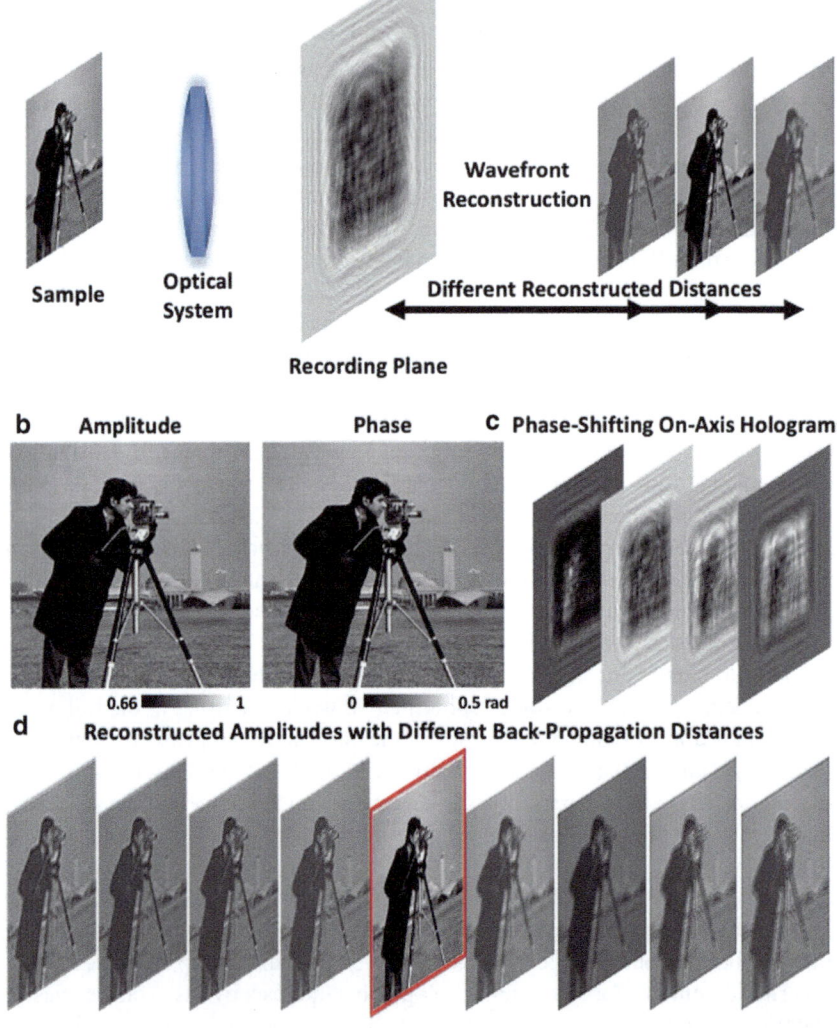

Fig. 5.25 Numerical simulations on refocusing in digital holography for in-focus image reconstruction. **a** Scheme; **b** setting amplitude and phase distributions; **c** phase-shifting on-axis holograms; **d** reconstructed intensity distributions corresponding to different propagation distances

autofocusing method for two-wavelength digital holographic microscopy according to the wavelength dependence of the diffraction process. In addition, deep learning has been also adopted in refocusing in digital holography [338]. The comparisons among some in-focus criteria can be referred in [333, 334, 339].

With the idea of refocusing in digital holography, many applications have been reported for high-quality observation of samples. Ferraro group analyzed bovine sperm head for morphometry analysis relying on refocusing in quantitative phase-contrast holographic microscopy [109]. Javidi group realized satisfied recognition of microorganism according to multi-focal images numerically reconstructed from single-shot in-line digital hologram using inverse Fresnel transformation [340]. Recently, the same group also obtained the in-focus phase image of cells using deep learning for focal plane determination [341]. Additionally, Lee group also adopted an electrically tunable lens to adjust the imaging plane according to the defocus distance calculated by digital holography [342], and this work provides a solution for focal plane tracking using digital holography. More information on refocusing in digital holography is available in reference [343].

Similar to digital holography, refocusing is also used in ptychography, especially in ptychographic iterative engine. Since the distance between the detector and the sample can hardly be measured precisely, the reconstructed sample information may suffer from the defocus blurs. In order to obtain the in-focus sample image, Dou et al. [344] introduced the numerical wavefront propagation into the retrieval method of the ptychographic iterative engine by introducing refocusing into the reconstruction iterations. Similarly, Loetgering et al. [345] also used such strategy to obtain in-focus sample information reconstruction: samples located in multiple slices can be reconstructed in high quality and their intervals can be precisely determined. It should be noted that it is different from the above mentioned tactics, through mathematical analysis, He et al. [346] found that when there exists axial (object to detector) distance error in ptychographic iterative engine, high-quality in-focus image can still be reconstructed using the lateral scanning position correction algorithms [347, 348] without correcting the axial distance during the iterations.

In addition to applications in digital holography and ptychography, refocusing is also used in the transport of intensity phase microscopy. The "in-focus" plane is often manually determined in phase imaging using transport of intensity phase microscopy. While an image can still be reconstructed if the central image deviates from the in-focus plane, the phase is no longer the in-focus one, and defocus aberration blurs the sample details. In order to compensate the defocus error, Wang group first designed the numerical in-focus compensation method used in transport of intensity phase microscopy based on refocusing [349]. Similar to the classical transport of intensity phase microscopy, a series of multi-focal images should be recorded. Due to the inevitable error in manual focusing, the central image position may not coincide with the actual in-focus plane. But via computing the intensity derivative and solving the Poisson equation, the phase distribution at the central plane can still be reconstructed. Combining with the captured intensity distribution, the wavefront at the central image plane can be obtained, which is actually a defocus wavefront. The in-focus wavefront can be numerically reconstructed from it through refocusing demonstrated in Fig. 5.26. First, multi-focal wavefronts are numerically computed from the defocus wavefront; then these wavefronts are evaluated using Tamura coefficient; and finally the in-focus wavefront can be determined. Verified by both simulations and experiments, not only the defocus distance can be determined, but also the in-focus wavefront can be obtained, proving the effectiveness of the numerical in-focus compensation method.

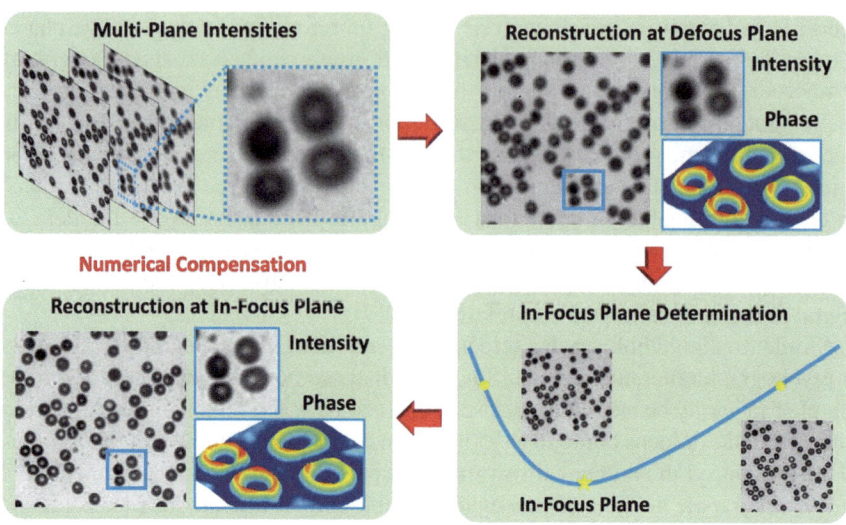

Fig. 5.26 Numerical in-focus compensation method used in transport of intensity phase microscopy

The numerical wavefront refocusing alone is unlikely to be able to bring the quality of the numerically compensated in-focus wavefront close to that of the actual one. In order to solve the problem, Wang group also proposed the mechanical in-focus compensation method [350]. First, the previous numerical in-focus compensation method is implemented. Then, with the determined in-focus position, multi-focal images including symmetric under- and over-focus images and the precise in-focus one can be recorded. Finally, the in-focus phase can be reconstructed by computing the intensity derivative and solving the Poisson equation. The principle of such method is demonstrated in Fig. 5.27. The refocusing performances of both numerical [349] and mechanical [350] in-focus compensation methods are also compared in Fig. 5.28. It is found that the mechanical in-focus compensation method performs better than the numerical one, especially in the conditions with large defocus distances.

Since the transport of intensity phase microscopy can be easily implemented using commercial microscopes, the wavefront refocusing can be extended as an autofocus method in microscopy. It is defined as the wavefront-sensing-based autofocusing [351] as demonstrated in Fig. 5.29. In this method, three multi-focal images are captured, and the defocus wavefront can be retrieved. Afterwards, the focal plane can be precisely determined via wavefront refocusing. When the defocus is not remarkable (it can be easily satisfied by manual focusing), the precise focal plane can be determined only using the wavefront-sensing-based autofocus method once. While the defocus is significant such as with the defocus distance over than 100 μm, the focal plane can still be precisely determined but using the method several times.

Compared to classical image-analysis-based autofocus methods, it only requires rather few multi-focal images, thus avoiding the numerous multi-focal image

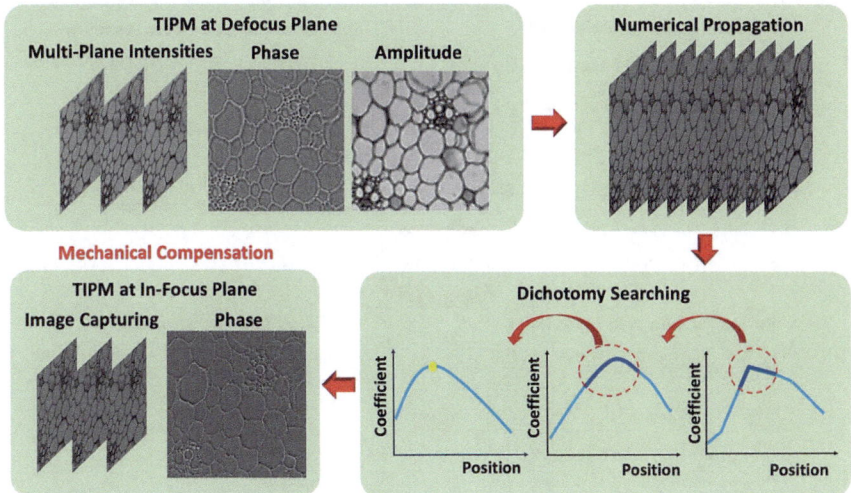

Fig. 5.27 Mechanical in-focus compensation method used in transport of intensity phase microscopy

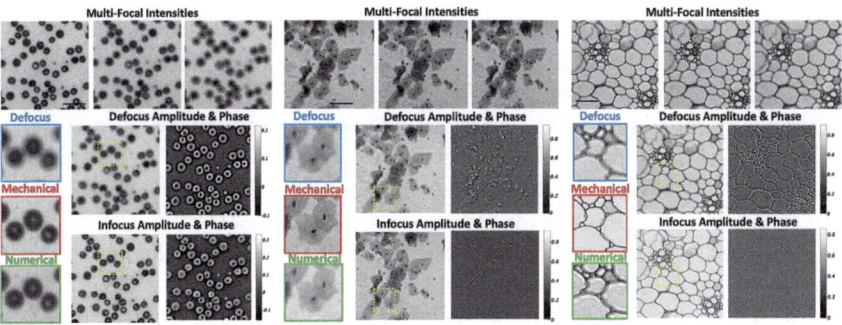

Fig. 5.28 Comparison on numerical and mechanical in-focus compensation method used in transport of intensity phase microscopy

recording through sample or micro-objective scanning. However, refocusing still consumes a lot of time during numerical wavefront propagation. In order to further accelerate the autofocus speed, Wang group also proposed GPU-aided [352] and pixel reduction-aided [353] wavefront-sensing-based autofocus methods, respectively. Since the numerical wavefront propagation and its in-focus evaluation can be carried out in parallel, GPU is adopted to remarkably accelerate the autofocus speed. Additionally, the refocusing speed greatly depends on the number of pixels in the image, with fewer pixels often processing faster. The pixel number of the retrieved wavefront is reduced before numerical wavefront propagation and in-focus evaluation by either downsampling as a frequency domain pixel reduction or by

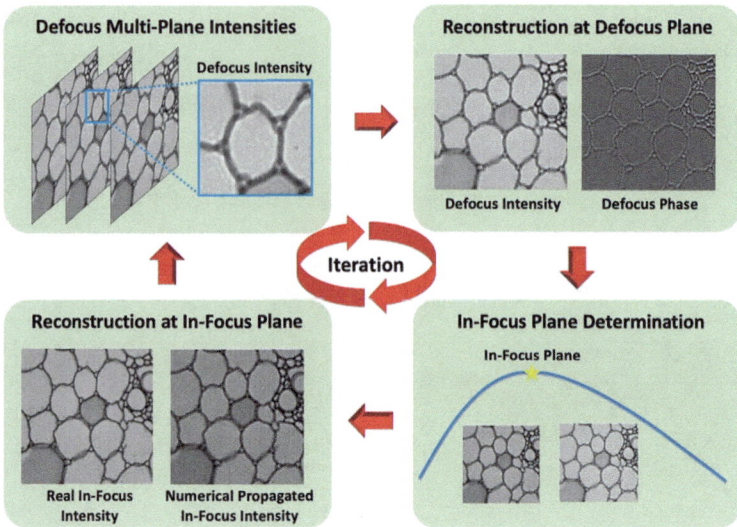

Fig. 5.29 Wavefront-sensing-based autofocusing

selecting the field of interest as a spatial domain pixel reduction. Wavefront-sensing-based autofocus can be accelerated significantly while maintaining high in-focus determination precision as shown in Fig. 5.30.

Essentially, refocusing is used in various phase imaging techniques, such as digital holography, ptychography, and transport of intensity phase microscopy, in order to obtain an in-focus image of high quality. Different from classical autofocusing which relies on sample or micro-objective scanning, numerical wavefront propagation is all that is required, thus avoiding complex operations. To achieve high-quality imaging, the in-focus image can be reconstructed based on various in-focus criteria such as intensity, phase, or both of them. In addition, wavefront-sensing-based autofocus is also developed as a new technique for autofocusing in various kinds of microscopy.

It is important to note that autofocusing only works for thin objects. For thick objects, it is not appropriate to define a focal plane according to the overall field of

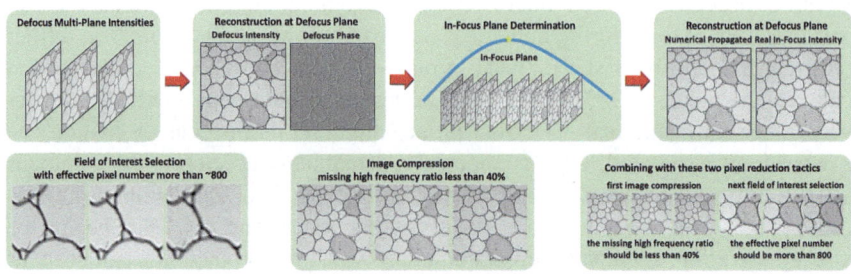

Fig. 5.30 Principle of the pixel reduction-aided wavefront-sensing-based autofocus method

view. In these conditions, the "overall focusing" is replaced by "local focusing", and the refocusing in optical phase imaging for depth of view extension is discussed in the next section.

5.5.3 Refocusing in Optical Phase Imaging for Depth of View Extension

The depth of view cannot be infinite due to limitations of the optical imaging system. In thick samples located over relatively large depths of view, only local regions are recorded in focus, while others are often defocused. While focal plane scanning can extend the depth of view, it requires complicated operations and has low temporal resolution. With numerical wavefront propagation, the wavefronts at different planes can be sequentially reconstructed for observations of local details inspired by optical phase imaging. Therefore, wavefront refocusing here can be treated as "local focusing", and provides a means of focusing on different planes within broad ranges in order to increase depth of field.

An example is provided as shown in Fig. 5.31a. Setting amplitude and phase distributions are shown in Fig. 5.31b, and the phase-shifting on-axis holograms is shown in Fig. 5.31c. Using phase-shifting phase retrieval algorithm, the wavefront distribution can be reconstructed. As a result of numerical wavefront propagation, Fig. 5.31d shows a series of reconstructed intensity distributions corresponding to different propagation distances. It shows that local fields of view become clearer when the propagation distance is closer to the actual defocus distance. With image fusion, the depth of view can be increased.

Works have been reported using refocusing in optical phase imaging for depth of view extension. Poon group obtained cross-sectional sample images in optical scanning holography [354]. Matoba group computed cross-sectional sample images combining with digital holographic microscopy and numerical wavefront propagation [355]. These works are examples of refocusing to specific planes for cross-sectional imaging especially for thick samples. Moreover, with multiple cross-sectional images, fused image can also be obtained as an all-in-focus image reconstruction [356]. Moreover, Ferraro group also extended the depths of view in digital holography using angular spectrum method especially for tilted samples [357–361]. In addition, Wang group adopted the similar idea in transport of intensity phase microscopy for depth of view extension [362] as demonstrated in Figs. 5.32 and 5.33. A little different from the above works, these works extended the depth of view through wavefront rotation rather than refocusing. However, both their aims are the same for depth of view extension, and wavefront rotation only fits for the tilted sample conditions.

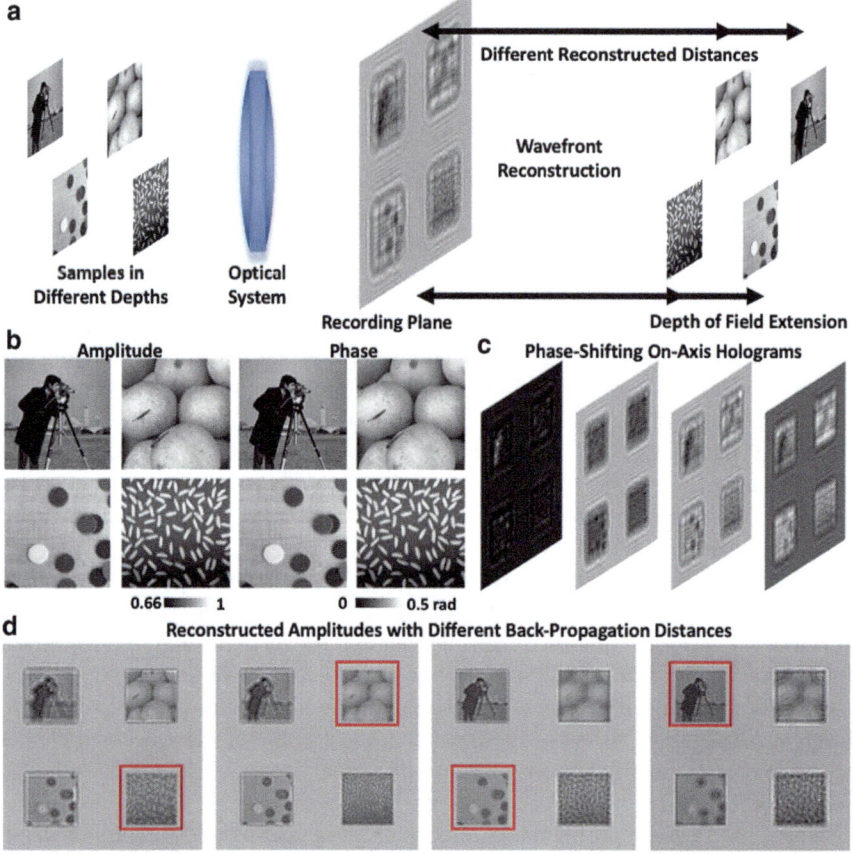

Fig. 5.31 Refocusing in optical phase imaging for depth of view extension. **a** Scheme; **b** setting amplitude and phase distributions; **c** phase-shifting on-axis holograms; **d** reconstructed intensity distributions corresponding to different propagation distances

5.5.4 Refocusing in Optical Phase Imaging for Three-Dimensional Particle Tracking

As previously illustrated, tracking is actually "specific focusing", which aims at three-dimensional localization of specific targets. Cell, virus, protein, mRNA, and other particles such as cells have been tracked for a long time, particularly in biological fields [363–371]. Most particle tracking studies can localize fluorescence-labeled targets in two or three dimensions [322]. However, label-free tracking not only simplifies sample preparation operations, but also has almost no impact on the targets. Thus, optical phase imaging-based tracking has been successfully used in various applications. In this part, particle image velocimetry (PIV) is introduced as an application of flow testing. Two applications in biological fields are then discussed. One of them is

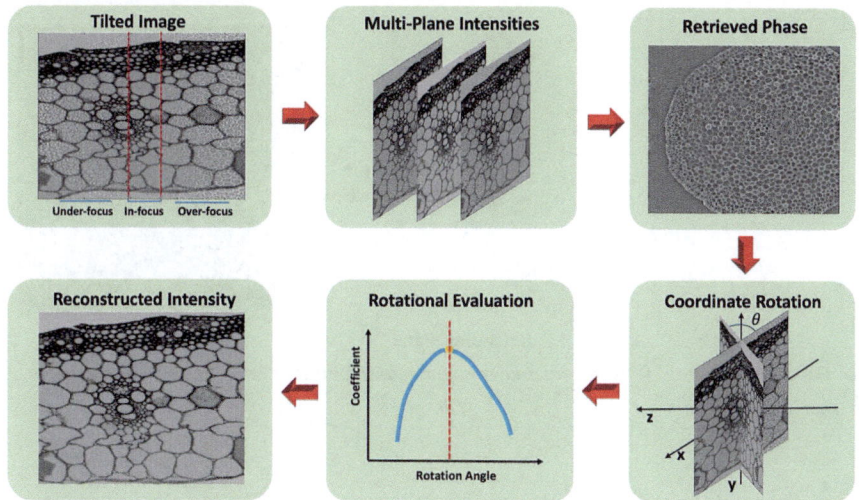

Fig. 5.32 Numerical tilting compensation in microscopy based on wavefront sensing using transport of intensity equation method

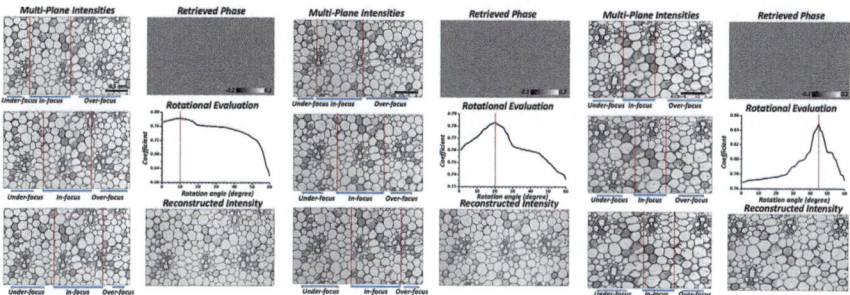

Fig. 5.33 Results of numerical tilting compensation in microscopy based on wavefront sensing using transport of intensity equation method in 10°, 30°, and 45° tilting conditions

sperm tracking as an example of cell tracking; another one is sub-diffraction sample tracking, which can be used to track viruses, proteins, and nanoparticles that are much smaller than the wavelengths.

Turbulent flow is a common physical phenomenon but is highly complex. In order to reflect the physics of complex flows, Particle Image Velocimetry (PIV) as a non-intrusive optical technique provides a solution for capturing velocity vectors via tracking multiple particles. If the tracer particles have enough flow following property, the motion of the tracer particles can truly reflect the turbulent flow field [372]. Various PIV techniques including stereo PIV [373], scanning PIV [374], plenoptic PIV [375], tomographic PIV [376], multi-projection PIV [377], rainbow PIV [378], and time-of-flight PIV [379] have been designed. Different from these PIV methods,

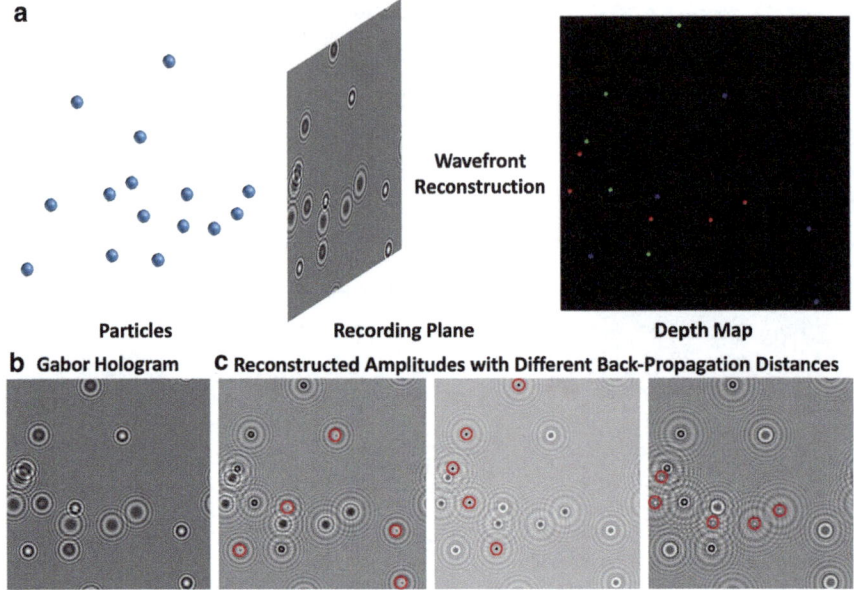

Fig. 5.34 Principle of holographic PIV. **a** Scheme and lateral positions of setting particles; **b** setting amplitude and phase distributions; **c** reconstructed intensity distributions

holographic PIV as shown in Fig. 5.34a provides another way for particle tracking [380, 381]. In holographic PIV, the lateral positions of particles can be easily determined from the hologram as revealed in Fig. 5.34b, while the axial positions of particles are often determined through refocusing via numerical wavefront propagation as shown in Fig. 5.34c. Compared to the above mentioned three-dimensional PIV techniques [109, 315–379], holographic PIV has two main advantages. Firstly, it often requires single camera, the optical system is still simple and cost-effective. Secondly, compared to stereoscopic imaging, the obtained wavefront can maintain an extremely high localization accuracy. Many early works focus on developing optimized in-focus criteria for particle position determination [377–391]. While many recent works improved the holographic PIV system and even updated the data processing algorithms. Grare et al. [392] proposed dual-wavelength three-dimensional holographic PIV to avoid the super-imposition of particle images. Katz group proposed dual-view holographic PIV to deal with the dense particle condition [393]. Koukourakis et al. [394] introduced wavefront shaping in holographic PIV relying on guide stars, and explained its potential on flow detection in biomedical applications. In addition, Masuda group have designed special purpose computer systems for holographic PIV to accelerate the processing speed [395, 396]. Moreover, in the data processing aspect, Hong group provided high fidelity digital in-line holographic method [397] and inverse problem method [398] for three-dimensional tracking both utilizing GPU processing, which significantly accelerated the processing speed. The same group also adopted machine learning in holographic PIV in order to obtain higher particle extraction rate, localization, and speed accuracy over a wide range of particle concentrations [399].

Digital holography, especially on-axis digital holography, is a rather suitable way for measuring three-dimensional velocity fields thus for PIV application. Compared to digital holography-based imaging, holographic PIV is more about particle tracking rather than particle imaging. Therefore, the on-axis digital holography is preferred for simplifying the system since only a coherent light source often a laser and an image recorder are required. The holographic PIV can hardly be considered as a recent technique, but with the development of processors and even deep learning methods, higher particle extraction rate, localization, and speed accuracy can be maintained even particle concentration is high.

Different from PIV in which particles negatively move with the flow, cells positively move in the liquid environment. Cellular motility is an important parameter to evaluate the activity of the cells, especially for sperms. Sperm quality detection is widely used in human and animal reproductive health assessment [400, 401]. According to the World Health Organization (WHO) [106], it is necessary to comprehensively evaluate the sperm quality from semen volume, appearance, pH value, liquefaction time and sperm concentration, viability, motility, morphology, and white blood cell content. Semen volume, appearance, pH value, and liquefaction time can be directly obtained by observing and measuring semen samples without any instrument. However, the sperm quality cannot be accurately evaluated only based on these macroscopic parameters. Sperm concentration combined with semen volume can quantitatively calculate the sperm number. Sperm motility can directly reflect sperm activity. Sperm morphology can be used to distinguish normal sperms, abnormal sperms, condensed sperms, immature sperms, and non-sperm cells including white blood cells. Therefore, compared to above macroscopic parameters, these microscopic parameters including sperm concentration, motility, and morphology measurement can evaluate sperm quality with higher accuracy.

Both sperm concentration and morphology can be easily measured and evaluated via classical microscopic imaging. It is worth noting that as a nearly transparent sample, the sperm imaging quality of the widely used bright-field microscopy is often poor. Though phase contrast and differential interference contrast microscopy can enhance the imaging contrast significantly, they can hardly reflect the actual sperm morphology quantitatively. To solve the problem, phase imaging is used in sperm morphology analysis through reconstructing the quantitative phase distributions of sperms. Especially using digital holographic microscopy, Ferraro group [107–109, 402] and Demirci group [112] have successfully analyzed sperm morphology according to sperm phase distributions. Moreover, Shaked group used quantitative interferometric microscopy to measure sperm morphology, compared it with the staining method recommended by WHO [106], and proved that the optical phase imaging can obtain high contrast images to accurately reflect sperm morphology without staining [104, 105]. However, most of these works only focus on static measurements, thus difficult for providing the solution for dynamic sperm analysis such as sperm motility testing.

In order to measure the sperm motility, many representative technologies have been proposed. Paper-based colorimetry directly determined sperm viability from the color of the test strip [403], however, only the sperm viability could be evaluated while not the sperm motility. Besides, photon correlation spectroscopy [404], laser scattering [405], and microfluidic [406, 407] methods have been designed for measuring sperm motility, as well as sperm size and concentration. Most of these methods can only reflect the "ensemble average" of sperms, however, they are impossible to analyze the "heterogeneity" of each sperm. For example, they can only measure the average speed of sperm movement in a population, but not the speed of each sperm. Therefore, in order to measure the sperm motility with higher accuracy, sperm motility testing method that is focusing on "single sperm level" is preferred. Compared to the above approaches [403–407], computer-assisted sperm analysis (CASA) [408–410] can track each sperm thus satisfying the "single sperm level" requirements, therefore it is a preferred way for sperm motility testing in high accuracy. It should be noted that CASA has the similar principle as PIV, both aiming at tracking particles. Though CASA has successfully used not only in sperm motility testing but also in sperm morphology observation, it still has some drawbacks. The most important one is that most CASA instruments can only image on a fixed plane, so only two-dimensional sperm tracking can be realized. Although algorithms have been proposed to calculate the three-dimensional motion trajectory from the two-dimensional information [411, 412], there are still inevitable errors. In addition, it has been reported that three-dimensional sperm tracking can be realized using fast focal plane scanning [413]. But this method remarkably reduced the recording frame rate, thus affecting the accuracy of sperm motility testing. Therefore, due to these drawbacks, the classical CASA is not a suitable method for sperm motility testing.

In order to update CASA for three-dimensional sperm tracking, digital holography was used [414]. Ferraro group investigated the use of an off-axis digital holographic microscope for detection and tracking of sperms [415]. The system is a classical Mach–Zehnder interferometer equipped with micro-objective for sample amplification. Sequential holograms were captured for reconstructing quantitative phase distributions of the field of view at different time. The target sperm positions in three dimensions were determined via refocusing the corresponding sperm. Therefore, four-dimensional (three-dimensions in space and one-dimension in time) tracking on sperm could be obtained. This method can track sperms in both high spatial and temporal resolutions relying on the equipped micro-objective and the high speed camera. However, due to the equipped micro-objective, its field of view is significantly reduced, thus only few sperms can be simultaneously tracked. Different from Ferraro group using off-axis digital holographic microscope equipped with micro-objective, Ozcan group provided three-dimensional sperm tracking based on dual-wavelength lensless digital holography [416–421]. Dual-wavelength light sources were used to irradiate sperm samples in different directions to obtain holograms with different projection angles. The transverse position of a single sperm can be directly obtained by identifying the sperm in the single-wavelength hologram, while the axial position can be obtained according to the position shift corresponding to dual-wavelength holograms. Compared with the traditional two-dimensional sperm tracking, this method can accurately locate the optical axis position of each sperm, so as to realize real three-dimensional tracking to accurately reflect sperm motility. Moreover, compared to that micro-objective equipped system [415], this lensless method can track sperm motility in three dimensions with both extremely large field of view and depth of field, and therefore, it can measure a large number of sperm samples.

Besides quantitative phase imaging, interference can also enhance the imaging contrast such as classical phase contrast microscopy and differential interference contrast microscopy. Besides these two classical and widely used techniques, various interference-based microscopy has been designed to imaging various targets in order to enhance the imaging contrast and sensitivity in a series of cellular observations [422–425]. Furthermore, the interference-based microscopy is also extended to sub-diffraction target imaging relying on confocal microscopy [426]. Recently, the interferometric scattering microscopy (iSCAT) was proposed for nanoscale particle imaging and tracking but only using wide-field microscopy [427]. In the perspective of optics, the collected signals of iSCAT microscopy are the interference between the back-scattered signal from the target and the reflection light from the slide. In other words, iSCAT can be considered as the on-axis digital holography in the reflection mode. It is worth noting that iSCAT is mainly adopted for imaging sensitivity enhancement, thus it can catch the sub-diffraction samples as nanoscale particles, proteins, and virus particles due to the enhanced imaging contrast and the signal to noise ratio. However, due to its single-shot property, it has also been adopted in single particle tracking, but only in two dimensions [428–435].

Since iSCAT also belongs to digital holography, it has the potentials in three-dimensional tracking [436]. The collected interference signal is coded by the phase, which includes the Gouy phase, the scattering phase, and the phase difference corresponding to the optical path difference between the sample and the light reflection interface. In them, the Gouy phase keeps $\pi/2$ for perfectly focused imaging condition, and the scattering phase can be pre-calculated from the refractive indexes of both the target and the environments. Therefore, the iSCAT signal is actually related to the distance between the target and the slide, therefore providing a way to measuring the axial positions of tracked particles. Sandoghdar group adopted iSCAT to study the electrostatic trapping in nanofluidic traps and slits [437]. By monitoring the iSCAT signals, the three-dimensional positions of the particles can be determined in nanometer precision. Besides iSCAT, Hsieh group also designed coherent brightfield (COBRI) microscopy in single particle tracking [438–440]. Different from iSCAT, in COBRI, the interference signal is actually from the forward scattered light and the transmission, which can be treated as the on-axis digital holography in transmission mode in the perspective of optics. Therefore, COBRI can also realize three-dimensional single particle tracking [438] relying the principle similar to that of iSCAT. Besides tracking, scattering-based interferometric microscopy can be applied in many other fields, such as imaging and sensing, and more information can be referred in the reviews on scattering-based interferometric microscopy [436, 441, 442].

These three examples of optical phase imaging-based tracking—PIV, sperm, and sub-diffraction targets—are discussed in this section. Unlike traditional tracking methods using fluorescence labeling and three-dimensional imaging tools, optical phase imaging-based tracking can achieve label-free three-dimensional particle tracking only with a simple wide-field imaging system. Optical phase imaging tracking offers two significant advantages. It can track targets without labeling. Secondly, optical phase imaging-based tracking does not require complicated optical systems for three-dimensional localization. These two advantages make label-free tracking an attractive solution for some applications.

5.5.5 Summary

Refocusing and tracking are discussed in this section, along with their applications in optical phase imaging. As with classical intensity-based imaging, optical phase imaging still requires "overall focusing" to reconstruct the in-focus phase distributions, so as to avoid blurring caused by defocusing. Using phase imaging, however, the depth of view can be extended by refocusing, which is actually based on "local focusing". In addition, single-shot phase imaging techniques can also be applied to particle tracking based on "specific focusing", as well as PIV, sperm, and sub-diffraction target tracking. Because the wavefront can be obtained completely via optical phase imaging, different regions of interest can be refocused through numerical wavefront propagation. Phase image refocusing and tracking can be applied to not only high-quality in-focus phase image reconstruction, depth of view extension, and single particle tracking discussed in this section, but also for other potential applications, such as all-in-focus imaging, cross-sectional imaging, and so on.

While the refocusing in optical phase imaging can reconstruct images at different planes, it is still regarded as cross-sectional imaging, rather than three-dimensional imaging or tomography. The refocused image can be viewed as the integral of the in-focus intensity at the refocused plane and the defocus intensity at other planes. In three-dimensional imaging, however, only the sample located within a thin layer, often smaller than the depth of view, can be reconstructed, while the sample intensity outside of such thin layers can be avoided. To promote two-dimensional imaging to three-dimensional imaging, the following section briefly reviews and discusses three-dimensional computational optical phase imaging.

5.6 Three-Dimensional (3D) Computational Phase Imaging

3D imaging is an interdisciplinary and longstanding topic in optics [443–447]. 3D imaging methods including confocal microscopy [448], stereoscopic imaging [449, 450], computed tomography (CT) [451], integral imaging [452, 453], magnetic resonant tomography (MRI) [454], and optical coherent tomography (OCT) [455] can provide details on 3D absorption or reflection properties of thick sample that are utilized in many applications in clinical diagnosis and biological research. Techniques of computational phase imaging including holography [456, 457], ptychography [458], and transport intensity Phase imaging [459] also can be applied to realize 3D imaging. In most of these methods, the sample is rotated about 180° for data acquisition and Radon transform [460] or scattering potential [461, 462] is applied for reconstruction. In this section, we present another approach to 3D phase imaging on thick samples that uses a multi-slice model [463–465]. We begin with an introduction on Classical 3D diffraction imaging using scattering potential to gain a deeper understanding of 3D imaging and take the discussion further to 3D multi-slice imaging.

5.6.1 Classical Optical Diffraction Tomography

3D phase imaging can be realized by using the optical setup shown in Fig. 5.35a. A 3D sample fixed on a rotational stage is illuminated by a parallel beam and can rotate along y-axis. If complex amplitudes of the illuminating light and scattering light are $U_i(x, y, z)$ and $U_s(x, y, z)$, respectively, the overall light field can be written as $U(x, y, z) = U_i(x, y, z) + U_s(x, y, z)$. When the scattered field $U_s(x, y, z)$ is much less than illuminating field $U_i(x, y, z)$, the scattered field can be written as [462]

$$U_s(x, y, z) = - \int G(x - x', y - y', z - z') F(x, y, z) U_i(x', y', z') dx' dy' dz' \tag{5.8}$$

where, $G(x - x', y - y', z - z')$ is Green function, and $F(x, y, z)$ is object function. By taking Fourier transform of both sides of Eq. (5.8), we obtain the following relation, known as the Fourier diffraction theorem

$$\tilde{F}(K_x, K_y, K_z) = \frac{ik_z}{\pi} \tilde{U}_s(k_x, k_y, z = 0) \tag{5.9}$$

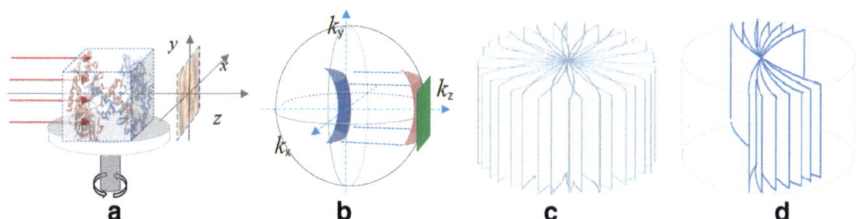

Fig. 5.35 Principle of 3D imaging using scattering potential. **a** Optical setup; **b** coordinate transform; **c** computed $\tilde{F}(K_x, K_y, K_z)$; **d** Obtained $\tilde{F}(K_x, K_y, K_z)$

where, k_x and k_y are spatial frequencies of scattered light, and K_x, K_y, and K_z are differences of k_x, k_y, and k_z to the spatial frequencies k_{x0}, k_{y0}, and k_{z0} of illuminated light, respectively. In experiments, the scattered field $U_s(x, y, z = 0)$ can be obtained by subtracting illuminating light $U_i(x, y, z = 0)$ from overall light $U(x, y, z = 0)$, and then $\tilde{U}_s(k_x, k_y, z = 0)$ in Eq. (5.9) was obtainable. Finally, $\tilde{F}(K_x, K_y, K_z)$ can be obtained by doing coordinate transform on $\tilde{U}_s(k_x, k_y, z = 0)$ in spatial frequency domain using $k_z = \sqrt{(nk_0)^2 - k_x^2 - k_y^2}$ and then shifting the transformed $\tilde{U}_s\left(k_x, k_y, \sqrt{(nk_0)^2 - k_x^2 - k_y^2}\right)$ with respect to illuminating vector of (k_{x0}, k_{y0}, k_{z0}), where n is the refraction index of sample, and $k_0 = 1/\lambda$. When the illumination is along z-axis, the illuminating vector is $(0, 0, nk_0)$, and the diffraction potential of $\tilde{F}(K_x, K_y, K_z)$ is $\frac{ik_z}{\pi}\tilde{U}_s\left(K_x, K_y, \sqrt{(nk_0)^2 - k_x^2 - k_y^2} - nk_0\right)$.

Figure 5.35b shows the physics of coordinate transform on $\tilde{U}_s(k_x, k_y, z = 0)$ to obtain $\tilde{U}_s\left(k_x, k_y, \sqrt{(nk_0)^2 - k_x^2 - k_y^2}\right)$, where the green rectangle indicates the $\tilde{U}_s(k_x, k_y, z = 0)$ computed from the scattered light field $U_s(x, y, z = 0)$, and by projecting it to the Ewald sphere we obtain the light red curved region that indicates the $\tilde{U}_s\left(k_x, k_y, \sqrt{(nk_0)^2 - k_x^2 - k_y^2}\right)$. Finally, when the light red region is shifted to the center of Ewald sphere we obtain the $\tilde{F}(K_x, K_y, K_z)$ that is indicated with the region of light blue in Fig. 5.35b. By rotating the sample along y-axis step by step and repeating above computations at each angle position, we can get the computed $\tilde{F}(K_x, K_y, K_z)$ shown in Fig. 5.35c. When the angle step size in sample rotation is fine enough, it will sufficiently fill the $K_x K_y K_z$ space, and the sample structure of $F(x, y, z)$ can be computed by doing inverse Fourier transform on the obtained $\tilde{F}(K_x, K_y, K_z)$ as shown in Fig. 5.35c.

The difference among specific diffraction tomographic methods lies on how to measure the scattered light field $U_s(x, y, z)$. Digital holographic tomography adopts an additional reference beam to form a hologram [456, 457], ptychography scans the sample to a raster of positions and records several frames of diffraction intensity at each angle [458], transport of intensity phase imaging measures the wave front of transmitted field by assuming that x-ray propagates along straight line and records two frames of images at slightly different axial positions [459]. In all these methods, tomographic imaging was realized by illuminating the sample from various angles relative to the sample. When illuminating angle is only in a small part of $[-90°, 90°]$ as in several experiments, the obtained $\tilde{F}(K_x, K_y, K_z)$ only occupies less part in Fig. 5.35d than in Fig. 5.35c and consequently, the reconstruction quality becomes lower. In Fourier ptychography, where a lens was applied to collect diffracted light, $\tilde{F}(K_x, K_y, K_z)$ was mainly limited by the lens aperture [443]. Except for rotating sample or illuminating beam, sectional phase imaging can also be realized for thick sample using multi-slice mode, where the sample is regarded as a stack of infinitely thin 2D slices, and the interval between successive slices is regarded as uniform with average refractive index of the sample material. The optical setup of these method is also quite simple since we need not rotate the sample and the illumination beam.

5.6.2   3D Imaging with Curved Illumination

A reconstructed image from digital holography can be refocused at any axial distance required, thereby revealing the three-dimensional structure of a thick sample. Since the image generated is always the result of interference between clear images of focused parts and blurred images of defocused parts, the 3D distinguishing capability of common digital holography is quite limited. The use of multi-slice mode in conjunction with digital holography in reflection optical geometry can enable the creation of real 3D imaging [467].

In Fig. 5.36a, a thick sample was sliced into N layers, where the nth slice is regarded as 2-D object with transmitting function of $T(x, y) = |T_n(x, y)|e^{i\phi(x,y)}$, the interval between the nth layer and the $(n + 1)$th layer was regarded as uniform. Under unit planar illumination, the light illuminated on the mth layer can be written as $\prod_{n=1}^{m-1} |T_n(x, y)|e^{i\phi(x,y)}$, and the reflection from the mth layer is

$$R_m(x, y) = \left(1 - |T_m(x, y)|\exp[i\phi_m(x, y)]\left[i \prod_{n=1}^{m-1} |T_n(x, y)|\exp[i\phi_n(x, y)]\right]\right)$$

(5.10)

The overall reflection from sample can be written as

$$R(x, y) = \sum_{m=1}^{N}\left(1 - |T_m(x, y)|\exp[i\phi_m(x, y)] \times \prod_{n=1}^{m-1} |T_n(x, y)|\exp[i\phi_n(x, y)]\right)$$

(5.11)

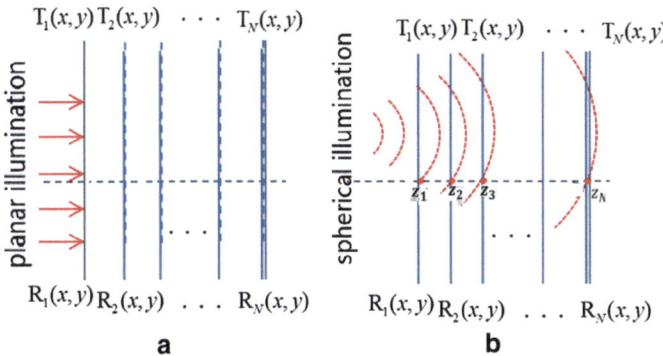

Fig. 5.36 Principle of depth resolved digital holography. **a** A thick sample was sliced into N layers; **b** a divergent illumination

If the sample is highly transparent the nth slice has transmitting modulus of $|T_n(x, y)| \approx 1$ and transmitting phase of $\phi_n(x, y) \ll \pi$ ($n = 1, 2, \ldots,$ N), and hence most of light transmitted through sample without any obvious alteration. This makes illuminations on all layers roughly the same and can still be approximated as unit planar illumination. Then the overall reflection from sample can be approximated as

$$R(x, y) = \sum_{m=1}^{N}(1 - |T_m(x, y)|\exp[i\phi_m(x, y)]) \tag{5.12}$$

Equation (5.12) is the interference among lights reflected from all layers, and $R(x, y)$ itself cannot reveal the 3D structure of sample. Three dimensional imaging can be realized using a divergent illumination as shown in Fig. 5.36b, where red dashed lines indicate spherical wave-fronts of equal phase, and the coordinate where these wave-fronts crosses the mth layer is $(x, y, z_{m=1,2,\ldots,N})$. For highly transparent sample with negligible absorption the complex field illuminating the mth layer can be approximated as $A\exp\left[i2\pi n_0\sqrt{x^2 + y^2 + (z_0 + z_m)^2}/\lambda\right]$. The reflected light can then be written as

$$R(x, y) = \sum_{m=1}^{N}\left(1 - |T_m(x, y)|\exp[i\phi_m(x, y)]\exp\left[i2\pi n_0\sqrt{x^2 + y^2 + (z_0 + z_m)^2}/\lambda\right]\right) \tag{5.13}$$

where Z_0 is the distance of the point light source to the upper surface of sample, it is obvious that the complex field $A\exp\left[i2\pi n_0\sqrt{x^2 + y^2 + (z_0 + z_m)^2}/\lambda\right]$ illuminating the mth layer can be found out when the position of the point source is decided. By dividing Eq. (5.10) with the complex field illumination of the kth layer $\exp\left[i2\pi n_0\sqrt{x^2 + y^2 + (z_0 + z_k)^2}/\lambda\right]$, $R(x, y)$ can be rewritten as

$$R'(x, y) = A(1 - |T_m(x, y)|\exp[i\phi_k(x, y)]) + A \sum_{m=1(m\neq k)}^{N} (1 - |T_m(x, y)|\exp[i\phi_k(x, y)])$$
$$\times \exp\left[i2\pi n_0\left(\frac{\sqrt{x^2 + y^2 + (z_0 + z_m)^2}}{\lambda} - \frac{\sqrt{x^2 + y^2 + (z_0 + z_k)^2}}{\lambda}\right)\right] \tag{5.14}$$

By changing the coordinate of illumination source from $(0, 0, z_0)$ to a series of positions $(0, 0, z_l = 0, 1, \ldots, L)$ and then adding all the corresponding $R'(x, y)$ together, we obtain the overall reflection as

$$R'(x, y) = AL(1 - |T_m(x, y)|) + \sum_{m=1(m\neq k)}^{N} A(1 - |T_m(x, y)|\exp[i\phi_k(x, y)])$$

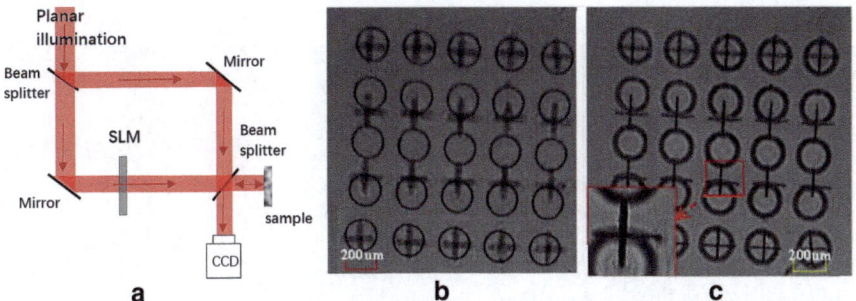

Fig. 5.37 **a** Experimental setup of depth resolved digital holography, **b** and **c** images of common wide field microscope

$$\times \sum_{k=1}^{L} \exp\left[i 2\pi n_0 \left(\frac{\sqrt{x^2 + y^2 + (z_0 + z_m)^2}}{\lambda} - \frac{\sqrt{x^2 + y^2 + (z_0 + z_k)^2}}{\lambda} \right) \right]$$

(5.15)

When the number L is large enough, the value of the second term on the right side of the Eq. (5.15) approaches to zero, and Eq. (5.15) is essentially the reflection function of the kth layer of the sample. The reflection function of other layers can also be reconstructed in the same way, realizing sectional imaging on thick samples.

Figure 5.37a is the schematic diagram of experimental setup to realize this kind of holographic sectional imaging, which is an off-axis holography in the form of Green-Twyman interferometer. The reflected light from the sample interfere with the reference light from a spatial light modulator (SLM) to form an off-axis hologram recorded by a CCD camera. The second beam splitter is located 80 mm from SLM, 10.6 mm from the sample, and 15.2 mm from the CCD. The curvature of the wavefront of the light illuminating the sample can be continuously adjusted with SLM. The sample is a piece of thin glass slide with shapes of small crosses fabricated on one side and circles on the other side. The thickness of glass slide is 1.2 mm. Figure 5.37b shows the image focused on the back surface of glass slide with small circles, and Fig. 5.37c shows the image focused on front surface of glass slide with small crosses. The diameter of these circles and the length of these crosses are both 200 μm, and the thickness of each line is a little less than 15 μm.

Figure 5.38a and b show two reconstructed images focusing on the front and back surfaces, respectively. Due to the interference between the reflected light from the front and back surfaces, the reconstructed image is superimposed with significant fringes, and the structure circles and crosses are almost impossible to discern.

By substituting a mirror for the sample in Fig. 5.37a and then recording the hologram formed, the complex amplitude leaving SLM can be reconstructed using the common holographic method. By writing the proper values to SLM, it is possible to adjust the curvature of the spherical wave-front incident on the sample surface. Figure 5.38c and d are expected light modulus reaching the front and back

(a)   (b)   (c)   (d)

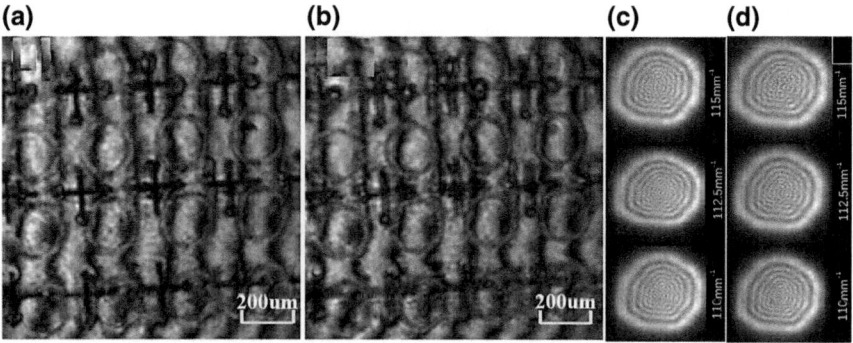

Fig. 5.38 Reconstructed image of common digital holography. **a** Reconstructed image focused on the front surface; **b** reconstructed image focused on back surface; **c** Illuminations on the front surface; **d** illuminations on the back surface

surface of the sample when wave-front curvatures are set to 115 mm^{-1}, 112.5 mm^{-1}, and 110 mm^{-1}, respectively.

250 digital holograms are recorded while the wavefront curvature of illumination is changed from 116.25 mm^{-1} to 110 mm^{-1} continuously, resulting in 250 reconstructions of reflected light. With the help of Eq. (5.11), we can add all these types of reconstructions together properly in order to obtain two reconstructions focused on the front surface and the back surface, respectively. In Fig. 5.39a and c, the reconstructed modulus and phase of the front surface of a glass slide have been shown; the small circles on the back surface have been completely excluded. The transverse resolution reached was about 5 μm. Figure 5.39b and d illustrate the reconstructed modulus and phase of the back surface of the glass slide where the structure of small crosses on the front surface was completely removed.

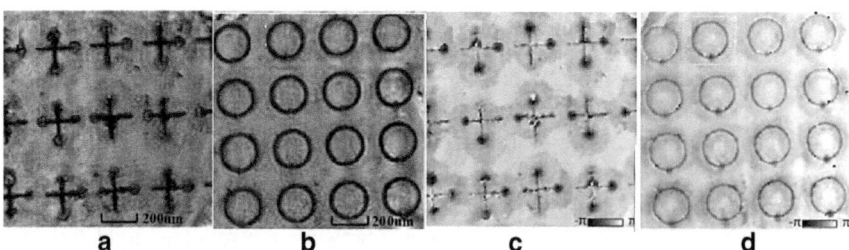

Fig. 5.39 Reconstruction results. Reconstructed amplitudes of front (**a**) and back (**b**) surfaces; reconstructed phases of front (**c**) and back (**d**) surfaces

5.6.3 3D Imaging with K-Domain Transform

Fourier components of a complex light field U(x, y, z) recorded at $z = z_0$ can be computed as $\tilde{U}\left(k_x, k_y\right)\Big|_{z=z_0}$. A coordinate transformation on $\tilde{U}\left(k_x, k_y\right)\Big|_{z=z_0}$ using $k_x = \sqrt{\lambda^{-2} - k_y^2 - k_z^2}$ results in $\tilde{U}\left(k_y, \sqrt{\lambda^{-2} - k_y^2 - k_z^2}\right)\Big|_{z=z_0}$, inverse Fourier transform which will provide the light field $U'(y, z)\big|_{y=y_0}$, changing the viewing plane from x–y plane to y–z plane. Thus, if the sample was illuminated by a light sheet in y–z plane, the image of illuminated slice can be obtained, and 3D imaging of a thick sample can be readily obtained by scanning the sample along x-axis [468].

The basic principle 3D imaging with k-domain can be illustrated with Fig. 5.40a. The light sheet propagates from left to right along the optical axis. Its complex amplitude A(x', y', z_0) on the plane (x', y', z_0) can be accurately recorded with common off-axis holography. The complex amplitude on other planes parallel to the plane (x', y', z_0) can be determined by calculating the Fresnel diffraction of A(x', y', z_0). This is the so-called "focus through" effect in digital holography, enabling the observation of the structure of a specimen at different depths. Since the clear image of focused layer is overlapped by blurred image of all other unfocused layers, common digital holography is not a real 3D imaging technique. However, if the complex amplitude B(y, z) on the (y, z) plane in Fig. 5.40a can be calculated from the reconstructed A(x', y', z_0), the structure of the sample slice illuminated by light sheet can be determined. Followed by a vertically scanning the sample through the light sheet, the 3D structure of sample can be clearly identified. This is the fundamental idea of K-domain transformation method.

If the complex light field A(x, y, z_0) on (x, y) plane in Fig. 5.40a is known, a (k_x, k_y) can be calculated as a (k_x, k_y) = FFT{A(x, y)}. The green rectangle in Fig. 5.40b indicates a (k_x, k_y), which is a 2-D matrix, and the value of its mnth element indicates the strength and phase of the

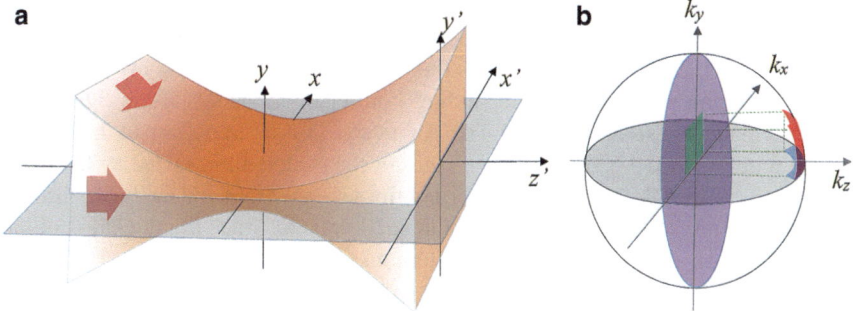

Fig. 5.40 Principle of 3D imaging with K-domain transform. **a** Schematic representation of the propagation of the light sheet from (x, y) plane to (x', y') plane. **b** Schematic representation of K-domain transform

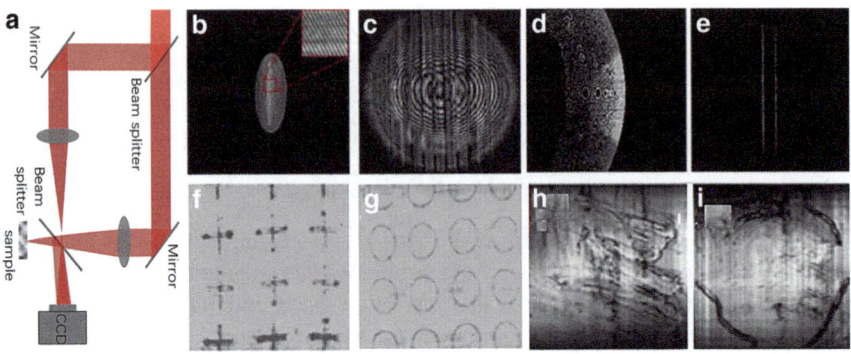

Fig. 5.41 Results of 3D imaging with K-domain transform. **a** Schematic representation of the optical setup; **b** Hologram of a piece of glass; **c** power spectra of reflected light on x–y plane; **d** power spectra of reflected light on y–z plane; **e** image reconstructed on y–z plane. **f** and **g** experimental results on a piece of glass with small crosses and circles etched on each side; **h** and **i** experimental results of a piece of glass with damaged optical films on both sides

Fourier component of $\exp\left[j\left(m\Delta k_x x + m\Delta k_y y + z\sqrt{\left(\tfrac{1}{\lambda}\right)^2 - m^2\Delta k_x^2 - n^2\Delta k_y^2} \right)\right]$.
By moving this element from the matrix position (m, n) to the position $\left(n, \sqrt{\left(\tfrac{1}{\lambda}\right)^2 - m^2\Delta k_x^2 - n^2\Delta k_y^2}/\Delta k_y \right)$, $b(k_y, k_z)$ shown with blue arc in Fig. 5.40b can be obtained. In other words, $b(k_y, k_z)$ was a rearrangement of matrix $a(k_x, k_y)$. Then, the light field $B(x, z, y|_{=0})$ on the (y, z) plane can be determined by making the inverse Fourier transform of $B(y, z) = FFT^{-1}\{b(k_y, k_z)\}$. The curved red region indicates $a(k_x, k_y, k_z)$.

Figure 5.41 shows experiments of K-domain transform 3D imaging in reflection geometry. Figure 5.41a shows the optical alignment applied, where a CW laser beam was used as illumination. Light reflected from sample interferes with reference beam and forms an off-axis hologram. The structure of the sample inside light-sheet can be calculated as described above. Sample is a glass slide with a thickness of 1.8 mm and the recorded hologram is shown in Fig. 5.41b, where fine interference fringes are clearly visible. Figure 5.41c shows the Fourier spectrum of reconstructed light on the recording plane and corresponding Fourier components in the light sheet plane calculated are shown in Fig. 5.41d. An inverse Fourier transform of the Fourier spectrum in Fig. 5.41d, generates the images of the sample illuminated by the light. It is shown in Fig. 6.7e, where the two bright lines are the front and the back surfaces of the glass, and the spacing between them is 100 pixels. Since the thickness of the glass is 1.8 mm, each pixel corresponds to a distance of 19 μm in Fig. 5.41e. The width of each line is 5 pixels in Fig. 5.41e, so the axial resolution reached is $18 \times 5 = 90$ μm. Figure 5.41f and g are experimental results on a piece of glass with small crosses and circles etched on each side. Figure 5.41h and i show a piece of glass with damaged optical film on both sides. Image on one side clearly excludes the image on the other side.

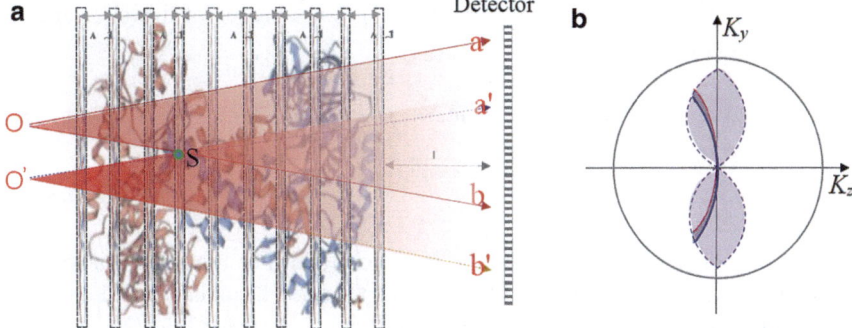

Fig. 5.42 Principle of 3D imaging with ptychography. **a** Scheme of 3D imaging with ptychography; **b** computed Fourier transform of scattering potential

5.6.4 3D Imaging with Ptychography

During the early stages of its development, Ptychographic Iterative Engine (PIE) only worked for very thin sample, whose transmission can be approximated as a 2-D function of $S(x, y)$, and its transmitted light $T(x, y)$ is approximated as the multiplication of transmission function $S(x, y)$ and illumination field $P(x, y)$. Thick sample cannot be directly imaged with common PIE since its transmission function keeps changing during its scanning through a localized illuminating probe [469]. 3PIE method was proposed by Andy Maiden in 2012 to realize sectional imaging [463]. The experimental setup and data acquisition of 3PIE are the same as that of common PIE, that is, the thick sample $S(x, y, z)$ is fixed on a translation stage, and when it was scanned through a localized illuminating probe $P(x, y)$ to a raster of positions of $(x-m\Delta x, y-n\Delta y)$, diffraction patterns $I_{mn}(x, y)$ formed were recorded by a detector placed at Fresnel or Fraunhofer distance. In iterative reconstruction, the thick sample was numerically sliced into N discrete slices $S_k(x, y)|_{k=1\cdots N}$ along z-axis as shown in Fig. 5.42, where the interval two successive slices are Δd, and the distance between the last sample slice to the detector is L.

In iterative reconstruction, the illumination on the first slice was assumed as $P_1(x, y)$, the transmitted field of this slice is computed as $T_1(x, y) = P_1(x, y) S_1(x-m\Delta x, y-n\Delta y)$, then by propagating $T_1(x, y)$ a distance of Δd to get the illumination incident on the second slice as $P_2(x, y) = \mathfrak{F}_{\Delta d}[T_1(x, y)]$ and the corresponding transmitted field as $T_2(x, y) = \mathfrak{F}_{\Delta d}[T_1(x, y)]S_2(x - m\Delta x, y - n\Delta y)$. By repeating this kind of computation at each slice we can get illumination field as $P_n(x, y) = \mathfrak{F}_{\Delta d}[T_{n-1}(x, y)]$ and corresponding transmitted field as $T_n(x, y) = P_n(x, y)S_n(x - m\Delta x, y - n\Delta y)$, and the light field reach the detector plane can be computed as $P_D(x, y) = \mathfrak{F}_D[T_N(x, y)] = |P_D(x, y)|e^{i\varphi_D(x, y)}$.

In backward propagation the computed light field on detected plane was assumed as $P'_D(x, y) = \sqrt{I_{mn}(x, y)}e^{i\varphi_D(x, y)}$. $\varphi_D(x, y)$ was a random initial guess. Then $P'_D(x, y)$ was back propagated to the last sample slice to get the transmitted light field as $T'_N(x, y) = \mathfrak{F}_{-L}\left[P'_D(x, y)\right]$, the transmission function of this slice and its illumination are updated as

$$S'_N(x - m\Delta x, y - n\Delta y) = S_N(x - m\Delta x, y - n\Delta y) + \frac{T'_N(x,y)-T_N(x,y)}{|P_n(x,y)|^2+\alpha} P^*(x, y)$$
$$P'_N(x, y) = P_N(x, y) + \frac{T'_N(x,y)-T_N(x,y)}{|S_N(x-m\Delta x, y-n\Delta y)|^2+\alpha} S^*_N(x - m\Delta x, y - n\Delta y)$$

(5.16)

By back propagating $P'_N(x, y)$ to the $N - 1$ sample slice to get the transmitted field as $T'_{N-1}(x, y) = \mathfrak{F}_{-\Delta}\left[P'_N(x, y)\right]$ and using the same formula pair of Eq. (5.16) we can get updated transmission function and illuminating field of the (N-1)th slice, and the transmission functions and illuminations of other slices can also be computed in the same way, realizing sectional imaging on thick sample after many rounds of iteration.

The physics for ptychography to realize 3D imaging can be illustrated with Fig. 5.42a. A divergent laser beam from a point laser source O illuminates a thick sample having weak diffraction characteristics. The dot S inside the sample was first illuminated by a light ray ob and followed by light ray $o'a'$, after the point laser source is shifted from O to O'. According to the principle of ODT, the Fourier transform of scattering potential of $\tilde{F}(K_x, K_y, K_z)$ can be known at the blue and red arcs in Fig. 5.42b after the complex transmitted light was reconstructed finally with ePIE using multi-slice algorithm. To avoid under-sampling of phase in iterative reconstruction the range of permitted divergent angle of illumination beam was limited, then these two arcs are quite close to each other, and thus the axial resolution capability reachable is quite low by using multi-slice PIE directly. For classic diffraction optical tomography, where the sample can be rotated by 180°, $\tilde{F}(K_x, K_y, K_z)$ is determined in the purple region surrounded by broken curves in Fig. 5.42b [466], thus higher axial resolution can be reached. With the aid of an objective to magnify the sample and subsequent 3D imaging with ptychography, however, multiple micrometers of axial resolution can be achieved [463].

For classic diffraction optical tomography, illumination within the sample volume is approximated by the original illumination when the sample is not present, so it can only be used to image very weak samples with transmission function $T(x, y, z) \approx 1.0$, and the reconstructed image $F(x, y, z)$ is essentially the high pass filtered $T(x, y, z)$. The 3D image reconstructed with ptychography was computed by updating both illumination and sample at each discrete slice with the same formula of ePIE. Since the illumination on each slice was the transmitted light of previous slice, assumption of weak diffraction is not required, and images reconstructed are directly the transmission function of $T(x, y, z)$ itself, which makes 3D ptychography suitable for a wider range of application than classical diffraction optical tomography.

Figure 5.43 shows a set of experimental results. Two layers of Stem sections of *Pachira macrocarpa* and Pumpkin with a spacing of 2.5 mm are taken as a sample, and a He–Ne laser beam with a divergent angle of 4.5° is used for illumination. The distance between the sample and the detector is 70 mm. The sample was then scanned by the illumination beam to 15 × 15 positions to get 225 diffraction patterns.

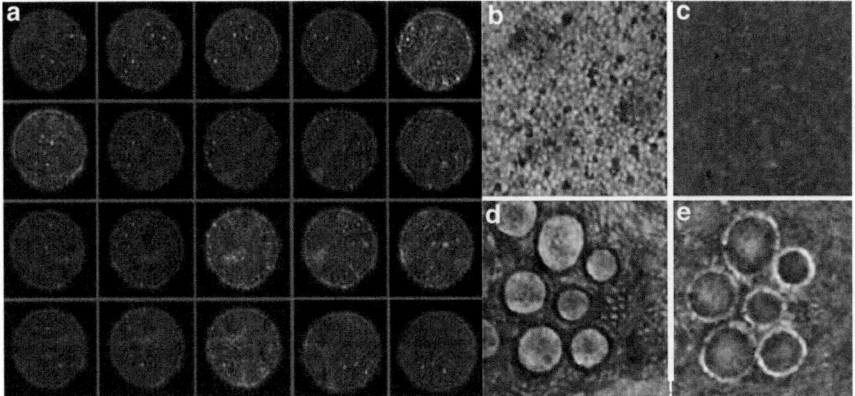

Fig. 5.43 3D imaging with ptychography. **a** Some recorded diffraction patterns; Reconstructed (**b**) amplitude and **c** phase of the first slice; reconstructed (**d**) amplitude and (**e**) phase of the second slide

Figure 5.43a shows some of the recorded diffraction patterns. Figure 5.43b and c show reconstructed modulus and phase of *Pachira macrocarpa* stem section, respectively. Figure 5.43d and e show reconstructed modulus and phase of *Pachira macrocarpa* stem section, respectively. We can find that there is no cross-talk between these reconstructed images and individual cells are clearly visible with a transverse resolution of about 11 μm. A single shot ptychography can also be used to achieve depth resolved imaging by illuminating a thick sample at different angles with a cluster of laser beams to generate a diffraction pattern array that can be recorded by a detector in one exposure [207, 470].

5.7 Summary

We conclude by discussing several typical applications of computational optical phase imaging, including stress and deformation measurement, optical testing, biomedical imaging, adaptive optics, refocusing, and tracking, as well as 3D imaging and sensing. Computational optical phase imaging can detect a wide variety of samples, ranging from macroscopic like optical elements and light waves to microscopic like tissues and cells. Due to their inherent advantages, these computational optical phase imaging techniques are applicable to a wide range of applications. Holography and interferometry, for example, have high sensitivity and accuracy, and are therefore widely used in optical shop testing and biomedical imaging. Ptychography has a large field of view and high spatial resolution, making it a preferred choice for static sample imaging and detection. The transmission of intensity phase imaging can be coupled with commercial microscopes; therefore, it is suitable for use in biomedical applications. The Shack-Hartmann wavefront sensor is particularly useful for detecting aberrations in adaptive optics. Computational optical phase imaging is not limited to only those applications mentioned above, and it can also be applied in many other areas as well. Computational optical phase imaging should, however, be considered according to the application requirements, such as spatial and temporal resolution, field of view, accuracy, etc.

References

1. Gdoutos, E.E.: Experimental Mechanics (An Introduction). Springer, Berlin (2022)
2. Doyle, K.B., Juergens, R.C., Genberg, V.L., Michels, G.J.: Numerical methods to compute optical errors due to stress birefringence. Proc. SPIE **34**, 34–42 (2002)
3. Dahmani, F., Lambropoulos, J.C., Schmid, A.W., Burns, S.J., Pratt, C.: Nanoindentation technique for measuring residual stress field around a laser-induced crack in fused silica. J. Mater. Sci. **33**, 4677–4685 (1998)
4. Neauport, J., Ambard, C., Cormont, P., Darbois, N., Destribats, J., Luitot, C., Rondeau, O.: Subsurface damage measurement of ground fused silica parts by HF etching techniques. Opt. Express **17**, 20448–20456 (2009)
5. Zhang, S., Xie, H., Zeng, X.T., Hing, P.: Residual stress characterization of diamond-like carbon coatings by an X-ray diffraction method. Surf. Coat. Tech. **122**, 219–224 (1999)
6. Zhang, Y., Xu, Y., You, C., Xu, D., Tang, J., Zhang, P., Dai, S.: Raman gain and femtosecond laser induced damage of Ge-As-S chalcogenide glasses. Opt. Express **25**, 8886–8895 (2017)
7. Gdoutos, E.E.: Matrix Theory of Photo-Elasticity. Springer-Verlag, Heidelberg (1979)
8. Narayanamurthy, C.S., Pedrini, G., Osten, W.: Digital holographic photoelasticity. Appl. Opt. **56**, F213–F217 (2017)
9. Griffin, D.W.: Phase-shifting interferometer. Opt. Lett. **26**, 140–141 (2001)
10. KrishnaKumar, V., Murukeshan, V.M., Asundi, A.: Opto-digital system for curvature measurement. Opt. Eng. **40**, 340–341 (2001)
11. Rastogi, P.K.: Measurement of curvature and twist of a deformed object by electronic speckle-shearing pattern interferometry. Opt. Lett. **21**, 905–907 (1996)
12. Moses, E.I.: The national ignition campaign: status and progress. Nucl. Fusion **53**, 104020 (2013)
13. Glass, A.J., Guenther, A.H.: Laser induced damage of optical elements-a status report. Appl. Opt. **12**, 637–649 (1973)
14. Sozet, M., Neauport, J., Lavastre, E., Roquin, N., Gallais, L., Lamaignère, L.: Laser damage growth with picosecond pulses. Opt. Lett. **41**, 2342–2345 (2016)
15. Cormont, P., Gallais, L., Lamaignère, L., Rullier, J.L., Combis, P., Hebert, D.: Impact of two CO_2 laser heatings for damage repairing on fused silica surface. Opt. Express **18**, 26068–26076 (2010)
16. Mendez, E., Nowak, K.M., Baker, H.J., Villarreal, F.J., Hall, D.R.: Localized CO_2 laser damage repair of fused silica optics. Appl. Opt. **45**, 5358–5367 (2006)
17. Kuske, A., Robertson, G.: Photoelastic Stress Analysis. Wiley-Interscience, London (1974)
18. Zhang, Z., Liu, H., Huang, J., Zhou, X., Ren, D., Cheng, X., Jiang, X., Wu, W., Zheng, W.: Residual stress near cracks of K and fused silica under 1064 nm nanosecond laser irradiation. Opt. Eng. **51**, 4201 (2012)
19. Wang, F., Zhang, Y.-P., Wang, H., Xu, W., Zhang, Y.-A., Li, C.-G.: Nondestructive evaluation of residual stress via digital holographic photoelasticity. J. Opt. **47**, 547–552 (2018)
20. Aben, H.K., Errapart, A., Ainola, L., Anton, J.: Photoelastic tomography for residual stress measurement in glass. Opt. Eng. **44**, 093601 (2005)

21. Gallais, L., Cormont, P., Rullier, J.L.: Investigation of stress induced by CO_2 laser processing of fused silica optics for laser damage growth mitigation. Opt. Express **17**, 23488–23501 (2009)

22. Doualle, T., Gallais, L., Cormont, P., Hébert, D., Combis, P., Rullier, J.-L.: Thermo-mechanical simulations of CO_2 laser-fused silica interactions. J. Appl. Phys. **119**, 113106 (2016)

23. Boyd, R.W.: Nonlinear Optics, 3rd edn. American Academic, Salt Lake City (2009)

24. Sakakura, M., Terazima, M.: Initial temporal and spatial changes of the refractive index induced by focused femtosecond pulsed laser irradiation inside a glass. Appl. Phys. B **71**, 024113 (2005)

25. Qi, N., Sun, S., Zhang, L., Yuan, X., Kong, Y., Veetil, S.P., Wang, S., Liu, C.: Microscopic three-dimensional inner stress measurement on laser induced damage. Opt. Express **28**, 24253–24261 (2020)

26. Jo, T., Kim, K., Kim, S., Pahk, H.: Thickness and surface measurement of transparent thin-film layers using white light scanning interferomei combined with reflectometry. J. Opt. Soc. Korea **18**, 236–243 (2014)

27. Kumar, R., Shakher, C.: Application of digital speckle pattern interferometry and wavelet transform in measurement of transverse vibrations in square plate. Opt. Laser Eng. **42**, 585–602 (2004)

28. Chen, R., Zhang, Q., He, W., Xie, H.M.: Orthogonal sampling moiré method and its application in microscale deformation field measurement. Opt. Laser Eng. **149** (2022)

29. Liu, C.: Simultaneous measurement of displacement and its spatial derivatives with a digital holographic method. Opt. Eng. **42**, 3443–3446 (2003)

30. Quiroga, J.A., Bernabeu, E.: Phase-unwrapping algorithm for noisy phase-map processing. Appl. Opt. **33**, 6725–6731 (1994)

31. Quiroga, J.A., González-Cano, A., Bernabeu, E.: Phase algorithm based on adaptive criterion. Appl. Opt. **34**, 2560–2563 (1995)

32. Charette, P.G., Hunter, I.W.: Robust phase-unwrapping method for phase images with high noise content. Appl. Opt. **35**, 3506–3513 (1996)

33. Haynam, C.A., Wegner, P.J., Auerbach, J.M., Bowers, M.W., Dixit, S.N., Erbert, G.V., Heestand, G.M., Henesian, M.A., Hermann, M.R., Jancaitis, K.S., Manes, K.R., Marshall, C.D., Mehta, N.C., Menapace, J., Moses, E., Murray, J.R., Nostrand, M.C., Orth, C.D., Patterson, R., Sacks, R.A., Shaw, M.J., Spaeth, M., Sutton, S.B., Williams, W.H., Wildmayer, C.C., White, R.K., Yang, S.T., van Wonterghem, B.M.: National ignition facility laser performance status. Appl. Opt. **46**, 3276–3303 (2007)

34. Williams, W.H., Auerbach, J.M., Henesian, M.A., Lawson, J.K., Hunt, J.T., Sacks, R.A., Widmayer, C.C.: Modeling characterization of the national ignition facility focal spot. Proc. SPIE **3264**, 93–104 (1998)

35. Homoelle, D., Bowers, M.W., Budge, T., Haynam, C., Heebner, J., Hermann, M., Jancaitis, K., Jarboe, J., LaFortune, K., Salmon, J.T., Schindler, T., Shaw, M.: Measurement of the repeatability of the prompt flashlamp-induced wavefront aberration on beamlines at the national ignition facility. Appl. Opt. **50**, 4382–4388 (2011)

36. Sutton, S.B., Marshall, C.D., Petty, C.S., Smith, L.K., van Wonterghem, B.M., Mills, S.: Thermal recovery of NIF amplifiers. Proc. SPIE **3047**, 560–570 (1996)

37. Wolfe, C.R., Lawson, J.K.: The measurement and analysis of wavefront structure from large aperture ICF optics. Proc. SPIE **2633**, 361–385 (1995)

38. Spaeth, M.L., Manes, K.R., Widmayer, C.C., Williams, W.H., Whitman, P.K., Henesian, M.A., Stowers, I.F., Honig, J.: National ignition facility wavefront requirements and optical architecture. Opt. Eng. **43**, 2854–2865 (2004)

39. Jiang, W., Li, H.: Hartmann-Shack wavefront sensing and control algorithm. Proc. SPIE **1271**, 82–93 (1990)

40. Cao, G., Yu, X.: Accuracy analysis of a Hartmann-Shack wavefront sensor operated with a faint object. Opt. Eng. **33**, 2331–2335 (1994)

41. Zavalova, V., Kudryashov, A.V.: Shack-Hartmann wave-front sensor for laser beam analyses. Proc. SPIE **4493**, 277–284 (2002)

42. Sato, S., Mori, T., Higashi, Y., Haya, S., Otsuka, M., Yamamoto, H.: A profilometer for synchrotron radiation mirrors. J. Electron Spectrosc. Relat. Phenom. **80**, 481–484 (1996)
43. Liu, X., Gao, Y., Chang, M.: A partial differential equation algorithm for wavefront reconstruction in lateral shearing interferometry. J. Opt. **11**, 045702 (2009)
44. Zhang, F., Rodenburg, J.M.: Phase retrieval based on wave-front relay and modulation. Phys. Rev. B **82**, 121104 (2010)
45. He, X., Liu, C., Gao, S., Wang, Y., Wang, J., Zhu, J.: Accurate focal spot diagnostics based on a single shot coherent modulation imaging. Laser Phys. Lett. **12**, 015005 (2015)
46. Tao, H., Veetil, S.P., Pan, X., Liu, C., Zhu, J.: Visualization of the influence of the air conditioning system to the high-power laser beam quality with the modulation coherent imaging method. Appl. Opt. **54**, 6632–6639 (2015)
47. Tyson, R.K. (ed.): Principles of Adaptive Optics. CRC, Boca Raton (2015)
48. Babcock, H.W.: Adaptive optics revisited. Science **249**, 253–257 (1990)
49. Ji, N., Milkie, D.E., Betzig, E.: Adaptive optics via pupil segmentation for high-resolution imaging in biological tissues. Nat. Methods **7**, 141–147 (2010)
50. Roorda, A., Romero-Borja, F., Donnelly, W.J., III., Queener, H., Hebert, T.J., Campbell, M.C.W.: Adaptive optics scanning laser ophthalmoscopy. Opt. Express **10**, 405–412 (2002)
51. Malacara, D. (ed.): Optical Shop Testing, 3rd edn. Wiley-Interscience, New Jersey (2007)
52. Taylor, H.D. (ed.): The Adjuetment and Testing of Telescope Objectives. CRC, Boca Raton (1947)
53. Zamkotsian, F., Dohlen, K., Lanzoni, P., Mazzanti, S., Michel, M.L., Buat, V., Burgarella, D.: Knife-edge test for characterization of subnanometer deformations in micro-optical surfaces. Proc. SPIE **3782** (1999)
54. Bruning, J.H., Herriott, D.R.: A versatile laser interferometer. Appl. Opt. **9**, 2180–2182 (1970)
55. Hariharan, P., Oreb, B.F., Eiju, T.: Digital phase-shifting interferometry: a simple error-compensating phase calculation algorithm. Appl. Opt. **26**, 2504–2506 (1987)
56. Bruning, J.H., Herriott, D.R., Gallagher, J.E., Rosenfeld, D.P., White, A.D., Brangaccio, D.J.: Digital wavefront measuring interferometer for testing optical surfaces and lenses. Appl. Opt. **13**, 2693–2703 (1974)
57. Rimmer, M.P., Wyant, J.C.: Evaluation of large aberrations using a lateral-shear interferometer having variable shear. Appl. Opt. **14**, 142–150 (1975)
58. Rimmer, M.P.: Method for evaluation lateral shearing interferograms. Appl. Opt. **13**, 623–629 (1974)
59. Wyant, J.C.: Use of an AC heterodyne lateral shear interferometer with real-time wave-front correction systems. Appl. Opt. **14**, 2622–2626 (1975)
60. Wang, Y., Chen, X., Cao, Z., Zhang, X., Liu, C., Mu, Q.: Gradient cross-correlation algorithm for scene-based Shack-Hartman wavefront sensing. Opt. Express **26**, 17549–17562 (2018)
61. Xia, F., Sinefeld, D., Li, B., Xu, C.: Two-photo Shack-Hartmann wavefront sensor. Opt. Lett. **42**, 1141–1144 (2017)
62. Rao, C., Jiang, W., Ling, N.: Measuring the power-law exponent of an atmospheric turbulence phase power spectrum with a Shack Hartmann wave-front sensor. Opt. Lett. **24**, 1008–1010 (1999)
63. Zheng, G., Horstmeyer, R., Yang, C.: Wide-field, high-resolution Fourier ptychographic microscopy. Nat. Photonics **7**, 739–745 (2013)
64. Ou, X., Zheng, G., Yang, C.: Embedded pupil function recovery for Fourier ptychographic microscopy. Opt. Express **22**, 4960–4972 (2014)
65. Gunjala, G., Sherwin, S., Shanker, A., Waller, L.: Aberration recovery by imaging a weak diffuser. Opt. Express **26**, 21054–21068 (2018)
66. Horstmeyer, R., Yang, C.: A phase space model of Fourier ptychographic microscopy. Opt. Express **22**, 338–358 (2014)
67. Ou, X., Horstmeyer, R., Yang, C., Zheng, G.: Quantitative phase imaging via Fourier ptychographic microscopy. Opt. Lett. **38**, 4845–4848 (2013)
68. Paxman, R.G., Schulz, T.J., Fienup, J.R.: Joint estimation of object and aberrations by using phase diversity. J. Opt. Soc. Am. A **9**, 1072–1085 (1992)

69. Gonsalves, R.A.: Phase retrieval and diversity in adaptive optics. Opt. Eng. **21**, 215829 (1982)
70. Raxman, R.G., Fienup, J.R.: Optical misalignment sensing and image reconstruction using phase diversity. J. Opt. Soc. Am. A **5**, 914–923 (1988)
71. Faulkner, H.M.L., Rodenburg, J.M.: Movable aperture lensless transmission microscopy: a novel phase retrieval algorithm. Phys. Rev. Lett. **93**, 023903 (2004)
72. Maiden, A.M., Rodenburg, J.M.: An improved ptychographical phase retrieval algorithm for diffractive imaging. Ultramicroscopy **109**, 1256–1262 (2009)
73. Pan, X., Veetil, S.P., Liu, C., Zhu, J.: High contrast imaging for weakly diffracting specimens with ptychographical iterative engine. Opt. Lett. **37**, 3348–3350 (2012)
74. Faulkner, H.M.L., Rodenburg, J.M.: Error tolerance of an iterative phase retrieval algorithm for moveable illumination microscopy. Ultramicroscopy **103**, 153–164 (2005)
75. Burdet, N., Shi, X., Parks, D., Clark, J.N., Huang, X., Kevan, S.D., Robinson, I.K.: Evaluation of partial coherence correction in X-ray ptychography. Opt. Express **23**, 5452–5467 (2015)
76. Rodenburg, J.M.: The phase problem, microdiffraction and wavelength-limited resolution-a discussion. Ultramicroscopy **27**, 413–422 (1989)
77. Zong, B., Luan, J., Jiang, Z., Kong, Y., Wang, S., Liu, C.: Quantitative aberration measurement with extended ptychographic iterative engine. Opt. Eng. **58**, 054102 (2019)
78. Moon, I., Javidi, B., Yi, F., Boss, F.D., Marquet, P.: Automated statistical quantification of three-dimensional morphology and mean corpuscular hemoglobin of multiple red blood cells. Opt. Express **20**, 10295–10309 (2012)
79. Yi, F., Moon, I., Lee, Y.H.: Three-dimensional counting of morphologically normal human red blood cells via digital holographic microscopy. J. Biomed. Opt. **20**, 016005 (2015)
80. Yi, F., Moon, I., Javidi, B.: Cell morphology-based classification of red blood cells using holographic imaging informatics. Biomed. Opt. Express **7**, 2385–2399 (2016)
81. Park, Y.K., Yamauchi, T., Choi, W., Dasari, R., Feld, M.S.: Spectroscopic phase microscopy for quantifying hemoglobin concentrations in intact red blood cells. Opt. Lett. **34**, 3668–3670 (2009)
82. Aknoun, S., Savatier, J., Bon, P., Galland, F., Abdeladim, L., Wattellier, B., Monneret, S.: Living cell dry mass measurement using quantitative phase imaging with quadriwave lateral shearing interferometry: an accuracy and sensitivity discussion. J. Biomed. Opt. **20**, 126009 (2015)
83. Moon, I., Yi, F., Lee, Y.H., Javidi, B., Boss, D., Marquet, P.: Automated quantitative analysis of 3D morphology and mean corpuscular hemoglobin in human red blood cells stored in different periods. Opt. Express **21**, 30947–30957 (2013)
84. Jaferzadeh, K., Moon, I.: Quantitative investigation of red blood cell three-dimensional geometric and chemical changes in the storage lesion using digital holographic microscopy. J. Biomed. Opt. **20**, 111218 (2015)
85. Cho, S., Kim, S., Kim, Y., Park, Y.K.: Optical imaging techniques for the study of malaria. Trends Biotechnol. **30**, 71–79 (2012)
86. Byeon, H., Ha, Y.-R., Lee, S.J.: Holographic analysis on deformation and restoration of malaria-infected red blood cells by antimalarial drug. J. Biomed. Opt. **20**, 115003 (2015)
87. Li, C., Chen, S., Klemba, M., Zhu, Y.: Integrated quantitative phase and birefringence microscopy for imaging malaria-infected red blood cells. J. Biomed. Opt. **21**, 090501 (2016)
88. Shaked, N.T., Satterwhite, L.L., Telen, M.J., Truskey, G.A., Wax, A.: Quantitative microscopy and nanoscopy of sickle red blood cells performed by wide field digital interferometry. J. Biomed. Opt. **16**, 030506 (2011)
89. Byun, H.S., Hillman, T.R., Higgins, J.M., Diez-Silva, M., Peng, Z., Dao, M., Dasari, R.R., Suresh, S., Park, Y.K.: Optical measurement of biomechanical properties of individual erythrocytes from a sickle cell patient. Acta Biomater. **8**, 4130–4138 (2012)
90. Javidi, B., Markman, A., Rawat, S., O'Connor, T., Anand, A., Andemariam, B.: Sickle cell disease diagnosis based on spatio-temporal cell dynamics analysis using 3D printed shearing digital holographic microscopy. Opt. Express **26**, 13614–13627 (2018)
91. Doblas, A., Roche, E., Ampudia-Blasco, F.J., Martinez-Corral, M., Saavedra, G., Garcia-Sucerquia, J.: Diabetes screening by telecentric digital holographic microscopy. J. Microsc. **261**, 285–290 (2016)

92. Roma, P.M.S., Siman, L., Amaral, F.T., Agero, U., Mesquita, O.N.: Total three-dimensional imaging of phase objects using defocusing microscopy: application to red blood cells. Appl. Phys. Lett. **104**, 251107 (2014)
93. Choi, W., Fang-Yen, C., Badizadegan, K., Oh, S., Lue, N., Dasari, R.R., Feld, M.S.: Tomographic phase microscopy. Nat. Methods **4**, 717–719 (2007)
94. Kim, Y., Shim, H., Kim, K., Park, H.J., Jang, S., Park, Y.K.: Profiling individual human red blood cells using common-path diffraction optical tomography. Sci. Rep. **4**, 6659 (2014)
95. Park, H.J., Ahn, T., Kim, K., Lee, S., Kook, S., Lee, D., Suh, I.B., Na, S., Park, Y.K.: Three-dimensional refractive index tomograms and deformability of individual human red blood cells from cord blood of newborn infants and maternal blood. J. Biomed. Opt. **20**, 111208 (2015)
96. Kim, K., Yoon, H.O., Diez-Silva, M., Dao, M., Dasari, R.R., Park, Y.K.: High-resolution three-dimensional imaging of red blood cells parasitized by Plasmodium falciparum and in situ hemozoin crystals using optical diffraction tomography. J. Biomed. Opt. **19**, 011005 (2014)
97. Park, H., Hong, S.-H., Kim, K., Cho, S.-H., Lee, W.-J., Kim, Y., Lee, S.-E., Park, Y.K.: Characterizations of individual mouse red blood cells parasitized by *Babesia microti* using 3-D holographic microscopy. Sci. Rep. **5**, 10827 (2015)
98. Kim, G., Jo, Y., Cho, H., Min, H., Park, Y.K.: Learning-based screening of hematologic disorders using quantitative phase imaging of individual red blood cells. Biosens. Bioelectron. **123**, 69–76 (2019)
99. Yoon, J., Kim, K., Park, H.J., Choi, C., Jang, S., Park, Y.K.: Label-free characterization of white blood cells by measuring 3D refractive index maps. Biomed. Opt. Express **6**, 3865–3875 (2015)
100. Schurmann, M., Scholze, J., Muller, P., Guck, J., Chan, C.J.: Cell nuclei have lower refractive index and mass density than cytoplasm. J. Biophotonics **9**, 1068–1076 (2016)
101. Steelman, Z.A., Eldridge, W.J., Weintraub, J.B., Wax, A.: Is the nuclear refractive index lower than cytoplasm? Validation of phase measurements and implications for light scattering technologies. J. Biophotonics **10**, 1714–1722 (2017)
102. Chung, J., Ou, X., Kulkarni, R.P., Yang, C.: Counting white blood cells from a blood smear using Fourier ptychographic microscopy. PLoS ONE **10**, e0133489 (2015)
103. Seo, S., Isikman, S.O., Sencan, I., Mudanyali, O., Su, T.W., Bishara, W., Erlinger, A., Ozcan, A.: High-throughput lens-free blood analysis on a chip. Anal. Chem. **82**, 4621–4627 (2010)
104. Balberg, M., Levi, M., Kalinowski, K., Barnea, I., Mirsky, S.K., Shaked, N.T.: Localized measurements of physical parameters within human sperm cells obtained with wide-field interferometry. J. Biophotonics **10**, 1305–1314 (2017)
105. Barnea, I., Karako, L., Mirsky, S.K., Levi, M., Balberg, M., Shaked, N.T.: Stain-free interferometric phase microscopy correlation with DNA fragmentation stain in human spermatozoa. J. Biophotonics **11**, e201800137 (2018)
106. World Health Organization: Laboratory Manual for the Examination of Human Semen, 5th edn. World Health Organization, Geneva (2010)
107. Di Caprio, G., Ferrara, M.A., Miccio, L., Merola, F., Memmolo, P., Ferraro, P., Coppola, G.: Holographic imaging of unlabelled sperm cells for semen analysis: a review. J. Biophotonics **8**, 779–789 (2015)
108. Miccio, L., Finizio, A., Puglisi, R., Balduzzi, D., Galli, A., Ferraro, P.: Dynamic DIC by digital holography microscopy for enhancing phase-contrast visualization. Biomed. Opt. Express **2**, 331–344 (2011)
109. Memmolo, P., Di Caprio, G., Distante, C., Paturzo, M., Puglisi, R., Balduzzi, D., Galli, A., Coppola, G., Ferraro, P.: Identification of bovine sperm head for morphometry analysis in quantitative phase-contrast holographic microscopy. Opt. Express **19**, 23215–23226 (2011)
110. Poola, P.K., Jayaraman, V., Chaithanya, K., Rao, D., John, R.: Quantitative label-free technique for morphological evaluation of human sperm-a promising tool in semen evaluation. OSA Continuum **1**, 1215–1225 (2018)

111. Su, T.-W., Erlinger, A., Tseng, D., Ozcan, A.: Compact and light-weight automated semen analysis platform using lensfree on-chip microscopy. Anal. Chem. **82**, 8307–8312 (2010)
112. Sobieranski, A.C., Inci, F., Tekin, H.C., Yuksekkaya, M., Comunello, E., Cobra, D., von Wangenheim, A., Demirci, U.: Portable lensless wide-field microscopy imaging platform based on digital inline holography and multi-frame pixel super-resolution. Light-Sci. Appl. **4**, e346 (2015)
113. Nygate, Y.N., Levi, M., Mirsky, S.K., Turko, N.A., Rubin, M., Barnea, I., Dardikman-Yoffe, G., Haifler, M., Shalev, A., Shaked, N.T.: Holographic virtual staining of individual biological cells. Proc. Natl. Sci. U.S.A. **117**, 9223–9231 (2020)
114. Majeed, H., Kandel, M.E., Han, K., Luo, Z., Macias, V., Tangella, K., Balla, A., Popescu, G.: Breast cancer diagnosis using spatial light interference microscopy. J. Biomed. Opt. **20**, 111210 (2015)
115. Majeed, H., Okoro, C., Kajdacsy-Balla, A., Toussaint, K.C., Popescu, G.: Quantifying collagen fiber orientation in breast cancer using quantitative phase imaging. J. Biomed. Opt. **22**, 046004 (2017)
116. Takabayashi, M., Majeed, H., Kajdacsy-Balla, A., Popescu, G.: Tissue spatial correlation as cancer marker. J. Biomed. Opt. **24**, 016502 (2019)
117. Sridharan, S., Macias, V., Tangella, K., Kajdacsy-Balla, A., Popescu, G.: Prediction of prostate cancer recurrence using quantitative phase imaging. Sci. Rep. **5**, 9976 (2015)
118. Sridharan, S., Macias, V., Tangella, K., Melamed, J., Dube, E., Kong, M.X., Kajdacsy-Balla, A., Popescu, G.: Prediction of prostate cancer recurrence using quantitative phase imaging: Validation on a general population. Sci. Rep. **6**, 33818 (2016)
119. Fanous, M., Keikhosravi, A., Kajdacsy-Balla, A., Eliceiri, K.W., Popescu, G.: Quantitative phase imaging of stromal prognostic markers in pancreatic ductal adenocarcinoma. Biomed. Opt. Express **11**, 1354–1364 (2020)
120. Ban, S., Min, E., Baek, S., Kwon, H.M., Popescu, G., Jung, W.: Optical properties of acute kidney injury measured by quantitative phase imaging. Biomed. Opt. Express **9**, 921–932 (2018)
121. Ban, S., Min, E., Ahn, Y., Popescu, G., Jung, W.: Effect of tissue staining in quantitative phase imaging. J. Biophotonics **11**, e201700402 (2018)
122. Horstmeyer, R., Ou, X., Zheng, G., Willems, P., Yang, C.: Digital pathology with Fourier ptychography. Comput. Med. Imag. Grap. **42**, 38–43 (2015)
123. Ozcan, A., McLeod, E.: Lensless imaging and sensing. Annu. Rev. Biomed. Eng. **18**, 77–102 (2016)
124. Zhang, Y., Greenbaum, A., Luo, W., Ozcan, A.: Wide-field pathology imaging using on-chip microscopy. Virchows Arch. **467**, 3–7 (2015)
125. Greenbaum, A., Akbari, N., Feizi, A., Luo, W., Ozcan, A.: Field-portable pixel super-resolution colour microscope. PLoS ONE **8**, e76475 (2013)
126. Greenbaum, A., Zhang, Y., Feizi, A., Chung, P.L., Luo, W., Kandukuri, S.R., Ozcan, A.: Wide-field computational imaging of pathology slides using lens-free on-chip microscopy. Sci. Transl. Med. **6**, 267ra175 (2014)
127. Rivenson, Y., Wu, Y., Wang, H., Zhang, Y., Feizi, A., Ozcan, A.: Sparsity-based multi-height phase recovery in holographic microscopy. Sci. Rep. **6**, 37862 (2016)
128. Luo, W., Greenbaum, A., Zhang, Y., Ozcan, A.: Synthetic aperture-based on-chip microscopy. Light-Sci. Appl. **4**, e261 (2015)
129. Rivenson, Y., Zhang, Y., Gnaydin, H., Teng, D., Ozcan, A.: Phase recovery and holographic image reconstruction using deep learning in neural networks. Light-Sci. Appl. **7**, 17141 (2018)
130. Wu, Y., Zhang, Y., Luo, W., Ozcan, A.: Demosaiced pixel super-resolution for multiplexed holographic color imaging. Sci. Rep. **6**, 28601 (2016)
131. Zhang, Y., Liu, T., Huang, Y., Teng, D., Bian, Y., Wu, Y., Rivenson, Y., Feizi, A., Ozcan, A.: Accurate color imaging of pathology slides using holography and absorbance spectrum estimation of histochemical stains. J. Biophotonics **12**, e201800335 (2019)
132. Zhang, Y., Shin, Y., Sung, K., Yang, S., Chen, H., Wang, H., Teng, D., Rivenson, Y., Kulkarni, R.P., Ozcan, A.: 3D imaging of optically cleared tissue using a simplified clarity method and on-chip microscopy. Sci. Adv. **3**, e1700553 (2017)

133. Jiang, S., Zhu, J., Song, P., Guo, C., Bian, Z., Wang, R., Huang, Y., Wang, S., Zhang, H., Zheng, G.: Wide-field, high-resolution lensless on-chip microscopy via near-field blind ptychographic modulation. Lab. Chip **20**, 1058–1065 (2020)
134. Godden, T.M., Suman, R., Humphry, M.J., Rodenburg, J.M., Maiden, A.M.: Ptychographic microscope for three-dimensional imaging. Opt. Express **22**, 12513–12523 (2014)
135. Li, P., Maiden, A.: Multi-slice ptychographic tomography. Sci. Rep. **8**, 2049 (2018)
136. Rivenson, Y., Liu, T., Wei, Z., Zhang, Y., de Haan, K., Ozcan, A.: PhaseStain: the digital staining of label-free quantitative phase microscopy images using deep learning. Light-Sci. Appl. **8**, 23 (2019)
137. Liu, T., Wei, Z., Rivenson, Y., de Haan, K., Zhang, Y., Wu, Y., Ozcan, A.: Deep learning-based color holographic microscopy. J. Biophotonics **12**, e201900107 (2019)
138. Wei, Q., Zhang, M., Yu, M., Xue, L., Liu, C., Vargas, J., Liu, F., Wang, S.: Rapid quantitative interferometric microscopy using fast Fourier transform and differential-integral based phase retrieval algorithm (FFT-DI-PRA). Opt. Commun. **456**, 124613 (2020)
139. Yu, W., Wang, S., Veetil, S., Gao, S., Liu, C., Zhu, J.: High-quality image reconstruction method for ptychography with partially coherent illumination. Phys. Rev. B **93**, 241105(R) (2016)
140. Sun, A., He, X., Kong, Y., Cui, H., Song, X., Xue, L., Wang, S., Liu, C.: Ultra-high speed digital micromirror device based ptychographic iterative engine method. Biomed. Opt. Express **8**, 3155–3162 (2017)
141. Ikeda, T., Popescu, G., Dasari. R.R., Feld, M.S.: Hilbert phase microscopy for investigating fast dynamics in transparent systems. Opt. Lett. **30**, 1165–1167 (2005)
142. Popescu, G., Ikeda, T., Best, C., Badizadegan, K., Dasari, R.R., Feld, M.S.: Erythrocyte structure and dynamics quantified by Hilbert phase microscopy. J. Biomed. Opt. **19**, 060503 (2005)
143. Popescu, G., Badizadegan, K., Dasari, R.R., Feld, M.S.: Observation of dynamic subdomains in red blood cells. J. Biomed. Opt. **11**, 040503 (2006)
144. Pham, H.V., Bhaduri, B., Tangella, K., Best-Popescu, C., Popescu, G.: Real time blood testing using quantitative phase imaging. PLoS ONE **8**, e55676 (2013)
145. Park, Y.K., Best, C.A., Badizadegan, K., Dasari, R.R., Feld, M.S., Kuriabova, T., Henle, M.L., Levine, A.J., Popescu, G.: Measurement of red blood cell mechanics during morphological changes. Proc. Natl. Acad. Sci. U.S.A. **107**, 6731–6736 (2010)
146. Park, Y.K., Best, C.A., Auth, T., Gov, N.S., Safran, S.A., Popescu, G., Suresh, S., Feld, M.S.: Metabolic remodeling of the human red blood cell membrane. Proc. Natl. Acad. Sci. U.S.A. **107**, 1289–1294 (2010)
147. Rappaz, B., Barbul, A., Hoffmann, A., Boss, D., Korenstein, R., Depeursinge, C., Magistretti, P.J., Marquet, P.: Spatial analysis of erythrocyte membrane fluctuations by digital holographic microscopy. Blood Cells Mol. Dis. **42**, 228–232 (2009)
148. Boss, D., Hoffmann, A., Rappaz, B., Depeursinge, C., Magistretti, P.J., van de Ville, D., Marquet, P.: Spatially-resolved eigen mode decomposition of red blood cells membrane fluctuations questions the role of ATP in flickering. PLoS ONE **7**, e40667 (2012)
149. Wang, S., Yan, K., Shan, Y., Xu, M., Liu, F., Xue, L.: Phase measurements of erythrocytes affected by metal ions with quantitative interferometric microscopy. Opt. Eng. **54**, 124105 (2015)
150. Basanta, B., Mikhail, K., Carlo, B., Krishna, T., Gabriel, P.: Optical assay of erythrocyte function in banked blood. Sci. Rep. **4**, 6211 (2014)
151. Jaferzadeh, K., Moon, I., Bardyn, M., Prudent, M., Tissot, J.-D., Rappaz, B., Javidi, B., Turcatti, G., Marquet, P.: Quantification of stored red blood cell fluctuations by time-lapse holographic cell imaging. Biomed. Opt. Express **9**, 4714–4729 (2018)
152. Kim, K., Kim, K.S., Park, H.J., Ye, J.C., Park, Y.K.: Real-time visualization of 3-D dynamic microscopic objects using optical diffraction tomography. Opt. Express **21**, 32269–32278 (2013)
153. Bhaduri, B., Wickland, D., Wang, R., Chan, V., Bashir, R., Popescu, G.: Cardiomyocyte imaging using real-time spatial light interference microscopy (SLIM). PLoS ONE **8**, e56930 (2013)

154. Rappaz, B., Moon, I., Yi, F., Javidi, B., Marquet, P., Turcatti, G.: Automated multi-parameter measurement of cardiomyocytes dynamics with digital holographic microscopy. Opt. Express **23**, 13333–13347 (2015)

155. Moon, I., Ahmadzadeh, E., Jaferzadeh, K., Kim, N.: Automated quantification study of human cardiomyocyte synchronization using holographic imaging. Biomed. Opt. Express **10**, 610–621 (2019)

156. Ahmadzadeh, E., Jaferzadeh, K., Shin, S., Moon, I.: Automated single cardiomyocyte characterization by nucleus extraction from dynamic holographic images using a fully convolutional neural network. Biomed. Opt. Express **11**, 1501–1516 (2020)

157. Mir, M., Wang, Z., Shen, Z., Bednarz, M., Bashir, R., Golding, I., Prasanth, S.G., Popescu, G.: Optical measurement of cycle-dependent cell growth. Proc. Natl. Acad. Sci. U.S.A. **108**, 13124–13129 (2011)

158. Mir, M., Bergamaschi, A., Katzenellenbogen, B.S., Popescu, G.: Highly sensitive quantitative imaging for monitoring single cancer cell growth kinetics and drug response. PLoS ONE **9**, e89000 (2014)

159. Girshovitz, P., Shaked, N.T.: Generalized cell morphological parameters based on interferometric phase microscopy and their application to cell life cycle characterization. Biomed. Opt. Express **3**, 1757–1773 (2012)

160. Kesavan, S.V., Navarro, F.P., Menneteau, M., Mittler, F., David-Watine, B., Dubrulle, N., Shorte, S.L., Chalmond, B., Dinten, J.-M., Allier, C.P.: Real-time label-free detection of dividing cells by means of lensfree video-microscopy. J. Biomed. Opt. **19**, 036004 (2014)

161. Kesavan, S.V., Momey, F., Cioni, O., David-Watine, B., Dubrulle, N., Shorte, S., Sulpice, E., Freida, D., Chalmond, B., Dinten, J.M., Gidrol, X., Allier, C.: High-throughput monitoring of major cell functions by means of lensfree video microscopy. Sci. Rep. **4**, 5942 (2014)

162. Rappaz, B., Cano, E., Colomb, T., Kuhn, J., Depeursinge, C., Simanis, V., Magistretti, P.J., Marquet, P.: Noninvasive characterization of the fission yeast cell cycle by monitoring dry mass with digital holographic microscopy. J. Biomed. Opt. **14**, 034049 (2009)

163. Pavillon, N., Kuhn, J., Moratal, C., Jourdain, P., Depeursinge, C., Magistretti, P.J., Marquet, P.: Early cell death detection with digital holographic microscopy. PLoS ONE **7**, e30912 (2012)

164. Khmaladze, A., Matz, R.L., Epstein, T., Jasensky, J., Holl, M.M.B., Chen, Z.: Cell volume changes during apoptosis monitored in real time using digital holographic microscopy. J. Struct. Biol. **178**, 270–278 (2012)

165. Ceballos, S., Kandel, M., Sridharan, S., Majeed, H., Monroy, F., Popescu, G.: Active intracellular transport in metastatic cells studied by spatial light interference microscopy. J. Biomed. Opt. **20**, 111209 (2015)

166. Ma, L., Rajshekhar, G., Wang, R., Bhaduri, B., Sridharan, S., Mir, M., Chakraborty, A., Iyer, R., Prasanth, S., Millet, L., Gillette, M.U., Popescu, G.: Phase correlation imaging of unlabeled cell dynamics. Sci. Rep. **6**, 32702 (2016)

167. Shaked, N.T., Rinehart, M.T., Wax, A.: Dual-interference-channel quantitative-phase microscopy of live cell dynamics. Opt. Lett. **34**, 767–769 (2009)

168. Bishitz, Y., Gabai, H., Girshovitz, P., Shaked, N.T.: Optical-mechanical signatures of cancer cells based on fluctuation profiles measured by interferometry. J. Biophotonics **7**, 624–630 (2014)

169. Rappaz, B., Marquet, P., Cuche, E., Emery, Y., Depeursinge, C., Magistretti, P.J.: Measurement of the integral refractive index and dynamic cell morphometry of living cells with digital holographic microscopy. Opt. Express **13**, 9361–9373 (2005)

170. Boss, D., Kuhn, J., Jourdain, P., Depeursinge, C., Magistretti, P.J., Marqueta, P.: Measurement of absolute cell volume, osmotic membrane water permeability, and refractive index of transmembrane water and solute flux by digital holographic microscopy. J. Biomed. Opt. **18**, 036007 (2013)

171. Jourdain, P., Pavillon, N., Moratal, C., Boss, D., Rappaz, B., Depeursinge, C., Marquet, P., Magistretti, P.J.: Determination of transmembrane water fluxes in neurons elicited by glutamate ionotropic receptors and by the cotransporters KCC2 and NKCC1: a digital holographic microscopy study. J. Neurosci. **31**, 11846–11854 (2011)

172. Jourdain, P., Boss, D., Rappaz, B., Moratal, C., Hernandez, M.-C., Depeursinge, C., Magistretti, P.J., Marquet, P.: Simultaneous optical recording in multiple cells by digital holographic microscopy of chloride current associated to activation of the ligand-gated chloride channel GABAA receptor. PLoS ONE 7, e51041 (2012)

173. Pache, C., Kuhn, J., Westphal, K., Toy, M.F., Parent, J., Buchi, O., Franco-Obregon, A., Depeursinge, C., Egli, M.: Digital holographic microscopy real-time monitoring of cytoarchitectural alterations during simulated microgravity. J. Biomed. Opt. 15, 026021 (2010)

174. Toy, M.F., Richard, S., Kuhn, J., Franco-Obregon, A., Egli, M., Depeursinge, C.: Enhanced robustness digital holographic microscopy for demanding environment of space biology. Biomed. Opt. Express 3, 313–326 (2012)

175. Alejandro, C., Martina, M., Lisa, M., Simonetta, G., Pietro, F.: Investigating fibroblast cells under "safe" and "injurious" blue-light exposure by holographic microscopy. J. Biophotonics 10, 919–927 (2017)

176. Mugnano, M., Memmolo, P., Miccio, L., Grilli, S., Merola, F., Calabuig, A., Bramanti, A., Mazzon, E., Ferraro, P.: In vitro cytotoxity evaluation of cadmium by label-free holographic microscopy. J. Biophotonics 11, e201800099 (2018)

177. Agero, U., Monken, C.H., Ropert, C., Gazzinelli, R.T., Mesquita, O.N.: Cell surface fluctuations studied with defocusing microscopy. Phys. Rev. E 67, 051904 (2003)

178. Neto, J.C., Agero, U., Oliveira, D.C.P., Gazzinelli, R.T., Mesquita, O.N.: Real-time measurements of membrane surface dynamics on macrophages and the phagocytosis of Leishmania parasites. Exp. Cell Res. 303, 207–217 (2005)

179. Neto, J.C., Agero, U., Gazzinelli, R.T., Mesquita, O.N.: Measuring optical and mechanical properties of a living cell with defocusing microscopy. Biophys. J. 91, 1108–1115 (2006)

180. Etcheverry, S., Gallardo, M.J., Solano, P., Suwalsky, M., Mesquita, O.N., Saavedra, C.: Real-time study of shape and thermal fluctuations in the echinocyte transformation of human erythrocytes using defocusing microscopy. J. Biomed. Opt. 17, 106013 (2012)

181. Li, Y., Di, J., Ma, C., Zhang, J., Zhong, J., Wang, K., Xi, T., Zhao, J.: Quantitative phase microscopy for cellular dynamics based on transport of intensity equation. Opt. Express 26, 586–593 (2018)

182. Li, Y., Di, J., Wu, W., Shang, P., Zhao, J.: Quantitative investigation on morphology and intracellular transport dynamics of migrating cells. Appl. Opt. 58, G162–G168 (2019)

183. Yu, W., Tian, X., He, X., Song, X., Xue, L., Liu, C., Wang, S.: Real time quantitative phase microscopy based on single-shot transport of intensity equation (ssTIE) method. Appl. Phys. Lett. 109, 071112 (2016)

184. Tian, X., Yu, W., Meng, X., Sun, A., Xue, L., Liu, C., Wang, S.: Real-time quantitative phase imaging based on transport of intensity equation with dual simultaneously recorded field of view. Opt. Lett. 41, 1427–1430 (2016)

185. Gong, Q., Wei, Q., Xu, J., Kong, Y., Jiang, Z., Qian, W., Zhu, Y., Xue, L., Liu, F., Liu, C., Wang, S.: Digital field of view correction combined dual-view transport of intensity equation method for real time quantitative imaging. Opt. Eng. 57, 063102 (2018)

186. Shan, Y., Gong, Q., Wang, J., Xu, J., Wei, Q., Liu, C., Xue, L., Wang, S., Liu, F.: Measurements on ATP induced cellular fluctuations using real-time dual view transport of intensity phase microscopy. Biomed. Opt. Express 10, 2337–2354 (2019)

187. Dannhauser, D., Rossi, D., Memmolo, P., Causa, F., Finizio, A., Ferraro, P., Netti, P.A.: Label-free analysis of mononuclear human blood cells in microfluidic flow by coherent imaging tools. J. Biophotonics 10, 683–689 (2017)

188. Gorthi, S.S., Schonbrun, E.: Phase imaging flow cytometry using a focus-stack collecting microscope. Opt. Lett. 37, 707–709 (2012)

189. Xue, L., Wang, S., Yan, K., Sun, N., Ferraro, P., Li, Z., Liu, F.: Gravity driven high throughput phase detecting cytometer based on quantitative interferometric microscopy. Opt. Commun. 316, 5–9 (2014)

190. Yan, K., Xue, L., Wang, S.: Field of view scanning based quantitative interferometric microscopic cytometers for cellular imaging and analysis. Microsc. Res. Tech. 81, 397–407 (2018)

191. Mahjoubfar, A., Chen, C., Niazi, K.R., Rabizadeh, S., Jalali, B.: Label-free high-throughput cell screening in flow. Biomed. Opt. Express **4**, 1618–1625 (2013)
192. Lau, A.K., Wong, T.T.W., Ho, K.K., Tang, M.Y.H., Chan, A.C.S., Wei, X., Lam, E.Y., Shum, H.C., Wong, K.K., Tsia, K.K.: Interferometric time-stretch microscopy for ultrafast quantitative cellular and tissue imaging. J. Biomed. Opt. **19**, 076001 (2014)
193. Guo, B., Lei, C., Kobayashi, H., Ito, T., Yalikun, Y., Jiang, Y., Tanaka, Y., Ozeki, Y., Goda, K.: High-throughput, label-free, single-cell, microalgal lipid screening by machine-learning-equipped optofluidic time-stretch quantitative phase microscopy. Cytom. Part A **91**, 494–502 (2017)
194. Guo, B., Lei, C., Wu, Y., Kobayashi, H., Ito, T., Yalikun, Y., Lee, S., Isozaki, A., Li, M., Jiang, Y., Yasumoto, A., Di Carlo, D., Tanaka, Y., Yatomi, Y., Ozeki, Y., Goda, K.: Optofluidic time-stretch quantitative phase microscopy. Methods **136**, 116–125 (2018)
195. Lei, C., Kobayashi, H., Wu, Y., Li, M., Isozaki, A., Yasumoto, A., Mikami, H., Ito, T., Nitta, N., Sugimura, T., Yamada, M., Yatomi, Y., Di Carlo, D., Ozeki, Y., Goda, K.: High-throughput imaging flow cytometry by optofluidic time-stretch microscopy. Nat. Protoc. **13**, 1603–1631 (2018)
196. Lee, K.C.M., Wang, M., Cheah, K.S.E., Chan, G.C.F., So, H.K.H., Wong, K.K.Y., Tsia, K.K.: Quantitative phase imaging flow cytometry for ultra-large-scale single-cell biophysical phenotyping. Cytom. Part A **95**, 510–520 (2019)
197. Wu, Y., Zhou, Y., Huang, C.-J., Kobayashi, H., Yan, S., Ozeki, Y., Wu, Y., Sun, C.-W., Yasumoto, A., Yatomi, Y., Lei, C., Goda, K.: Intelligent frequency-shifted optofluidic time-stretch quantitative phase imaging. Opt. Express **28**, 519–532 (2020)
198. Yan, H., Wu, Y., Zhou, Y., Xu, M., Paie, P., Lei, C., Sheng Yan, Goda, K.: Virtual optofluidic time-stretch quantitative phase imaging. APL Photonics **5**, 046103 (2020)
199. Chen, C., Lu, Y.-N., Huang, H., Yan, K., Jiang, Z., He, X., Kong, Y., Liu, C., Liu, F., Xue, L., Wang, S.: PhaseRMiC: phase real-time microscope camera for live cell imaging. Biomed. Opt. Express **12**, 5261–5271 (2021)
200. Sidorenko, P., Cohen, O.: Single-shot ptychography. Optica **3**, 9–14 (2016)
201. Chen, B.K., Sidorenko, P., Lahav, O., Peleg, O., Cohen, O.: Multiplexed single-shot ptychography. Opt. Lett. **43**, 5379–5382 (2018)
202. Wengrowicz, O., Peleg, O., Zahavy, T., Loevsky, B., Cohen, O.: Deep neural networks in single-shot ptychography. Opt. Express **28**, 17511–17520 (2020)
203. He, X., Liu, C., Zhu, J.: Single-shot Fourier ptychography based on diffractive beam splitting. Opt. Lett. **43**, 214–217 (2018)
204. He, X., Veetil, S.P., Pan, X., Sun, A., Liu, C., Zhu, J.: High-speed ptychographic imaging based on multiple-beam illumination. Opt. Express **26**, 25869–25879 (2018)
205. He, X., Liu, C., Zhu, J.: Single-shot aperture-scanning Fourier ptychography. Opt. Express **26**, 28187–28196 (2018)
206. He, X., Pan, X., Liu, C., Zhu, J.: Single-shot phase retrieval based on beam splitting. Appl. Opt. **57**, 4832–4838 (2018)
207. Chang, C., Pan, X., Tao, H., Liu, C., Veetil, S.P., Zhu, J.: Single-shot ptychography with highly tilted illuminations. Opt. Express **28**, 28441–28451 (2020)
208. Park, Y.K., Popescu, G., Badizadegan, K., Dasari, R.R., Feld, M.S.: Diffraction phase and fluorescence microscopy. Opt. Express **14**, 8263–8268 (2006)
209. Eldridge, W.J., Hoballah, J., Wax, A.: Molecular and biophysical analysis of apoptosis using a combined quantitative phase imaging and fluorescence resonance energy transfer microscope. J. Biophotonics **11**, e201800126 (2018)
210. Nicolas, P., Alexander, B., Daniel, B., Corinne, M., Jonas, K., Pascal, J., Christian, D., Pierre, J.M., Pierre, M.: Cell morphology and intracellular ionic homeostasis explored with a multimodal approach combining epifluorescence and digital holographic microscopy. J. Biophotonics **3**, 432–436 (2010)
211. Nygate, Y.N., Singh, G., Barnea, I., Shaked, N.T.: Simultaneous off-axis multiplexed holography and regular fluorescence microscopy of biological cells. Opt. Lett. **43**, 2587–2590 (2018)

212. Kumar, M., Quan, X., Awatsuji, Y., Cheng, C., Hasebe, M., Tamada, Y., Matoba, O.: Common-path multimodal three-dimensional fluorescence and phase imaging system. J. Biomed. Opt. **25**, 032010 (2020)
213. de Kernier, I., Ali-Cherif, A., Rongeat, N., Cioni, O., Morales, S., Savatier, J., Monneret, S., Blandin, P.: Large field-of-view phase and fluorescence mesoscope with microscopic resolution. J. Biomed. Opt. **24**, 036501 (2019)
214. Coskun, A.F., Su, T.-W., Ozcan, A.: Wide field-of-view lens-free fluorescent imaging on a chip. Lab. Chip **10**, 824–827 (2010)
215. Liao, J., Wang, Z., Zhang, Z., Bian, Z., Guo, K., Nambiar, A., Jiang, Y., Jiang, S., Zhong, J., Choma, M., Zheng, G.: Dual light-emitting diode-based multichannel microscopy for whole-slide multiplane, multispectral and phase imaging. J. Biophotonics **11**, e201700075 (2018)
216. Zheng, J., Zuo, C., Gao, P., Nienhaus, G.U.: Dual-mode phase and fluorescence imaging with a confocal laser scanning microscope. Opt. Lett. **43**, 5689–5692 (2018)
217. Kang, J.W., Lue, N., Kong, C.-R., Barman, I., Dingari, N.C., Goldfless, S.J., Niles, J.C., Dasari, R.R., Feld, M.S.: Combined confocal Raman and quantitative phase microscopy system for biomedical diagnosis. Biomed. Opt. Express **2**, 2484–2492 (2011)
218. Kim, K., Park, W.S., Na, S., Kim, S., Kim, T., Heo, W.D., Park, Y.K.: Correlative three-dimensional fluorescence and refractive index tomography: bridging the gap between molecular specificity and quantitative bioimaging. Biomed. Opt. Express **8**, 5688–5697 (2017)
219. Shin, S., Kim, D., Kim, K., Park, Y.K.: Super-resolution three-dimensional fluorescence and optical diffraction tomography of live cells using structured illumination generated by a digital micromirror device. Sci. Rep. **8**, 9183 (2018)
220. Dong, D., Huang, X., Li, L., Mao, H., Mo, Y., Zhang, G., Zhang, Z., Shen, J., Liu, W., Wu, Z., Liu, G., Liu, Y., Yang, H., Gong, Q., Shi, K., Chen, L.: Super-resolution fluorescence-assisted diffraction computational tomography reveals the three-dimensional landscape of the cellular organelle interactome. Light-Sci. Appl. **9**, 11 (2020)
221. Zhang, Q., Zhong, L., Tang, P., Yuan, Y., Liu, S., Tian, J., Lu, X.: Quantitative refractive index distribution of single cell by combining phase-shifting interferometry and AFM imaging. Sci. Rep. **7**, 2532 (2017)
222. Eldridge, W.J., Ceballos, S., Shah, T., Park, H.S., Steelman, Z.A., Zauscher, S., Wax, A.: Shear modulus measurement by quantitative phase imaging and correlation with atomic force microscopy. Biophys. J. **117**, 696–705 (2019)
223. Ding, H., Wang, Z., Nguyen, F., Boppart, S.A., Popescu, G.: Fourier transform light scattering of inhomogeneous and dynamic structures. Phys. Rev. Lett. **101**, 238102 (2008)
224. Ding, H., Berl, E., Wang, Z., Millet, L.J., Gillette, M.U., Liu, J., Boppart, M., Popescu, G.: Fourier transform light scattering of biological structure and dynamics. IEEE Trans. Nucl. Sci. **16**, 909–918 (2010)
225. Ding, H., Nguyen, F., Boppart, S.A., Popescu, G.: Optical properties of tissues quantified by Fourier-transform light scattering. Opt. Lett. **34**, 1372–1374 (2009)
226. Ding, H., Millet, L.J., Gillette, M.U., Popescu, G.: Actin-driven cell dynamics probed by Fourier transform light scattering. Biomed. Opt. Express **1**, 260–267 (2010)
227. Kim, K., Park, Y.K.: Fourier transform light scattering angular spectroscopy using digital inline holography. Opt. Lett. **37**, 4161–4163 (2012)
228. Yu, H.S., Park, H.J., Kim, Y., Kim, M.W., Park, Y.K.: Fourier-transform light scattering of individual colloidal clusters. Opt. Lett. **37**, 2577–2579 (2012)
229. Kim, Y., Park, Y.K., Higgins, J.M., Dasari, R.R., Suresh, S.: Anisotropic light scattering of individual sickle red blood cells. J. Biomed. Opt. **17**, 040501 (2012)
230. Jo, Y.J., Jung, J.H., Lee, J.W., Shin, D., Park, H.J., Nam, K.T., Park, J.-H., Park, Y.K.: Angle-resolved light scattering of individual rod-shaped bacteria based on Fourier transform light scattering. Sci. Rep. **4**, 5090 (2014)
231. Jo, Y.J., Jung, J.H., Kim, M., Park, H.J., Kang, S.-J., Park, Y.K.: Label-free identification of individual bacteria using Fourier transform light scattering. Opt. Express **23**, 15792–15805 (2015)

232. Jang, Y., Jang, J., Park, Y.K.: Dynamic spectroscopic phase microscopy for quantifying hemoglobin concentration and dynamic membrane fluctuation in red blood cells. Opt. Express **20**, 9673–9681 (2012)
233. Pham, H., Bhaduri, B., Ding, H., Popescu, G.: Spectroscopic diffraction phase microscopy. Opt. Lett. **37**, 3438–3440 (2012)
234. Turko, N.A., Roitshtain, D., Blum, O., Kemper, B., Shaked, N.T.: Dynamic measurements of flowing cells labeled by gold nanoparticles using full-field photothermal interferometric imaging. J. Biomed. Opt. **22**, 066012 (2017)
235. Turko, N.A., Barnea, I., Blum, O., Korenstein, R., Shaked, N.T.: Detection and controlled depletion of cancer cells using photothermal phase microscopy. J. Biophotonics **8**, 755–763 (2018)
236. Zhu, M., Baffou, G., Meyerbroker, N., Polleux, J.: Micropatterning thermoplasmonic gold nanoarrays to manipulate cell adhesion. ACS Nano **6**, 7227–7233 (2012)
237. Samira, K., Pierre, B., Dominique, V., Elizabeth, G., Niall, M., David, M., Serge, M., Guillaume, B.: Optical imaging and characterization of graphene and other 2D materials using quantitative phase microscopy. ACS Photonics **4**, 3130–3139 (2017)
238. Khadir, S., Andren, D., Chaumet, P.C., Monneret, S., Bonod, N., Kall, M., Sentenac, A., Baffou, G.: Full optical characterization of single nanoparticles using quantitative phase imaging. Optica **7**, 243–248 (2020)
239. Oh, J.T., Lee, G.-H., Rho, J., Shin, S., Lee, B.J., Nam, Y., Park, Y.K.: Optical measurements of three-dimensional microscopic temperature distributions around gold nanorods excited by localized surface plasmon resonance. Phys. Rev. Appl. **11**, 044079 (2019)
240. Park, Y.K., Depeursinge, C., Popescu, G.: Quantitative phase imaging in biomedicine. Nat. Photonics **12**, 578–589 (2018)
241. Majeed, H., Sridharan, S., Mir, M., Ma, L., Min, E., Jung, W., Popescu, G.: Quantitative phase imaging for medical diagnosis. J. Biophotonics **10**, 177–205 (2017)
242. Lee, K.R., Kim, K., Jung, J., Heo, J.H., Cho, S., Lee, S., Chang, G., Jo, Y.J., Park, H., Park, Y.K.: Quantitative phase imaging techniques for the study of cell pathophysiology: from principles to applications. Sensors **13**, 4170–4191 (2013)
243. Jung, J.H., Kim, K., Yu, H.S., Lee, K.R., Lee, S.E., Nahm, S.H., Park, H.J., Park, Y.K.: Biomedical applications of holographic microspectroscopy. Appl. Opt. **53**, G111–G122 (2014)
244. Platt, B.C., Shack, R.: History and principles of Shack-Hartmann wavefront sensing. J. Refract. Surg. **17**, S573–S577 (2001)
245. Beverage, J.L., Shack, R.V., Descour, M.R.: Measurement of the three-dimensional microscope point spread function using a Shack-Hartmann wavefront sensor. J. Microsc. **205**, 61–75 (2002)
246. Colomb, T., Montfort, F., Kuhn, J., Aspert, N., Cuche, E., Marian, A., Charriere, F., Bourquin, S., Marquet, P., Depeursinge, C.: Numerical parametric lens for shifting, magnification, and complete aberration compensation in digital holographic microscopy. J. Opt. Soc. Am. A **23**, 3177–3190 (2006)
247. Zheng, G., Ou, X., Horstmeyer, R., Yang, C.: Characterization of spatially varying aberrations for wide field-of-view microscopy. Opt. Express **21**, 15131–15143 (2013)
248. Booth, M.J.: Adaptive optical microscopy: the ongoing quest for a perfect image. Light-Sci. Appl. **3**, e165 (2014)
249. Debarre, D., Booth, M.J., Wilson, T.: Image based adaptive optics through optimisation of low spatial frequencies. Opt. Express **15**, 8176–8190 (2007)
250. Debarre, D., Botcherby, E.J., Watanabe, T., Srinivas, S., Booth, M.J., Wilson, T.: Image-based adaptive optics for two-photon microscopy. Opt. Lett. **34**, 2495–2497 (2009)
251. Antonello, J., Verhaegen, M., Fraanje, R., van Werkhoven, T., Gerritsen, H.C., Keller, C.U.: Semidefinite programming for model-based sensorless adaptive optics. J. Opt. Soc. Am. A **29**, 2428–2438 (2012)
252. Scrimgeour, J., Curtis, J.E.: Aberration correction in wide-field fluorescence microscopy by segmented-pupil image interferometry. Opt. Express **20**, 14534–14541 (2012)

253. Milkie, D.E., Betzig, E., Ji, N.: Pupil-segmentation-based adaptive optical microscopy with full-pupil illumination. Opt. Lett. **36**, 4206–4208 (2011)
254. Park, J.-H., Kong, L., Zhou, Y., Cui, M.: Large-field-of-view imaging by multi-pupil adaptive optics. Nat. Methods **14**, 581–583 (2017)
255. Rahman, S.A., Booth, M.J.: Direct wavefront sensing in adaptive optical microscopy using backscattered light. Appl. Opt. **52**, 5523–5532 (2013)
256. Tao, X., Norton, A., Kissel, M., Azucena, O., Kubby, J.: Adaptive optical two-photon microscopy using autofluorescent guide stars. Opt. Lett. **38**, 5075–5078 (2013)
257. Aviles-Espinosa, R., Andilla, J., Porcar-Guezenec, R., Olarte, O.E., Nieto, M., Levecq, X., Artigas, D., Loza-Alvarez, P.: Measurement and correction of in vivo sample aberrations employing a nonlinear guide-star in two-photon excited fluorescence microscopy. Biomed. Opt. Express **2**, 3135–3149 (2011)
258. Azucena, O., Crest, J., Kotadia, S., Sullivan, W., Tao, X., Reinig, M., Gavel, D., Olivier, S., Kubby, J.: Adaptive optics wide-field microscopy using direct wavefront sensing. Opt. Lett. **36**, 825–827 (2011)
259. Vermeulen, P., Muro, E., Pons, T., Loriette, V., Fragola, A.: Adaptive optics for fluorescence wide-field microscopy using spectrally independent guide star and markers. J. Biomed. Opt. **16**, 076019 (2011)
260. Tao, X., Crest, J., Kotadia, S., Azucena, O., Chen, D.C., Sullivan, W., Kubby, J.: Live imaging using adaptive optics with fluorescent protein guide-stars. Opt. Express **20**, 15969–15982 (2012)
261. Feierabend, M., Ruckel, M., Denk, W.: Coherence-gated wave-front sensing in strongly scattering samples. Opt. Lett. **29**, 2255–2257 (2004)
262. Rueckel, M., Mack-Bucher, J.A., Denk, W.: Adaptive wavefront correction in two-photon microscopy using coherence-gated wavefront sensing. Proc. Natl. Acad. Sci. U.S.A. **103**, 17137–17142 (2006)
263. Tuohy, S., Podoleanu, A.G.: Depth-resolved wavefront aberrations using a coherence-gated Shack-Hartmann wavefront sensor. Opt. Express **18**, 3458–3476 (2010)
264. Wang, J., Leger, J.-F., Binding, J., Boccara, A.C., Gigan, S., Bourdieu, L.: Measuring aberrations in the rat brain by coherence-gated wavefront sensing using a Linnik interferometer. Biomed. Opt. Express **3**, 2510–2525 (2012)
265. Yoon, S., Kim, M., Jang, M., Choi, Y., Choi, W., Kang, S., Choi, W.: Deep optical imaging within complex scattering media. Nat. Rev. Phys. **2**, 141–158 (2020)
266. Horstmeyer, R., Ruan, H., Yang, C.: Guidestar-assisted wavefront-shaping methods for focusing light into biological tissue. Nat. Photonics **9**, 563–571 (2015)
267. Mosk, A.P., Lagendijk, A., Lerosey, G., Fink, M.: Controlling waves in space and time for imaging and focusing in complex media. Nat. Photonics **6**, 283–292 (2012)
268. Vellekoop, I.M., Lagendijk, A., Mosk, A.P.: Exploiting disorder for perfect focusing. Nat. Photonics **4**, 320–322 (2010)
269. Vellekoop, I.M., Mosk, A.P.: Focusing coherent light through opaque strongly scattering media. Opt. Lett. **32**, 2309–2311 (2007)
270. Bertolotti, J., van Putten, E.G., Blum, C., Lagendijk, A., Vos, W.L., Mosk, A.P.: Non-invasive imaging through opaque scattering layers. Nature **491**, 232–234 (2012)
271. Judkewitz, B., Horstmeyer, R., Vellekoop, I.M., Papadopoulos, I.N., Yang, C.: Translation correlations in anisotropically scattering media. Nat. Phys. **11**, 684–689 (2015)
272. Cua, M., Zhou, E., Yang, C.: Imaging moving targets through scattering media. Opt. Express **25**, 3935–3945 (2017)
273. Hong, P., Ojambati, O.S., Lagendijk, A., Mosk, A.P., Vos, W.L.: Three-dimensional spatially resolved optical energy density enhanced by wavefront shaping. Optica **5**, 844–849 (2018)
274. Ruan, H., Haber, T., Liu, Y., Brake, J., Kim, J., Berlin, J.M., Yang, C.: Focusing light inside scattering media with magnetic-particle-guided wavefront shaping. Optica **4**, 1337–1343 (2017)
275. Si, K., Fiolka, R., Cui, M.: Fluorescence imaging beyond the ballistic regime by ultrasound-pulse-guided digital phase conjugation. Nat. Photonics **6**, 657–661 (2012)

276. Xu, X., Liu, H., Wang, L.V.: Time-reversed ultrasonically encoded optical focusing into scattering media. Nat. Photonics **5**, 154–157 (2011)

277. Judkewitz, B., Wang, Y.M., Horstmeyer, R., Mathy, A., Yang, C.: Speckle-scale focusing in the diffusive regime with time reversal of variance-encoded light (TROVE). Nat. Photonics **7**, 300–305 (2013)

278. Popoff, S.M., Lerosey, G., Carminati, R., Fink, M., Boccara, A.C., Gigan, S.: Measuring the transmission matrix in optics: an approach to the study and control of light propagation in disordered media. Phys. Rev. Lett. **104**, 100601 (2010)

279. Popoff, S., Lerosey, G., Fink, M., Boccara, A.C., Gigan, S.: Image transmission through an opaque material. Nat. Commun. **1**, 81 (2010)

280. Dremeau, A., Liutkus, A., Martina, D., Katz, O., Schulke, C., Krzakala, F., Gigan, S., Daudet, L.: Reference-less measurement of the transmission matrix of a highly scattering material using a DMD and phase retrieval techniques. Opt. Express **23**, 11898–11911 (2015)

281. Yoon, J., Lee, K.R., Park, J., Park, Y.K.: Measuring optical transmission matrices by wavefront shaping. Opt. Express **23**, 10158–10167 (2015)

282. Park, J.-H., Park, C.H., Yu, H., Cho, Y.-H., Park, Y.K.: Active spectral filtering through turbid media. Opt. Lett. **37**, 3261–3263 (2012)

283. Park, J., Park, J.-H., Yu, H., Park, Y.K.: Focusing through turbid media by polarization modulation. Opt. Lett. **40**, 1667–1670 (2015)

284. Lee, K.R., Lee, J., Park, J.-H., Park, J.-H., Park, Y.K.: One-wave optical phase conjugation mirror by actively coupling arbitrary light fields into a single-mode reflector. Phys. Rev. Lett. **115**, 153902 (2015)

285. Park, J.-H., Park, C., Yu, H.S., Park, J., Han, S., Shin, J., Ko, S.H., Nam, K.T., Cho, Y.-H., Park, Y.K.: Subwavelength light focusing using random nanoparticles. Nat. Photonics **7**, 454–458 (2013)

286. Park, C., Park, J.-H., Rodriguez, C., Yu, H.S., Kim, M., Jin, K., Han, S., Shin, J., Ko, S.H., Nam, K.T., Lee, Y.-H., Cho, Y.-H., Park, Y.K.: Full-field subwavelength imaging using a scattering superlens. Phys. Rev. Lett. **113**, 113901 (2014)

287. Bianchi, S., Leonardo, R.D.: A multi-mode fiber probe for holographic micromanipulation and microscopy. Lab Chip **12**, 635–639 (2012)

288. Bianchi, S., Rajamanickam, V.P., Ferrara, L., Fabrizio, E.D., Liberale, C., Leonardo, R.D.: Focusing and imaging with increased numerical apertures through multimode fibers with micro-fabricated optics. Opt. Lett. **38**, 4935–4938 (2013)

289. Cizmar, T., Dholakia, K.: Exploiting multimode waveguides for pure fibre-based imaging. Nat. Commun. **3**, 1027 (2012)

290. Cizmar, T., Dholakia, K.: Shaping the light transmission through a multimode optical fibre: complex transformation analysis and applications in biophotonics. Opt. Express **19**, 18871–18884 (2011)

291. Ploschner, M., Tyc, T., Cizmar, T.: Seeing through chaos in multimode fibres. Nat. Photonics **9**, 529–535 (2015)

292. Ohayon, S., Caravaca-Aguirre, A., Piestun, R., DiCarlo, J.J.: Minimally invasive multimode optical fiber microendoscope for deep brain fluorescence imaging. Biomed. Opt. Express **9**, 1492–1509 (2018)

293. Caravaca-Aguirre, A.M., Piestun, R.: Single multimode fiber endoscope. Opt. Express **25**, 1656–1665 (2017)

294. Turtaev, S., Leite, I.T., Altwegg-Boussac, T., Pakan, J.M.P., Rochefort, N.L., Cizmar, T.: High-fidelity multimode fibre-based endoscopy for deep brain in vivo imaging. Light-Sci. Appl. **7**, 92 (2018)

295. Loterie, D., Farahi, S., Papadopoulos, I., Goy, A., Psaltis, D., Moser, C.: Digital confocal microscopy through a multimode fiber. Opt. Express **23**, 23845–23858 (2015)

296. Morales-Delgado, E.E., Psaltis, D., Moser, C.: Two-photon imaging through a multimode fiber. Opt. Express **23**, 32158–32170 (2015)

297. Flaes, D.E.B., Stopka, J., Turtaev, S., de Boer, J.F., Tyc, T., Cizmar, T.: Robustness of light-transport processes to bending deformations in graded-index multimode waveguides. Phys. Rev. Lett. **120**, 233901 (2018)

298. Lee, K.R., Park, Y.K.: Exploiting the speckle-correlation scattering matrix for a compact reference-free holographic image sensor. Nat. Commun. **7**, 13359 (2016)
299. Park, J., Cho, J.-Y., Park, C., Lee, K.R., Lee, H., Cho, Y.-H., Park, Y.K.: Scattering optical elements: stand-alone optical elements exploiting multiple light scattering. ACS Nano **10**, 6871–6876 (2016)
300. Yu, H., Lee, K.R., Park, J., Park, Y.K.: Ultrahigh-definition dynamic 3D holographic display by active control of volume speckle fields. Nat. Photonics **11**, 186–192 (2017)
301. Choi, Y., Yang, T.D., Fang-Yen, C., Kang, P., Lee, K.J., Dasari, R.R., Feld, M.S., Choi, W.: Overcoming the diffraction limit using multiple light scattering in a highly disordered medium. Phys. Rev. Lett. **107**, 023902 (2011)
302. Choi, Y., Kim, M., Yoon, C., Yang, T.D., Lee, K.J., Choi, W.: Synthetic aperture microscopy for high resolution imaging through a turbid medium. Opt. Lett. **36**, 4263–4265 (2011)
303. Choi, Y., Yoon, C., Kim, M., Yang, T.D., Fang-Yen, C., Dasari, R.R., Lee, K.J., Choi, W.: Scanner-free and wide-field endoscopic imaging by using a single multimode optical fiber. Phys. Rev. Lett. **109**, 203901 (2012)
304. Choi, Y., Yoon, C., Kim, M., Choi, W., Choi, W.: Optical imaging with the use of a scattering lens. IEEE J. Sel. Top. Quant. **20**, 6800213 (2014)
305. Kim, D., Moon, J., Kim, M., Yang, T.D., Kim, J., Chung, E., Choi, W.: Toward a miniature endomicroscope: pixelation-free and diffraction-limited imaging through a fiber bundle. Opt. Lett. **39**, 1921–1924 (2014)
306. Kim, M., Choi, W., Choi, Y., Yoon, C., Choi, W.: Transmission matrix of a scattering medium and its applications in biophotonics. Opt. Express **23**, 12648–12668 (2015)
307. Yoon, C., Kang, M., Hong, J.H., Yang, T.D., Xing, J., Yoo, H., Choi, Y., Choi, W.: Removal of back-reflection noise at ultrathin imaging probes by the single-core illumination and widefield detection. Sci. Rep. **7**, 6524 (2017)
308. Kang, S., Jeong, S., Choi, W., Ko, H., Yang, T.D., Joo, J.H., Lee, J.-S., Lim, Y.-S., Park, Q.-H., Choi, W.: Imaging deep within a scattering medium using collective accumulation of single-scattered waves. Nat. Photonics **9**, 253–258 (2015)
309. Kang, S., Kang, P., Jeong, S., Kwon, Y., Yang, T.D., Hong, J.H., Kim, M., Song, K.-D., Park, J.H., Lee, J.H., Kim, M.J., Kim, K.H., Choi, W.: High-resolution adaptive optical imaging within thick scattering media using closed-loop accumulation of single scattering. Nat. Commun. **8**, 2157 (2017)
310. Kim, M., Jo, Y., Hong, J.H., Kim, S., Yoon, S., Song, K.-D., Kang, S., Lee, B., Kim, G.H., Park, H.-C., Choi, W.: Label-free neuroimaging in vivo using synchronous angular scanning microscopy with single-scattering accumulation algorithm. Nat. Commun. **10**, 3152 (2019)
311. Popoff, S.M., Aubry, A., Lerosey, G., Fink, M., Boccara, A.C., Gigan, S.: Exploiting the time-reversal operator for adaptive optics, selective focusing, and scattering pattern analysis. Phys. Rev. Lett. **107**, 263901 (2011)
312. Kim, M., Choi, Y., Yoon, C., Choi, W., Kim, J., Park, Q.-H., Choi, W.: Maximal energy transport through disordered media with the implementation of transmission eigenchannels. Nat. Photonics **6**, 581–585 (2012)
313. Choi, Y., Hillman, T.R., Choi, W., Lue, N., Dasari, R.R., So, P.T.C., Choi, W., Yaqoob, Z.: Measurement of the time-resolved reflection matrix for enhancing light energy delivery into a scattering medium. Phys. Rev. Lett. **111**, 243901 (2013)
314. Jeong, S., Lee, Y.-R., Choi, W., Kang, S., Hong, J.H., Park, J.-S., Lim, Y.-S., Park, H.-G., Choi, W.: Focusing of light energy inside a scattering medium by controlling the time-gated multiple light scattering. Nat. Photonics **12**, 277–283 (2018)
315. Firestone, L., Cook, K., Culp, K., Talsania, N., Preston, K.: Comparison of autofocus methods for automated microscopy. Cytometry **12**, 195–206 (1991)
316. Sun, Y., Duthaler, S., Nelson, B.J.: Autofocusing in computer microscopy: selecting the optimal focus algorithm. Microsc. Res. Tech. **65**, 139–149 (2004)
317. Guo, K., Liao, J., Bian, Z., Heng, X., Zheng, G.: InstantScope: a low-cost whole slide imaging system with instant focal plane detection. Biomed. Opt. Express **6**, 3210–3216 (2015)

318. Liao, J., Bian, L., Bian, Z., Zhang, Z., Patel, C., Hoshino, K., Eldar, Y.C., Zheng, G.: Single-frame rapid autofocusing for brightfield and fluorescence whole slide imaging. Biomed. Opt. Express **7**, 4763–4768 (2016)

319. Jiang, S., Liao, J., Bian, Z., Guo, K., Zhang, Y., Zheng, G.: Transform- and multi-domain deep learning for single-frame rapid autofocusing in whole slide imaging. Biomed. Opt. Express **9**, 1601–1612 (2018)

320. Pinkard, H., Phillips, Z., Babakhani, A., Fletcher, D.A., Waller, L.: Deep learning for single-shot autofocus microscopy. Optica **6**, 794–797 (2019)

321. Dastidar, T.R., Ethirajan, R.: Whole slide imaging system using deep learning-based automated focusing. Biomed. Opt. Express **11**, 480–491 (2020)

322. von Diezmann, L., Shechtman, Y., Moerner, W.E.: Three-dimensional localization of single molecules for super-resolution imaging and single-particle tracking. Chem. Rev. **117**, 7244–7275 (2017)

323. Mir, M., Bhaduri, B., Wang, R., Zhu, R., Popescu, G.: Quantitative phase imaging. Prog. Opt. **57**, 133–217 (2012)

324. Gillespie, J., King, R.A.: The use of self-entropy as a focus measure in digital holography. Pattern Recogn. Lett. **9**, 19–25 (1989)

325. Liebling, M., Unser, M.: Autofocus for digital Fresnel holograms by use of a Fresnelet-sparsity criterion. J. Opt. Soc. Am. A **21**, 2424–2430 (2004)

326. Dubois, F., Schockaert, C., Callens, N., Yourassowsky, C.: Focus plane detection criteria in digital holography microscopy by amplitude analysis. Opt. Express **14**, 5895–5908 (2006)

327. Dubois, F., Mallahi, A.E., Dohet-Eraly, J., Yourassowsky, C.: Refocus criterion for both phase and amplitude objects in digital holographic microscopy. Opt. Lett. **39**, 4286–4289 (2014)

328. Dohet-Eraly, J., Yourassowsky, C., Dubois, F.: Fast numerical autofocus of multispectral complex fields in digital holographic microscopy with a criterion based on the phase in the Fourier domain. Opt. Lett. **41**, 4071–4074 (2016)

329. Li, W., Loomis, N.C., Hu, Q., Davis, C.S.: Focus detection from digital in-line holograms based on spectral l_1 norms. J. Opt. Soc. Am. A **24**, 3054–3062 (2007)

330. Langehanenberg, P., Kemper, B., Dirksen, D., von Bally, G.: Autofocusing in digital holographic phase contrast microscopy on pure phase objects for live cell imaging. Appl. Opt. **47**, D176–D182 (2008)

331. Memmolo, P., Distante, C., Paturzo, M., Finizio, A., Ferraro, P., Javidi, B.: Automatic focusing in digital holography and its application to stretched holograms. Opt. Lett. **36**, 1945–1947 (2011)

332. Memmolo, P., Paturzo, M., Javidi, B., Netti, P.A., Ferraro, P.: Refocusing criterion via sparsity measurements in digital holography. Opt. Lett. **39**, 4719–4722 (2014)

333. Lyu, M., Yuan, C., Li, D., Situ, G.: Fast autofocusing in digital holography using the magnitude differential. Appl. Opt. **56**, F152–F157 (2017)

334. Zhang, Y., Wang, H., Wu, Y., Tamamitsu, M., Ozcan, A.: Edge sparsity criterion for robust holographic autofocusing. Opt. Lett. **42**, 3824–3827 (2017)

335. Ren, Z., Chen, N., Lam, E.Y.: Automatic focusing for multisectional objects in digital holography using the structure tensor. Opt. Lett. **42**, 1720–1723 (2017)

336. Ou, H., Wu, Y., Lam, E.Y., Wang, B.-Z.: New autofocus and reconstruction method based on a connected domain. Opt. Lett. **43**, 2201–2203 (2018)

337. Gao, P., Yao, B., Rupp, R., Min, J., Guo, R., Ma, B., Zheng, J., Lei, M., Yan, S., Dan, D., Ye, T.: Autofocusing based on wavelength dependence of diffraction in two-wavelength digital holographic microscopy. Opt. Lett. **37**, 1172–1174 (2012)

338. Ren, Z., Xu, Z., Lam, E.Y.: Learning-based nonparametric autofocusing for digital holography. Optica **5**, 337–344 (2018)

339. Fonseca, E.S.R., Fiadeiro, P.T., Pereira, M., Pinheiro, A.: Comparative analysis of autofocus functions in digital in-line phase-shifting holography. Appl. Opt. **55**, 7663–7674 (2016)

340. Javidi, B., Moon, I., Yeom, S., Carapezza, E.: Three-dimensional imaging and recognition of microorganism using single-exposure on-line (SEOL) digital holography. Opt. Express **13**, 4492–4506 (2005)

341. Jaferzadeh, K., Hwang, S.-H., Moon, I., Javidi, B.: No-search focus prediction at the single cell level in digital holographic imaging with deep convolutional neural network. Biomed. Opt. Express **10**, 4276–4289 (2019)

342. Kim, J.W., Lee, B.H.: Autofocus tracking system based on digital holographic microscopy and electrically tunable lens. Curr. Opt. Photon. **3**, 27–32 (2019)

343. Memmolo, P., Miccio, L., Paturzo, M., Di Caprio, G., Coppola, G., Netti, P.A., Ferraro, P.: Recent advances in holographic 3D particle tracking. Adv. Opt. Photon. **7**, 713–755 (2015)

344. Dou, J., Gao, Z., Ma, J., Yuan, C., Yang, Z., Wang, L.: Iterative autofocusing strategy for axial distance error correction in ptychography. Opt. Laser Eng. **98**, 56–61 (2017)

345. Loetgering, L., Du, M., Eikema, K.S.E., Witte, S.: zPIE: an autofocusing algorithm for ptychography. Opt. Lett. **45**, 2030–2033 (2020)

346. He, X.: Analysis of influence of object-detector distance error on the reconstructed object and probe in ptychographic imaging. In Preparation

347. Maiden, A.M., Humphry, M.J., Sarahan, M.C., Kraus, B., Rodenburg, J.M.: An annealing algorithm to correct positioning errors in ptychography. Ultramicroscopy **120**, 64–72 (2012)

348. Zhang, F., Peterson, I., Vila-Comamala, J., Diaz, A., Berenguer, F., Bean, R., Chen, B., Menzel, A., Robinson, I.K., Rodenburg, J.M.: Translation position determination in ptychographic coherent diffraction imaging. Opt. Express **21**, 13592–13606 (2013)

349. Tian, X., Meng, X., Yu, W., Song, X., Xue, L., Liu, C., Wang, S.: In-focus quantitative intensity and phase imaging with the numerical focusing transport of intensity equation method. J. Opt. **18**, 105302 (2016)

350. Meng, X., Tian, X., Kong, Y., Sun, A., Yu, W., Qian, W., Song, X., Cui, H., Xue, L., Liu, C., Wang, S.: Rapid in-focus corrections on quantitative amplitude and phase imaging using transport of intensity equation method. J. Microsc. **266**, 253–262 (2017)

351. Xu, J., Tian, X., Meng, X., Kong, Y., Gao, S., Cui, H., Liu, F., Xue, L., Liu, C., Wang, S.: Wavefront-sensing-based autofocusing in microscopy. J. Biomed. Opt. **22**, 086012 (2017)

352. Jiang, Z., Kong, Y., Liu, F., Liu, C., Wang, S.: Graphics processing unit (GPU) aided wavefront-based autofocusing in microscopy. AIP Adv. **8**, 105328 (2018)

353. Xu, J., Kong, Y., Jiang, Z., Gao, S., Xue, L., Liu, F., Liu, C., Wang, S.: Accelerating wavefront-sensing-based autofocusing using pixel reduction in spatial and frequency domains. Appl. Opt. **58**, 3003–3012 (2019)

354. Kim, T., Poon, T.-C.: Autofocusing in optical scanning holography. Appl. Opt. **48**, H153–H159 (2009)

355. Kumar, M., Quan, X., Awatsuji, Y., Tamada, Y., Matoba, O.: Digital holographic multimodal cross-sectional fluorescence and quantitative phase imaging system. Sci. Rep. **10**, 7580 (2020)

356. Tang, M., Liu, C., Wang, X.P.: Autofocusing and image fusion for multi-focus plankton imaging by digital holographic microscopy. Appl. Opt. **59**, 333–345 (2020)

357. Ferraro, P., Grilli, S., Alfieri, D., De Nicola, S., Finizio, A., Pierattini, G., Javidi, B., Coppola, G., Striano, V.: Extended focused image in microscopy by digital holography. Opt. Express **13**, 6738–6749 (2005)

358. De Nicola, S., Finizio, A., Pierattini, G., Ferraro, P., Alfieri, D.: Angular spectrum method with correction of anamorphism for numerical reconstruction of digital holograms on tilted planes. Opt. Express **13**, 9935–9940 (2005)

359. Paturzo, M., Ferraro, P.: Creating an extended focus image of a tilted object in Fourier digital holography. Opt. Express **17**, 20546–20552 (2009)

360. Matrecano, M., Paturzo, M., Finizio, A., Ferraro, P.: Enhancing depth of focus in tilted microfluidics channels by digital holography. Opt. Lett. **38**, 896–898 (2013)

361. Matrecano, M., Paturzo, M., Ferraro, P.: Tilted objects EFI extracted at once by 3D output of the angular spectrum method. Opt. Laser Eng. **51**, 1353–1359 (2013)

362. Hu, J., Meng, X., Wei, Q., Kong, Y., Jiang, Z., Xue, L., Liu, F., Liu, C., Wang, S.: Numerical tilting compensation in microscopy based on wavefront sensing using transport of intensity equation method. J. Opt. **20**, 035301 (2018)

363. Shen, H., Tauzin, L.J., Baiyasi, R., Wang, W., Moringo, N., Shuang, B., Landes, C.F.: Single particle tracking: from theory to biophysical applications. Chem. Rev. **117**, 7331–7376 (2017)

364. Rabut, G., Ellenberg, J.: Automatic real-time three-dimensional cell tracking by fluorescence microscopy. J. Microsc. **216**, 131–137 (2004)
365. Gadelha, H., Hernandez-Herrera, P., Montoya, F., Darszon, A., Corkidi, G.: Human sperm uses asymmetric and anisotropic flagellar controls to regulate swimming symmetry and cell steering. Sci. Adv. **6**, eaba5168 (2020)
366. Lakadamyali, M., Rust, M.J., Babcock, H.P., Zhuang, X.: Visualizing infection of individual influenza viruses. Proc. Natl. Acad. Sci. U.S.A. **100**, 9280–9285 (2003)
367. Brandenburg, B., Zhuang, X.: Virus trafficking-learning from single-virus tracking. Nat. Rev. Microbiol. **5**, 197–208 (2007)
368. Yildiz, A., Forkey, J.N., McKinney, S.A., Ha, T., Goldman, Y.E., Selvin, P.R.: Myosin V walks hand-over-hand: single fluorophore imaging with 1.5-nm localization. Science **300**, 2061–2065 (2003)
369. Toprak, E., Selvin, P.R.: New fluorescent tools for watching nanometer-scale conformational changes of single molecules. Annu. Rev. Biophys. Biomol. Struct. **36**, 349–369 (2007)
370. Zimyanin, V.L., Belaya, K., Pecreaux, J., Gilchrist, M.J., Clark, A., Davis, I., Johnston, D.S.: In vivo imaging of Oskar mRNA transport reveals the mechanism of posterior localization. Cell **134**, 843–853 (2008)
371. Vargas, D.Y., Raj, A., Marras, S.A.E., Kramer, F.R., Tyagi, S.: Mechanism of mRNA transport in the nucleus. Proc. Natl. Acad. Sci. U.S.A. **102**, 17008–17013 (2005)
372. Westerweel, J., Elsinga, G.E., Adrian, R.J.: Particle image velocimetry for complex and turbulent flows. Annu. Rev. Fluid Mech. **45**, 409–436 (2013)
373. Prasad, A.K., Adrian, R.J.: Stereoscopic particle image velocimetry applied to liquid flows. Exp. Fluids **15**, 49–60 (1993)
374. Hori, T., Sakakibara, J.: High-speed scanning stereoscopic PIV for 3D vorticity measurement in liquids. Meas. Sci. Technol. **15**, 1067–1078 (2004)
375. Fahringer, T.W., Lynch, K.P., Thurow, B.S.: Volumetric particle image velocimetry with a single plenoptic camera. Meas. Sci. Technol. **26**, 115201 (2015)
376. Elsinga, G.E., Scarano, F., Wieneke, B., van Oudheusden, B.W.: Tomographic particle image velocimetry. Exp. Fluids **41**, 933–947 (2006)
377. Zhao, J., Liu, H., Cai, W.: Numerical and experimental validation of a single-camera 3D velocimetry based on endoscopic tomography. Appl. Opt. **58**, 1363–1373 (2019)
378. Xiong, J., Idoughi, R., Aguirre-Pablo, A.A., Aljedaani, A.B., Dun, X., Fu, Q., Thoroddsen, S.T., Heidrich, W.: Rainbow particle imaging velocimetry for dense 3D fluid velocity imaging. ACM T. Graphic. **36**, 36 (2017)
379. Paciaroni, M.E., Chen, Y., Lynch, K.P., Guildenbecher, D.R.: Backscatter particle image velocimetry via optical time-of-flight sectioning. Opt. Lett. **43**, 312–315 (2018)
380. Hinsch, K.D.: Holographic particle image velocimetry. Meas. Sci. Technol. **13**, R61–R72 (2002)
381. Katz, J., Sheng, J.: Applications of holography in fluid mechanics and particle dynamics. Annu. Rev. Fluid Mech. **42**, 531–555 (2010)
382. Malek, M., Allano, D., Coetmellec, S., Lebrun, D.: Digital in-line holography: influence of the shadow density on particle field extraction. Opt. Express **12**, 2270–2279 (2004)
383. Yang, Y., Kang, B.S., Choo, Y.J.: Application of the correlation coefficient method for determination of the focal plane to digital particle holography. Appl. Opt. **47**, 817–824 (2008)
384. Yang, Y., Kang, B.S.: Experimental validation for the determination of particle positions by the correlation coefficient method in digital particle holography. Appl. Opt. **47**, 5953–5960 (2008)
385. Wormald, S.A., Coupland, J.: Particle image identification and correlation analysis in microscopic holographic particle image velocimetry. Appl. Opt. **48**, 6400–6407 (2009)
386. Wu, Y., Wu, X., Wang, Z., Grehan, G., Chen, L., Cen, K.: Measurement of microchannel flow with digital holographic microscopy by integrated nearest neighbor and cross-correlation particle pairing. Appl. Opt. **50**, H297–H305 (2011)
387. Guildenbecher, D.R., Reu, P.L., Stuaffacher, H.L., Grasser, T.: Accurate measurement of out-of-plane particle displacement from the cross correlation of sequential digital in-line holograms. Opt. Lett. **38**, 4015–4018 (2013)

388. Seo, K.W., Lee, S.J.: High-accuracy measurement of depth-displacement using a focus function and its cross-correlation in holographic PTV. Opt. Express **22**, 15542–15553 (2014)
389. Wu, Y., Wu, X., Yang, J., Wang, Z., Gao, X., Zhou, B., Chen, L., Qiu, K., Grehan, G., Cen, K.: Wavelet-based depth-of-field extension, accurate autofocusing, and particle pairing for digital inline particle holography. Appl. Opt. **53**, 556–564 (2014)
390. Hesseling, C., Homeyer, T., Peinke, J., Gulker, G.: Particle depth position detection by 2D correlation in digital in-line holography. Opt. Lett. **41**, 4947–4950 (2016)
391. Hasegawa, S., Miaki, T.: Whole phase curvature-based particle positioning and size determination by digital holography. Appl. Opt. **59**, 7201–7210 (2020)
392. Grare, S., Allano, D., Coetmellec, S., Perret, G., Corbin, F., Brunel, M., Grehan, G., Lebrun, D.: Dual-wavelength digital holography for 3D particle image velocimetry: experimental validation. Appl. Opt. **55**, A49–A53 (2016)
393. Gao, J., Katz, J.: Self-calibrated microscopic dual-view tomographic holography for 3D flow measurements. Opt. Express **26**, 16708–16725 (2018)
394. Koukourakis, N., Fregin, B., Konig, J., Buttner, L., Czarske, J.W.: Wavefront shaping for imaging-based flow velocity measurements through distortions using a Fresnel guide star. Opt. Express **24**, 22074–22087 (2016)
395. Masuda, N., Ito, T., Kayama, K., Kono, H., Satake, S., Kunugi, T., Sato, K.: Special purpose computer for digital holographic particle tracking velocimetry. Opt. Express **14**, 587–592 (2006)
396. Abe, Y., Masuda, N., Wakabayashi, H., Kazo, Y., Ito, T., Satake, S., Kunugi, T., Sato, K.: Special purpose computer system for flow visualization using holography technology. Opt. Express **16**, 7686–7692 (2008)
397. Toloui, M., Hong, J.: High fidelity digital inline holographic method for 3D flow measurements. Opt. Express **23**, 27159–27173 (2015)
398. Mallery, K., Hong, J.: Regularized inverse holographic volume reconstruction for 3D particle tracking. Opt. Express **27**, 18069–18084 (2019)
399. Shao, S., Mallery, K., Kumar, S.S., Hong, J.: Machine learning holography for 3D particle field imaging. Opt. Express **28**, 2987–2999 (2020)
400. Eisenberg, M.L., Li, S., Behr, B., Pera, R.R., Cullen, M.R.: Relationship between semen production and medical comorbidity. Fertil. Steril. **103**, 66–71 (2015)
401. Rodriguez-Martinez, H.: State of the art in farm animal sperm evaluation. Reprod. Fertil. Dev. **19**, 91–101 (2007)
402. Bianco, V., Mandracchia, B., Marchesano, V., Pagliarulo, V., Olivieri, F., Coppola, S., Paturzo, M., Ferraro, P.: Endowing a plain fluidic chip with micro-optics: a holographic microscope slide. Light-Sci. Appl. **6**, e17055 (2017)
403. Nosrati, R., Gong, M.M., Gabriel, M.C.S., Pedraza, C.E., Zini, A., Sinton, D.: Paper-based quantification of male fertility potential. Clin. Chem. **62**, 458–465 (2016)
404. Frost, J., Cummins, H.Z.: Motility assay of human sperm by photon correlation spectroscopy. Science **212**, 1520–1522 (1981)
405. Earnshaw, J.C., Munroe, G., Thompson, W., Traub, A.I.: Automated laser light scattering system for assessment of sperm motility. Med. Biol. Eng. Comput. **23**, 263–268 (1985)
406. Kanakasabapathy, M.K., Sadasivam, M., Singh, A., Preston, C., Thirumalaraju, P., Venkataraman, M., Bormann, C.L., Draz, M.S., Petrozza, J.C., Shafiee, H.: An automated smartphone-based diagnostic assay for point-of-care semen analysis. Sci. Transl. Med. **9**, eaai7863 (2017)
407. Chen, Y.A., Huang, Z.W., Tsai, F.S., Chen, C.Y., Lin, C.M., Wo, A.M.: Analysis of sperm concentration and motility in a microfluidic device. Microfluid. Nanofluid. **10**, 59–67 (2011)
408. Mortimer, S.T.: CASA-practical aspects. J. Androl. **21**, 515–524 (2000)
409. Amann, R.P., Katz, D.F.: Reflections on CASA after 25 years. J. Androl. **25**, 317–325 (2004)
410. Amann, R.P., Waberski, D.: Computer-assisted sperm analysis (CASA): capabilities and potential developments. Theriogenology **81**, 5–17 (2014)
411. Crenshaw, H.C., Ciampaglio, C.N., McHenry, M.: Analysis of the three-dimensional trajectories of organisms: estimates of velocity, curvature and torsion from positional information. J. Exp. Biol. **203**, 961–982 (2000)

412. Gurarie, E., Grunbaum, D., Nishizaki, M.T.: Estimating 3D movements from 2D observations using a continuous model of helical swimming. Bull. Math. Biol. **73**, 1358–1377 (2011)

413. Corkidi, G., Taboada, B., Wood, C.D., Guerrero, A., Darszon, A.: Tracking sperm in three-dimensions. Biochem. Biophys. Res. Commun. **373**, 125–129 (2008)

414. Sheng, J., Malkiel, E., Katz, J., Adolf, J., Belas, R., Place, A.R.: Digital holographic microscopy reveals prey-induced changes in swimming behavior of predatory dinoflagellates. Proc. Natl. Acad. Sci. U.S.A. **104**, 17512–17517 (2007)

415. Di Caprio, G., Mallahi, A.E., Ferraro, P., Dale, R., Coppola, G., Dale, B., Coppola, G., Dubois, F.: 4D tracking of clinical seminal samples for quantitative characterization of motility parameters. Biomed. Opt. Express **5**, 690–700 (2014)

416. Su, T.-W., Xue, L., Ozcan, A.: High-throughput lensfree 3D tracking of human sperms reveals rare statistics of helical trajectories. Proc. Natl. Acad. Sci. U.S.A. **109**, 16018–16022 (2012)

417. Su, T.-W., Choi, I., Feng, J., Huang, K., McLeod, E., Ozcan, A.: Sperm trajectories form chiral ribbons. Sci. Rep. **3**, 1664 (2013)

418. Daloglu, M.U., Luo, W., Shabbir, F., Lin, F., Kim, K., Lee, I., Jiang, J.Q., Cai, W.J., Ramesh, V., Yu, M.-Y., Ozcan, A.: Label-free 3D computational imaging of spermatozoon locomotion, head spin and flagellum beating over a large volume. Light-Sci. Appl. **7**, 17121 (2018)

419. Su, T.W., Choi, I., Feng, J., Huang, K., Ozcan, A.: High-throughput analysis of horse sperms' 3D swimming patterns using computational on-chip imaging. Anim. Reprod. Sci. **169**, 45–55 (2016)

420. Daloglu, M.U., Ozcan, A.: Computational imaging of sperm locomotion. Biol. Reprod. **97**, 182–188 (2017)

421. Daloglu, M.U., Lin, F., Chong, B., Chien, D., Veli, M., Luo, W., Ozcan, A.: 3D imaging of sex-sorted bovine spermatozoon locomotion, head spin and flagellum beating. Sci. Rep. **8**, 15650 (2018)

422. Curtis, A.S.: The mechanism of adhesion of cells to glass: a study by interference reflection microscopy. J. Cell Biol. **20**, 199–215 (1964)

423. Gingell, D., Todd, I.: Interference reflection microscopy. A quantitative theory for image interpretation and its application to cell-substratum separation measurement. Biophys. J. **26**, 507–526 (1979)

424. Prins, F.A., van Diemen-Steenvoorde, R., Bonnet, J., Velde, I.C.: Reflection contrast microscopy of ultrathin sections in immunocytochemical localization studies: a versatile technique bridging electron microscopy with light microscopy. Histochemistry **99**, 417–425 (1993)

425. Limozin, L., Sengupta, K.: Quantitative reflection interference contrast microscopy (RICM) in soft matter and cell adhesion. ChemPhysChem **10**, 2752–2768 (2009)

426. Lindfors, K., Kalkbrenner, T., Stoller, P., Sandoghdar, V.: Detection and spectroscopy of gold nanoparticles using supercontinuum white light confocal microscopy. Phys. Rev. Lett. **93**, 037401 (2004)

427. Kukura, P., Ewers, H., Muller, C., Renn, A., Helenius, A., Sandoghdar, V.: High-speed nanoscopic tracking of the position and orientation of a single virus. Nat. Methods **6**, 923–927 (2009)

428. Jacobsen, V., Stoller, P., Brunner, C., Vogel, V., Sandoghdar, V.: Interferometric optical detection and tracking of very small gold nanoparticles at a water-glass interface. Opt. Express **14**, 405–414 (2006)

429. Lin, Y.H., Chang, W.L., Hsieh, C.L.: Shot-noise limited localization of single 20 nm gold particles with nanometer spatial precision within microseconds. Opt. Express **22**, 9159–9170 (2014)

430. Spillane, K.M., Ortega-Arroyo, J., de Wit, G., Eggeling, C., Ewers, H., Wallace, M.I., Kukura, P.: High-speed single-particle tracking of GM1 in model membranes reveals anomalous diffusion due to interleaflet coupling and molecular pinning. Nano Lett. **14**, 5390–5397 (2014)

431. Andrecka, J., Ortega-Arroyo, J., Takagi, Y., de Wit, G., Fineberg, A., MacKinnon, L., Young, G., Sellers, J.R., Kukura, P.: Structural dynamics of myosin 5 during processive motion revealed by interferometric scattering microscopy. eLife **4**, e05413 (2015)

432. Andrecka, J., Takagi, Y., Mickolajczyk, K.J., Lippert, L.G., Sellers, J.R., Hancock, W.O., Goldman, Y.E., Kukura, P.: Interferometric scattering microscopy for the study of molecular motors. Methods Enzymol. **581**, 517–539 (2016)

433. Spindler, S., Ehrig, J., Konig, K., Nowak, T., Piliarik, M., Stein, H.E., Taylor, R.W., Garanger, E., Lecommandoux, S., Alves, I.D., Sandoghdar, V.: Visualization of lipids and proteins at high spatial and temporal resolution via interferometric scattering (iSCAT) microscopy. J. Phys. D: Appl. Phys. **49**, 274002 (2016)

434. Wu, H.M., Lin, Y.H., Yen, T.C., Hsieh, C.L.: Nanoscopic substructures of raft-mimetic liquid-ordered membrane domains revealed by high-speed single-particle tracking. Sci. Rep. **6**, 20542 (2016)

435. de Wit, G., Albrecht, D., Ewers, H., Kukura, P.: Revealing compartmentalized diffusion in living cells with interferometric scattering microscopy. Biophys. J. **114**, 2945–2950 (2018)

436. Ortega-Arroyo, J., Kukura, P.: Interferometric scattering microscopy (iSCAT): new frontiers in ultrafast and ultrasensitive optical microscopy. Phys. Chem. Chem. Phys. **14**, 15625–15636 (2012)

437. Krishnan, M., Mojarad, N., Kukura, P., Sandoghdar, V.: Geometry-induced electrostatic trapping of nanometric objects in a fluid. Nature **467**, 692–695 (2010)

438. Huang, Y.F., Zhuo, G.Y., Chou, C.Y., Lin, C.H., Chang, W., Hsieh, C.L.: Coherent brightfield microscopy provides the spatiotemporal resolution to study early stage viral infection in live cells. ACS Nano **11**, 2575–2585 (2017)

439. Huang, Y.F., Zhuo, G.Y., Chou, C.Y., Lin, C.H., Hsieh, C.L.: Label-free, ultrahigh-speed, 3D observation of bidirectional and correlated intracellular cargo transport by coherent brightfield microscopy. Nanoscale **9**, 6567–6574 (2017)

440. Cheng, C.Y., Liao, Y.H., Hsieh, C.L.: High-speed imaging and tracking of very small single nanoparticles by contrast enhanced microscopy. Nanoscale **11**, 568–577 (2019)

441. Hsieh, C.L.: Label-free, ultrasensitive, ultrahigh-speed scattering-based interferometric imaging. Opt. Commun. **422**, 69–74 (2018)

442. Young, G., Kukura, P.: Interferometric scattering microscopy. Annu. Rev. Phys. Chem. **70**, 301–322 (2019)

443. Ambs, P., Bigue, L., Fainman, Y., Binet, R., Collineau, J., Lehureau, J.C., Huignard, J.P.: Image reconstruction using electrooptic holography. Paper presented at the 16th Annual Meeting of the IEEE Lasers and Electro-Optics Society, Tucson, 27–28 October (2003)

444. Yamaguchi, I., Zhang, T.: Phase-shifting digital holography. Opt. Lett. **22**, 1268–1270 (1997)

445. Xiao, X., Javidi, B., Martinez-Corral, M., Stern, A.: Advances in three-dimensional integral imaging: sensing, display, and applications. Appl. Opt. **52**, 546–560 (2013)

446. Chen, C.C., Zhu, C., White, E.R., Chiu, C.Y., Scott, M.C., Regan, B.C., Marks, L.D., Huang, Y., Miao, J.: Three-dimensional imaging of dislocations in a nanoparticle at atomic resolution. Nature **496**, 74–77 (2013)

447. Golab, A., Ward, C.R., Permana, A., Lennox, P., Botha, P.: High-resolution three-dimensional imaging of coal using microfocus X-ray computed tomography, with special reference to modes of mineral occurrence. Int. J. Coal Geol. **113**, 97–108 (2013)

448. Carlsson, K., Aslund, N.: Confocal imaging for 3-D digital microscopy. Appl. Opt. **26**, 3232–3238 (1987)

449. Radfar, E., Jang, W.H., Freidoony, L., Park, J., Kwon, K., Jung, B.: Single-channel stereoscopic video imaging modality based on transparent rotating deflector. Opt. Express **23**, 27661–27671 (2015)

450. Yano, S., Ide, S., Mitsuhashi, T., Thwaites, H.: A study of visual fatigue and visual comfort for 3D HDTV/HDTV images. Displays **23**, 191–201 (2002)

451. Kalender, W.A.: X-ray computed tomography. Phys. Med. Biol. **51**, 29–43 (2006)

452. Martinez-Corral, M., Javidi, B., Martinez-Cuenca, R., Saavedra, G.: Integral imaging with improved depth of field by use of amplitude-modulated microlens arrays. Appl. Opt. **43**, 5806–5813 (2004)

453. Martinez-Cuenca, R., Saavedra, G., Martinez-Corral, M., Javidi, B.: Enhanced depth of field integral imaging with sensor resolution constraints. Opt. Express **12**, 5237–5242 (2004)

454. Dong, Z.C., Andrews, T., Xie, C.M., Yokoo, T.: Advances in MRI Techniques and Applications. Biomed Res. Int. **2015**, 139043 (2015)
455. Huang, D., Swanson, E.A., Lin, C.P., Schuman, J.S., Schuman, J.S., Chang, W., Hee, M.R., Flotte, T., Gregory, K., Puliafito, C.A., Fujimoto, J.G.: Optical coherence tomography. Science **254**, 1178–1181 (1991)
456. Lin, Y., Cheng, C., Poon, T.: Tomographic imaging of a digital holographic microscope. Paper presented at the 2011 Digital Holography and Three-Dimensional Imaging, Tokyo, 9–11 May 2011
457. Lin, Y.C., Chen, H.C., Tu, H.Y., Liu, C.Y., Cheng, C.J.: Optically driven full-angle sample rotation for tomographic imaging in digital holographic microscopy. Opt. Lett. **42**, 1321–1324 (2017)
458. Dierolf, M., Menzel, A., Thibault, P., Schneider, P., Kewish, C.M., Wepf, R., Bunk, O., Pfeiffer, F.: Ptychographic X-ray computed tomography at the nanoscale. Nature **467**, 436–439 (2010)
459. Sheppard, C.J.R.: Three-dimensional phase imaging with the intensity transport equation. Appl. Opt. **41**, 5951–5955 (2002)
460. Ambartsoumian, G., Quinto, E.T.: Generalized Radon transforms and applications in tomography. Inverse Probl. **36**, 020301 (2020)
461. Streibl, N.: Three-dimensional imaging by a microscope. J. Opt. Soc. Am. A **2**, 121–127 (1985)
462. Wolf, E.: Three-dimensional structure determination of semi-transparent objects from holographic data. Opt. Commun. **1**, 153–156 (1969)
463. Maiden, A.M., Humphry, M.J., Rodenburg, J.M.: Ptychographic transmission microscopy in three dimensions using a multi-slice approach. J. Opt. Soc. Am. A **29**, 1606–1614 (2012)
464. Godden, T.M., Suman, R., Humphry, M.J., Rodenburg, J.M., Maiden, A.M.: Ptychographic microscope for three-dimensional imaging. Opt. Exp. **22**, 12513–12523 (2014)
465. Suzuki, A., Furutaku, S., Shimomura, K., Yamauchi, K., Kohmura, Y., Ishikawa, T., Takahashi, Y.: High-resolution multislice x-ray ptychography of extended thick objects. Phys. Rev. Lett. **112**, 053903 (2014)
466. Horstmeyer, R., Chung, J., Ou, X.Z., Zheng, G., Yang, C.H.: Diffraction tomography with Fourier ptychography. Optica **3**, 827–835 (2016)
467. Jiang, Z.L., Veetil, S.P., Liu, C., Zhu, J.Q.: Depth resolved imaging by digital holography with an illumination of constantly changing curvature. Opt. Lett. **40**, 3001–3004 (2015)
468. Zhang, X.D., Liu, C., Zhu, J.Q.: K-domain transform based three-dimensional microscopy. Appl. Phys. Lett. **113**, 221106 (2018)
469. Liu, C., Walther, T., Rodenburg, J.M.: Influence of thick crystal effects on ptychographic image reconstruction with moveable illumination. Ultramicroscopy **109**, 1263–1275 (2009)
470. Chang, C.C., Pan, X.C., Tao, H., Liu, C., Veetil, S.P., Zhu, J.Q.: 3D single-shot ptychography with highly tilted illuminations. Opt. Express **29**, 30878–30891 (2021)

Chapter 6
Recent Trends in Computational Optical Phase Imaging

In recent years, phase imaging has experienced tremendous success and has opened up new exciting frontiers. As an example, deep learning algorithms are capable of learning patterns and rules from a series of data sets in order to perform specific tasks. Computational optical phase imaging records light intensity $I(x, y)$ like any other optical imaging. When the detector is perfectly ideal in both spatial resolution and dynamic range, the data acquisition can be mathematically written as Eq. (6.1), in which $O(x_0, y_0, z_0)$ is the complex amplitude of object and A is the operating matrix decided by the optical system.

$$I(x, y) = |AO(x_0, y_0, z_0)|^2 \tag{6.1}$$

As an example, A is the Fresnel diffraction integral of $O(x_0, y_0, z_0)$ in coherent diffraction imaging, while acting as the imaging function of $O(x_0, y_0, z_0)$ in transport of intensity phase imaging. The goal of computational phase imaging is essentially to find a corresponding matrix B to obtain $O(x_0, y_0, z_0)$ as Eq. (6.2).

$$O(x_0, y_0, z_0) = B|AO(x_0, y_0, z_0)|^2 \tag{6.2}$$

B is available for holography, interferometry, and transport of intensity phase imaging with proper approximations and the phase of the object $O(x_0, y_0, z_0)$ can be obtained analytically from recorded intensity $I(x, y)$. However, this is not the case with coherent diffraction imaging where the phase of object $O(x_0, y_0, z_0)$ is to be numerically reconstructed using a time-consuming iterative approach. Through deep learning, it is possible to generate a numerical estimate of B based on training data sets, thus enabling a faster computation of the target phase.

The rapid improvements in consumer electronics and wireless communications could allow smartphones to be developed into low-cost and portable optical imaging devices and can provide simpler, smarter, and more robust computational phase imaging. As a result, primary health care providers will be able to improve their screening and detection practices. In this chapter we discuss recent advances in phase imaging in conjunction with deep learning and point-of-care microscopy.

6.1 Deep Learning in Computational Optical Phase Imaging

Recently, deep learning approaches have been employed in computational optical phase imaging and its applications as illustrated in Fig. 6.1. These methods are complementary to many classical ones. In this short section, we review works related to deep learning in computational optical phase imaging. This short section is intended not only as a new tutorial for deep learning-based computational optical phase imaging, but also to inform readers that deep learning-based imaging is an essential part of imaging and photography.

6.1.1 Deep Learning Used in Phase Retrieval

Phase retrieval is one application of deep learning in computational optical phase imaging. Different optical phase imaging techniques have been reported, including coherent diffraction imaging, digital holography, transport of intensity phase imaging, Shack-Hartmann sensing, and even ghost imaging. The Barbastathis group introduced deep neural networks to lensless phase imaging by solving end-to-end inverse problems in computational imaging [1]. Further, they considered the spatial frequency of training database in order to enhance the spatial resolution of the deep learning-based phase retrieval [2]. Most importantly, they proved that training is

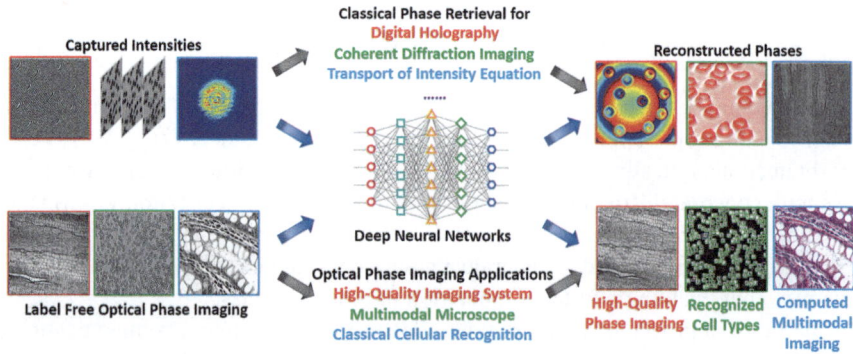

Fig. 6.1 Deep learning used in computational optical phase imaging and its applications

important to deep learning-based computational optical phase imaging and a well-trained neural network can learn the underlying physical model [3], which validates that deep learning-based phase retrieval can be used in practice.

Ozcan group adopted deep learning in digital holographic microscopy for phase and amplitude reconstruction while significantly eliminating twin-image and self-interference-related spatial artifacts [4–6]. Lam group proposed learning frameworks for holographic reconstruction [7, 8] and capsule network to deal with limited computational conditions [9]. Situ group proposed learning-based end-to-end approach for in-line digital holographic reconstruction [10]. Moon group employed both convolutional neural network [11] and conditional generative adversarial model [12] for holographic reconstruction. Zhao group adopted Y-Net in digital holographic microscopy for both phase and amplitude reconstruction [13, 14]. Lee group reported deep learning-based holographic reconstruction for high-fidelity, low-light quantitative phase imaging [15]. Deep learning-based phase retrieval has been successfully experimented in coherent diffraction imaging [16–18]. A combination of deep learning and ptychography enhances the quality of computational optical phase imaging. For example, in ptychographical iterative engine, Cohen group proposed a deep learning-based single-shot ptychography reconstruction method [19], and Barbastathis group reported deep learning based-coherent modulation imaging [20]. Moreover, in Fourier ptychographic microscopy, Zheng group first modeled the Fourier ptychographic microscopy using convolutional neural network and recovered the complex object information using trained network [21]. Moreover, several groups [22–26] proposed neural network models for Fourier ptychographic microscopic reconstruction. Deep learning has been experimented in transport of intensity phase microscopy [27], Shack-Hartmann sensing [28–30], illumination coded imaging [31], and ghost phase imaging with a point detector [32]. Moreover, Tian group designed Bayesian convolutional neural network which can quantify the uncertainty of deep learning predictions [33]. Waller group reported the untrained deep neural network based on quantitative phase imaging [34]. Deep learning has been further extended to phase unwrapping [35–37].

6.1.2 Deep Learning Used in Computational Optical Phase Imaging Applications

Besides phase retrieval, deep learning has also been employed in many computational optical phase imaging applications. A direct application is specimen recognition and classification. Ozcan group introduced deep learning techniques into digital holography platform for label-free detections on parasites in bodily fluids [38], bioaerosol [39], rare cells [40] and particle aggregation [41]. Besides, based on the obtained optical phase images, several groups [42–45] recognized and classified cells from phase images using deep learning. Moreover, combining quantitative phase imaging and machine learning, it was not only possible to discern healthy B cells from lymphoblasts, but also to classify stages of B cell in acute lymphoblastic leukemia [46]. Popescu group recognized head, midpiece, and tail of a sperm from its phase images [47]. Deep learning-based approaches were very useful in diagnosing several diseases from red blood cell phase images [48–51].

Deep learning can also enhance the performance of computational optical phase imaging. For example, with limited projection angles, Barbastathis group still obtained high-resolution phase tomography of dense layered objects using deep neural networks [52] and Park group used cycle generative adversarial network to reduce coherent noise in three-dimensional quantitative phase imaging [53]. Barbastathis group constructed deep neural networks to recover phase in weak light illumination conditions [54–56]. Additionally, retrieved wavefront images from computational optical phase imaging can also be numerically transformed to images in other modes such as bright field, fluorescence, polarization, and chemical stain imaging through deep learning. Thus, the microscope with a single imaging mode can be extended to multi-modal imaging, allowing it to significantly improve its imaging capabilities. Ozcan group first provided the idea of PhaseStain [57], in which digital H&E image can be computed from specimen phase images. With Deep learning, the same group has also obtained bright field [58, 59], color [60], and polarization images [61] from holographic reconstructions. Besides, Shaked group obtained virtually stained images from holographic reconstructed label-free isolated biological cells [62]; Zheng group obtained virtual bright-field and fluorescence staining images from Fourier ptychographic microscopy reconstruction [63]; and Popescu group presented phase imaging with computational specificity as a combination of quantitative phase imaging and deep learning, which provides high specificity for unlabeled live cells [64].

6.1.3 Discussions

More and more works are published in the last few years on the combination of computational optical phase imaging and deep learning techniques [65–68]. These works can be roughly classified into two categories: one is deep learning-based phase retrieval, and the other is deep learning applied to computational optical phase imaging. In these fields, the function of deep learning is to construct a relationship between input and output. These fields rely on deep learning methods to determine the relationship between the input distributions as phase distributions and the output distributions as intensity distributions. Moreover, deep learning methods can be applied to computational optical phase imaging applications such as target recognition and classification to retrieve the relation between retrieved phase distributions and the recognized and classified results; and, in the application of imaging mode transformation, deep learning methods can determine the relation between different imaging modes.

Deep learning has been shown to perform well in computational optical phase imaging, but it does have some drawbacks. Its main disadvantage is that it requires high-quality training data. Reconstruction quality is directly affected by it, and it is often difficult to obtain high-quality training data. The training process itself is lengthy and complex. Secondly, it often has a poor generalization owing to the fact that the reconstructed relation between input and output is specific to the systems that generated the training data. In contrast to most conventional phase retrieval methods that do provide analytical and iterative solutions for a definite relation between phase and intensity using matured algorithms, deep learning methods require massive training data collection and heavy training that are not as cost-effective to provide the same quality as classical methods. Nevertheless, deep learning is useful in certain settings, for instance, it can suppress twin images in holographic reconstruction while providing a better performance than classical approaches. Deep learning can offer specimen phase in satisfactory quality even at low signal to noise ratios. Further, it is suitable for target recognition and classification, especially in medical imaging. In imaging mode transformation, deep learning is preferable since relationships between various imaging modes are difficult to establish, a problem that can be solved by using massive training data. Deep learning is an advantage in computational optical phase imaging when the relationship between input and output is hard to reconstruct. Deep learning-based imaging is an important part of computational imaging and photography, but not all of it.

The applications of deep learning to computational optical phase imaging are examined in this short section. This technique is often used in phase retrieval, and it has been used in almost all computational optical phase imaging techniques. Furthermore, deep learning in computational optical phase imaging applications has been shown to enhance optical phase imaging performance in several examples. In conclusion, both pros and cons of deep learning-based computational optical phase imaging are discussed. However, deep learning may not be a suitable solution in many phase retrieval applications, not only due to the fact that it relies on large volume of high-quality training data, but also because of the existing phase retrieval algorithms, both analytical and iterative, are already capable of accurate phase reconstruction. However, deep learning tactics still have their place when it is hard to establish the relations between input and output.

6.2 Point-of-Care Computational Optics Phase Imaging

The ability of microscopes to provide direct observations and quantitative measurements on various kinds of samples in high resolution and contrast has always made them widely used in biological research and medical diagnostics. Unfortunately, due to their large sizes and expensive costs, commercial microscopes are often too large and expensive for on-site use. To expand the application range, various kinds of compact microscopes have been designed, and most of them are capable of satisfying the sample-in-analysis-out requirement, thus supporting point-of-care applications. For example, Zheng group constructed an open-source, fully-automated microscope with trans- and epi-illumination, polarization contrast, and fluorescence

[69], Collins et al. built a microscope with trans- and epi-illumination, polarization contrast, and florescence [70], and Tkaczyk group constructed small, inexpensive, and high-throughput fluorescence microscopes [71, 72]. In addition to these traditional compact microscopes, smartphone microscopy is also a potential solution for building portable or even handheld point-of-care imaging devices. Recent smartphones are not only equipped with high-quality CMOS sensors for image recording, but also with powerful processors for displaying and processing images, depending on the sample-in-analysis-out scenario.

A typical smartphone microscope design utilizes a micro-objective for image magnification, and an eyepiece connects the micro-objective to the lens of the smartphone camera. Using such configurations, Fletcher proposed Cellscopes successfully used in mobile health [73–75]. The Chan group used the design to develop a smartphone fluorescence microscope to diagnose patients who had human immunodeficiency virus (HIV) or hepatitis B virus (HBV) [76]. Wang and Liu et al. also adapted the configuration for smartphone microscopes to capture bright-field and fluorescence images for cell counting [77] and single molecule mercury ion imaging [78], respectively. In addition, more compact lenses were used instead of the pair of micro-objective and eyepiece to further reduce the size. As an example, Fletcher group improved their Cellscope by reversing the smartphone camera lens [79, 80]. Moreover, the Ozcan group has developed a series of smartphone fluorescence microscopes using such configurations to image cells [81], pathogens [82–85], viruses [86] and even DNA [87, 88]. Smartphone microscopes were developed that are suitable for various situations. A few examples are oil droplet-based smartphone microscope [89], camera flash and ambient light-based smartphone microscope [90], smartphone hyper-spectral microscope [91], handheld smartphone confocal microscope [92], and plasmonics enhanced smartphone fluorescence microscope [93]. Deep learning was also used to significantly improve the imaging performance of smartphone microscope [94].

A new generation of compact chip-size microscopes has also been developed. A series of chip-size microscopes designed by Yang group are lens-free [95–104]. Using pinholes in an array, these chip-size optofluidic microscopes can achieve extremely high resolution due to the fact that the pinhole size determines the imaging resolution rather than the pixel size. Furthermore, the pinhole array can also be replaced with a Fresnel zone plate array [99]. While in their updated version, optofluidic scanning and pinhole/Fresnel zone plate array can be replaced by illuminations in multiple directions to achieve sub-pixel imaging resolution [100] based on the pixel super resolution algorithm [101, 102]. The chip-size microscope not only effectively reduces the sizes of microscopes, but also achieves satisfactory imaging resolution for multiple applications [103–105]. The same tactic is applied in a smartphone microscope, but the camera lens should be removed [106]. A point-of-care microscope from the Prakash group is different from previous point-of-care microscopes because it has a paper-based configuration and is relatively cheap [107, 108], while maintaining its many imaging modes and high imaging quality with 140× magnification and 2 μm resolution.

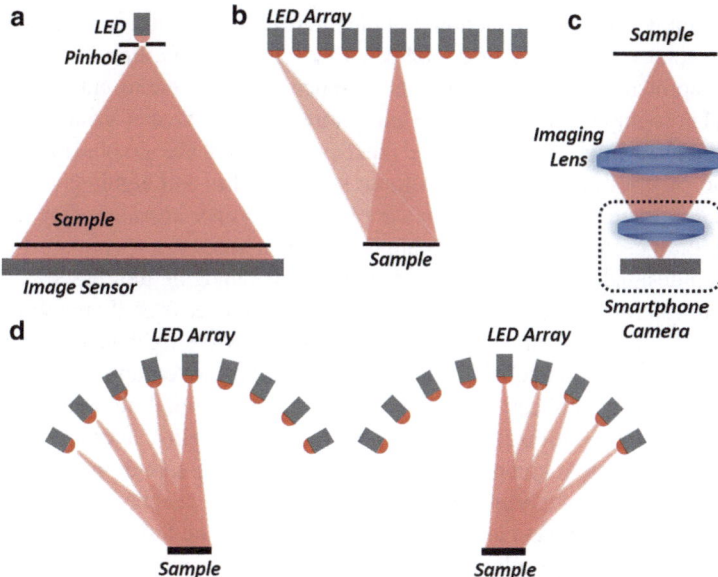

Fig. 6.2 Classical schemes in point-of-care computational optical phase microscopes. **a** Point-of-care digital holographic microscopy; **b** point-of-care ptychographical microscopy; **c** point-of-care transport of intensity phase microscopy; **d** point-of-care differential phase contrast microscopy

The point-of-care microscopes listed above all use bright-/dark-field and fluorescence imaging. In contrast with amplitude imaging, phase imaging often provides higher contrast, especially in label-free situations, so computational optical phase microscopy is preferable to amplitude imaging in such cases. A variety of point-of-care computational optical phase microscopes are discussed in this section, including those based on digital holography, ptychography, transport of intensity phase microscopy, and differential phase contrast microscopy. Figure 6.2 lists the classical schemes in point-of-care computational optical phase microscopes. Each of these point-of-care computational optical phase microscopes provides quantitative phase images with simple and portable structures; more importantly, they can support sample-in-analysis-out detection systems.

6.2.1   Point-of-Care Digital Holographic Microscopy

Among various phase imaging techniques, digital holographic microscopy is the mostly widely used since it has pretty high accuracy but only requires relatively simple devices and operations. Various interferometric systems such as on-axis, off-axis and even slightly off-axis configurations and different phase retrieval algorithms such as phase shifting, spatial filtering, and iteration ones have been designed.

However, many classical digital holographic microscopes are still constructed based on commercial microscopes or complicated systems, still limiting their potentials in on-site applications. Several attempts have been made in the recent past to make the digital holographic microscopy systems more compact. Shaked group designed τ interferometers, which are portable and inexpensive digital holographic microscopes in on- and off-axis geometries and without strict stability and highly coherent illumination conditions [109, 110]. Both sample and reference arms are integrated in a compact volume, and the reference beam is easily generated using a pinhole filtering. Park group designed a white-light quantitative phase imaging unit based on lateral shearing interferometry, providing a cost-effective and user-friendly phase imaging tool [111]. However, both the τ interferometers and the white-light quantitative phase imaging unit still rely on commercial microscopes, thus not completely solving the problem for on-site application. A typical point-of-care digital holographic microscopes should integrate the systems of illumination, imaging, interferometry, and recording system into a compact structure. Many point-of-care digital holographic microscopy designs have been described, and they can be roughly divided into two categories: lenses-based and lens-free designs. Although the former has a high resolution while the latter has a large field of view, both are compact, inexpensive, and have been successfully used for a variety of purposes.

Several lens-based point-of-care digital holographic microscopes have been proposed by the Javidi group [50, 112–115]. Their work [112] followed the classical Mach–Zehnder interferometer configuration but compacted the optical system from coherent source to image recorder, resulting in a portable digital holographic microscope. The system contains a commercial micro-objective, but in order to further enhance the resolution, an extra microsphere is also adopted as the secondary micro-objective to pursue higher resolution. Unfortunately, the system is still slightly large and complicated. In order to further compact the digital holographic microscopes, Javidi group reported several lateral shearing-based ones [50, 113–115], which are quite different from the traditional digital holographic microscopy design. These designs divide the magnified wavefront into two, but with lateral shearing by using a glass plate, these two wavefronts can interfere to produce a shearing hologram, which is captured by a camera [113]. The common path design not only avoids the extra setting of the reference arm, but also remarkably compacts the optical system. When samples (such as cells) are compressed distributed in the FoV, the sample phase can be directly obtained thus avoiding the extra integral computation since in these conditions, the shearing interference is actually generated by sample and background. With its micro-objective, the lateral shearing holographic microscope not only offers higher resolution and a capacity for high-quality cell imaging, but also supports live cell dynamic imaging due to the single-shot phase retrieval capability. With its satisfactory spatio-temporal resolution, live red blood cells were observed dynamically, and the spatio-temporal information obtained from the point-of-care digital holographic microscope supports the diagnosis of sickle cell disease successfully [114]. Moreover, besides integrating with a CCD/CMOS camera, the lateral shearing configuration is also extended to a point-of-care smartphone digital holographic microscopy system with an extra eyepiece [115]. Additionally, deep learning is also adopted in this system for cell identification and disease diagnosis to pursue high efficiency and accuracy [50].

The introduction of a micro-objective can improve the resolution of imaging, but it inevitably limits the field of view at the same time. To observe more samples with a large field of view, lensless digital holographic microscopy is preferred. Ozcan group is a pioneer in this area and has already developed many designs, methods, and applications. While the optical configuration of these lensfree digital holographic microscopes is rather simple as schemed in Fig. 6.2a, the monochromator and other independent optical elements make the system large and complicated, thus making them unsuitable for point-of-care applications. With the goal of further compressing the system, Ozcan group has proposed a series of point-of-care designs [116, 117]: the partially coherent source is often an LED filtered by a pinhole to obtain both high temporal and spatial coherence, the sample is directly set above a CMOS camera chip, and all of these elements are integrated using 3D printing; furthermore, with the mature wavefront reconstruction methods, both sample amplitude and phase distributions can be retrieved. In addition, such lensfree design has been extended to smartphone digital holographic microscopes [118] and tomographic microscopes [119]. Due to the lack of a lens in the system, imaging aberration is completely avoided, plus the field of view is remarkably enlarged to more than 20 mm^2, but the resolution is still limited. Ozcan group has proposed a range of solutions to improve resolution. With enough iterations, the sample amplitude and phase can be reconstructed by back and forth propagation between the captured multi-height holograms. A rotating axial shift stage is employed in the design of digital holographic microscopes for point-of-care use [120, 121] in order to change the distance between the sample and the CMOS camera chip for multi-height hologram recording. Another improvement suggests reconstruction of super-resolution hologram from multiple low resolution hologram but with sub-pixel shifting reported earlier [100]. Using multiple fiber-optic waveguides from LED arrays to generate shifted illuminations, low resolution holograms shifted at sub-pixel level is recorded, and the super-resolution hologram is reconstructed for super-resolution amplitude and phase retrieval [122]. Another approach uses a nanolens as the near-field micro-objective to collect higher spatial frequency light for resolution enhancement [123], which shares the similar idea reported earlier [112]. Moreover, these point-of-care digital holographic microscopes have also been successfully applied in several applications, such as parasite detection [116], cell analysis [117, 124, 125], bacterial sensing [126], and bioaerosol testing [39]. Besides, Im et al. have also designed Raspberry Pi [127] and smartphone [128]-based point of care digital holographic microscopes for the detections of lymphoma cells. Moreover, Allier group has designed point-of-care digital holographic microscopes focusing on dynamic cell observations [129–131], and these video microscopes have been successfully used in cell monitoring inside the incubator.

In addition to their simplicity and high accuracy, these point-of-care digital holographic microscopes have the same advantages as the digital holograph. Additionally, compact systems with micro-objectives can often achieve rather high resolutions in wavefront reconstruction, while lensfree configurations offer extremely large fields of view. Point-of-care digital holographic microscopes have been successfully used in a variety of applications, including cell imaging, showing their potential in biological and medical fields.

6.2.2   Point-of-Care Ptychographic Microscopy

(Point-of-care) digital holographic microscopes are still unable to achieve high resolution and wide field of view simultaneously. Since ptychographic microscopy can extract sample phase distributions not only in high resolution but also with large fields of view, it is often used in conditions that require both. It has been widely used in various applications, including biological and medical observations, but it relies on commercial microscopes or scanning devices, which are huge, expensive, and complex, making it unsuitable for point-of-care use. Three point-of-care ptychographic microscopes were developed, including miniatures of a Fourier ptychographic microscope, a ptychographic iterative engine microscope, and a structured illumination microscope, and are briefly described here.

Through the use of multiple illuminations in different directions, Fourier ptychographic microscopy can reconstruct phase images with a resolution that goes beyond the limits of the numerical aperture. Based on this principle, Zheng group designed a FPscope [132] by using an LED array similar to classical Fourier ptychographic microscope, while replacing the often used micro-objective with a cellphone lens. The LED array, the cellphone lens, and the image recorder are integrated in a 3D printed structure. Using such a low-cost and simple design, the system could reach sub-micron resolution and extremely large FoV of 0.88 mm^2. Harvey group improved the FPscope by integrating with a Raspberry Pi single-board computer for camera and illumination LED array control and performed autonomous data acquisition [133]. Moreover, LED position calibration and aberration correction are also considered in order to pursue higher accuracy and resolution. The improved FPscope can reach a resolution of 780 nm with a FoV of 4 mm^2. Moreover, Yang group also designed a point-of-care ptychographic microscope with a resolution of 1.26 μm and a FoV of ~24 mm^2 and used in incubator called EmSight [134]. Different from classical ptychographic microscopes that provide amplitude and phase images, EmSight also offers fluorescence image to extract more details of the cells during long-time incubation; it successfully tracked movements of dopaminergic neurons over a 21-day period using different imaging modes. All these designs use LED array as the illumination system as schemed in Fig. 6.2b.

In contrast to Fourier ptychographic microscopy based on iterations in the frequency domain, ptychographic iterative engine relies on iterations in the spatial domain and can reach diffraction limit and extremely large FoV. Various works have been proposed on ptychographic iterative engine including reconstruction algorithm optimizations, optical system improvements, and increasing applications. However, most of these works still rely on huge and complicated systems limiting the on-site applications. Following the classical ptychographic iterative engine system, Zheng group simplified it and designed a point-of-care ptychographic iterative engine microscope [135]. It still has a lensless configuration composed of coherent light source, an image recorder and a two-dimensional scanning system, all integrated by a 3D printing structures. Through recording a series of diffraction patterns via scanning the sample, phase image with a resolution of 780 nm and the extremely large FoV that is limited by the scanning system can be reconstructed. Based on the fundamental functions on wavefront imaging, the proposed point-of-care ptychographic iterative engine microscope can perform effective cell segmentation and obtain digital refocusing on thick biological samples, showing a potential for whole slide and thick sample imaging.

Similar to Fourier ptychographic microscopy, structure illumination microscopy can also enhance the imaging resolution by extending the sample spatial frequency distribution. Zheng group designed a lensless point-of-care ptychographic microscope using speckle illumination [136]. The speckle illumination light is obtained from a laser diode passing through a diffuser. A cost-effective galvo scanner deflects the speckle illumination in different directions to generate multiple illuminations on the sample. The sample is directly located above the image recorder chip which is used to record a series of images corresponding to different speckle illuminations relying on galvo scanning. Using the ptychographic iterative engine-based reconstruction algorithm, both the speckle illuminations and sample can be reconstructed through iterations in the spatial domain. Since no lens is required in the system, the point-of-care structure illumination microscope not only avoids the imaging aberrations, but also has a pretty large FoV.

A main advantage of these point-of-care ptychographic microscopes is their ability to achieve high resolution (even close to the diffraction limit) while maintaining a wide field of view at the same time. For this reason, they are preferred in applications such as slide imaging requiring both imaging resolution and field of view. As scanning and recording is required for accurate imaging, these point-of-care ptychographic microscopes require precision systems for accurate scanning and a stable imaging environment; therefore, they are not suitable for imaging applications requiring high speed or dynamic observations.

6.2.3 Point-of-Care Transport of Intensity Phase Microscopy

The principle of transport of intensity phase imaging entails taking a series of multi-focal images, computing the intensity derivative, then solving the Poisson equation to obtain a quantitative phase distribution. Transport of intensity phase imaging was earlier used in short-wavelength (X-ray) imaging. Since transport of intensity phase imaging can be combined with commercial optical microscopes, especially with the development of real-time transport of intensity phase imaging, it has found several applications in cell and tissue observations. There have been few reports of point-of-care transport of intensity phase microscopy, unlike digital holography and ptychography, which have many designs. This could be because it is rather difficult to capture high-quality multi-focal images using only simple and compact systems. However, two of the works reported point-of-care transport of intensity phase microscopes both relying on smartphones but using different tactics for multi-focal image recording.

Wang group designed a point-of-care transport of intensity phase microscope [137] with a configuration similar to smartphone bright-field microscope that is composed of transmitted-illumination LED light source, a micro-objective, an eyepiece, and a smartphone. Figure 6.2c reveals such scheme. However, more than a smartphone-based bright-field microscope that records image at a fixed focal position, the proposed smartphone transport of intensity phase microscope records multi-focal images by using the manual focusing function of the smartphone for intensity derivative computation. Therefore, the quantitative sample phase can be reconstructed by solving the Poisson equation using the smartphone application. However, there are still two problems in manual focusing-based point-of-care transport of intensity phase microscope; the relation between the manual focusing of the smartphone application and the actual focal shifting of the system is unknown, and there exists scaling error among multi-focal images induced by smartphone camera lens during focal adjustment. In order to solve the first problem, the relation between the manual focusing and the actual focal shifting needs to be calibrated. This is done by shifting a sample (often a resolution test chart) by placing it on a precision translation stage along the imaging axis to simulate different imaging planes. By shifting the sample to specific imaging plane, the in-focus image can be obtained by adjusting the manual focusing to a specific value. Shifting the sample equidistantly to different imaging planes, its corresponding manual focusing values can be obtained for fitting a relation between the manual focusing of the smartphone application and the actual focal shifting of the system. During this calibration process, the scaling is also analyzed according to the regular pattern size changes. With calibrated defocus relation and scaling, it is then possible to retrieve high quality image from recorded and compensated multi-focal images.

Different from the above design, Zhan group proposed a single-shot-based point-of-care transport of intensity phase microscope [138]. The system is rather simple without the requirement of focal adjustment. Its main configuration is a bright-field smartphone microscope with a single lens instead of using a pair of micro-objective and eyepiece. Moreover, a distorted micro-grating is inserted into the imaging system, which aims at providing multi-focal images on one imaging plane. This smartphone can capture three images, an in-focus image in the center, and two under- and over-focus images on either side. By solving the Poisson equation, sample phase image can be reconstructed from single-shot recorded multi-focal images. This single-shot point-of-care transport of intensity phase microscope can realize real-time phase imaging, however, it suffers from small FoV.

The point-of-care transport of intensity phase microscopes can be used for biological and medical observations, such as blood testing, which are both discussed in these works. When compared to the previously introduced point-of-care digital holographic and ptychographic microscopes, the best advantage with point-of-care transport of intensity phase microscopy is that it has the rapid and simple phase retrieval speed, without phase unwrapping or wavefront iteration, therefore, it can provide phase images in fast speed even when using low-end processors. Its accuracy, however, is slightly lower than the previous introduced devices. Phase imaging accuracy can be improved by capturing more multi-focal images or by using iterations, but they complicate the system design and increase data processing load. Point-of-care phase microscopy is suitable when fast imaging is required.

6.2.4 Point-of-Care Differential Phase Contrast Microscopy

Differential phase contrast microscopy, a newly developed quantitative phase imaging technique, can reconstruct sample phase based on image intensity with multi-angle illuminations created by sequentially patterning the LED array source [139–141]. Waller's group optimized the Cellscope by using a domed LED array to compact differential phase contrast microscopy [142]. Quantitative phase can be reconstructed by solving the inverse problem as a simple deconvolution in Fourier domain.

The point-of-care differential phase contrast microscope is only an updated version of classical bright-field microscope by using programmable LED array instead of Kohler illumination as schemed in Fig. 6.2d. Therefore, the system even becomes simpler. Though multiple image recording is still required, the captured image number is significantly less than those required in point-of-care ptychographic microscopes. The optical system is fixed without any mechanical shifting, successfully avoiding the FoV mismatch in point-of-care transport of intensity phase microscopes. Moreover, contrast enhanced images can also be computed from captured images with multi-angle illuminations, providing more sample perspectives. Considering these advantages, the differential phase contrast microscope is suited for point-of-care device development and can also be extended for quasi-real-time phase imaging. Moreover, point-of-care differential phase contrast microscope can also share the same system as point-of-care Fourier ptychographic microscope, and the phase obtained by differential phase contrast can be used as the initial guess in Fourier ptychographic microscopy in order to accelerate the reconstruction speed [143].

6.2.5 Discussions

Different types of point-of-care computational optical phase microscopes are discussed in this section, including digital holographic microscopes, ptychographic microscopes, transport of intensity phase microscopes, and differential phase contrast microscopes. Due to their compact design, the systems can be efficiently reduced in size and price, and can provide high-quality quantitative phase imaging in a sample-in-analysis-out mode. Thus, on-site optical phase imaging with point-of-care phase microscopy expands the range of applications.

Today, point-of-care microscopes can be created using portable or even hand-held electronic devices (image recorders, light sources, microcomputers, and smartphones). The point-of-care microscopes are equipped with image recorders such as smartphone or industrial cameras that are connected to computers, allowing for further processing of the image to improve the picture quality (resolution and contrast enhancement) and reflect more sample details (cell recognition, segmentation and counting) and provide more sample perspectives (phase and fluorescence imaging) with multiple computational imaging techniques. In this fashion, these point-of-care microscopes are often "smart" microscopes, referring to the compact design as well as the extensive imaging capabilities, both of which support high-quality biological and medical observations.

These "smart" point-of-care microscopes are promising diagnostic tools due to their cost-effective prices, compact designs, and high-quality imaging performance, especially in underdeveloped community hospitals unable to afford expensive commercial microscopes. Even though it provides high-quality images in different perspectives, they cannot provide medical diagnostics, which require a trained physician. While mobile health, such as telemedicine and telediagnosis, can partially solve the problem, a remote professional doctor is still necessary. A possible solution in medical diagnostics is artificial intelligence/machine learning, which can significantly reduce labor costs. Therefore, when these "smart" point-of-care microscopes are upgraded to "intelligent" point-of-care microscopes equipped with artificial intelligence diagnostic algorithms, they can be used for disease diagnostics in underdeveloped community hospitals. A future direction in this field may be to combine "smart" point-of-care microscopes and artificial intelligence/machine learning as "intelligent" point-of-care microscopes for on-site medical diagnosis.

References

1. Sinha, A., Lee, J., Li, S., Barbastathis, G.: Lensless computational imaging through deep learning. Optica **4**, 1117–1125 (2017)
2. Li, S., Barbastathis, G.: Spectral pre-modulation of training examples enhances the spatial resolution of the phase extraction neural network (PhENN). Opt. Express **26**, 29340–29352 (2018)
3. Deng, M., Li, S., Zhang, Z., Kang, I., Fang, N.X., Barbastathis, G.: On the interplay between physical and content priors in deep learning for computational imaging. Opt. Express **28**, 24152–24170 (2020)
4. Rivenson, Y., Zhang, Y., Gunaydin, H., Teng, D., Ozcan, A.: Phase recovery and holographic image reconstruction using deep learning in neural networks. Light-Sci. Appl. **7**, 17141 (2018)
5. Wu, Y., Rivenson, Y., Zhang, Y., Wei, Z., Gunaydin, H., Lin, X., Ozcan, A.: Extended depth-of-field in holographic imaging using deep-learning-based autofocusing and phase recovery. Optica **5**, 704–710 (2018)
6. Rivenson, Y., Wu, Y., Ozcan, A.: Deep learning in holography and coherent imaging. Light-Sci. Appl. **8**, 85 (2019)
7. Ren, Z., Xu, Z., Lam, E.Y.: Learning-based nonparametric autofocusing for digital holography. Optica **5**, 337–344 (2018)

8. Ren, Z., Xu, Z., Lam, E.Y.: End-to-end deep learning framework for digital holographic reconstruction. Adv. Photonics **1**, 016004 (2019)

9. Zeng, T., So, H.K.H., Lam, E.Y.: RedCap: residual encoder-decoder capsule network for holographic image reconstruction. Opt. Express **28**, 4876–4887 (2020)

10. Wang, H., Lyu, M., Situ, G.: eHoloNet: a learning-based end-to-end approach for in-line digital holographic reconstruction. Opt. Express **26**, 22603–22614 (2018)

11. Jaferzadeh, K., Hwang, S.H., Moon, I., Javidi, B.: No-search focus prediction at the single cell level in digital holographic imaging with deep convolutional neural network. Biomed. Opt. Express **10**, 4276–4289 (2019)

12. Moon, I., Jaferzadeh, K., Kim, Y., Javidi, B.: Noise-free quantitative phase imaging in Gabor holography with conditional generative adversarial network. Opt. Express **28**, 26284–26301 (2020)

13. Wang, K., Dou, J., Qian, K., Di, J., Zhao, J.: Y-Net: a one-to-two deep learning framework for digital holographic reconstruction. Opt. Lett. **44**, 4765–4768 (2019)

14. Wang, K., Qian, K., Di, J., Zhao, J.: Y4-Net: a deep learning solution to one-shot dual-wavelength digital holographic reconstruction. Opt. Lett. **45**, 4220–4223 (2020)

15. Zhang, Z., Zheng, Y., Xu, T., Upadhya, A., Lim, Y.J., Mathews, A., Xie, L., Lee, W.M.: Holo-UNet: hologram-to-hologram neural network restoration for high fidelity low light quantitative phase imaging of live cells. Biomed. Opt. Express **11**, 5478–5487 (2020)

16. Bai, C., Zhou, M., Min, J., Dang, S., Yu, X., Zhang, P., Peng, T., Yao, B.: Robust contrast-transfer-function phase retrieval via flexible deep learning networks. Opt. Lett. **44**, 5141–5144 (2019)

17. Wang, F., Bian, Y., Wang, H., Lyu, M., Pedrini, G., Osten, W., Barbastathis, G., Situ, G.: Phase imaging with an untrained neural network. Light-Sci. Appl. **9**, 77 (2020)

18. Cherukara, M.J., Zhou, T., Nashed, Y., Enfedaque, P., Hexemer, A., Harder, R.J., Holt, M.V.: AI-enabled high-resolution scanning coherent diffraction imaging. Appl. Phys. Lett. **117**, 044103 (2020)

19. Wengrowicz, O., Peleg, O., Zahavy, T., Loevsky, B., Cohen, O.: Deep neural networks in single-shot ptychography. Opt. Express **28**, 17511–17520 (2020)

20. Kang, I., Zhang, F., Barbastathis, G.: Phase extraction neural network (PhENN) with coherent modulation imaging (CMI) for phase retrieval at low photon counts. Opt. Express **28**, 21578–21600 (2020)

21. Jiang, S., Guo, K., Liao, J., Zheng, G.: Solving Fourier ptychographic imaging problems via neural network modeling and TensorFlow. Biomed. Opt. Express **9**, 3306–3319 (2018)

22. Strohl, F., Jadhav, S., Ahluwalia, B.S., Agarwal, K., Prasad, D.K.: Object detection neural network improves Fourier ptychography reconstruction. Opt. Express **28**, 37199–37208 (2020)

23. Zhang, J., Xu, T., Shen, Z., Qiao, Y., Zhang, Y.: Fourier ptychographic microscopy reconstruction with multiscale deep residual network. Opt. Express **27**, 8612–8625 (2019)

24. Zhang, J., Xu, T., Li, X., Zhang, Y., Chen, Y., Wang, X., Wang, S., Wang, C.: High-throughput deep learning microscopy using multi-angle super-resolution. IEEE Photonics J. **12**, 6900914 (2020)

25. Sun, M., Chen, X., Zhu, Y., Li, D., Mu, Q., Xuan, L.: Neural network model combined with pupil recovery for Fourier ptychographic microscopy. Opt. Express **27**, 24161–24174 (2019)

26. Zhang, J., Tao, X., Yang, L., Wu, R., Sun, P., Wang, C., Zheng, Z.: Forward imaging neural network with correction of positional misalignment for Fourier ptychographic microscopy. Opt. Express **28**, 23164–23175 (2020)

27. Wang, K., Di, J., Li, Y., Ren, Z., Qian, K., Zhao, J.: Transport of intensity equation from a single intensity image via deep learning. Opt. Laser Eng. **134**, 106233 (2020)

28. Hu, L., Hu, S., Gong, W., Si, K.: Learning-based Shack-Hartmann wavefront sensor for high-order aberration detection. Opt. Express **27**, 33504–33517 (2019)

29. Hu, L., Hu, S., Gong, W., Si, K.: Deep learning assisted Shack-Hartmann wavefront sensor for direct wavefront detection. Opt. Lett. **45**, 3741–3744 (2020)

30. DuBose, T.B., Gardner, D.F., Watnik, A.T.: Intensity-enhanced deep network wavefront reconstruction in Shack-Hartmann sensors. Opt. Lett. **45**, 1699–1702 (2020)
31. Li, X., Qi, H., Jiang, S., Song, P., Zheng, G., Zhang, Y.: Quantitative phase imaging via a cGAN network with dual intensity images captured under centrosymmetric illumination. Opt. Lett. **44**, 2879–2882 (2019)
32. Komuro, K., Nomura, T., Barbastathis, G.: Deep ghost phase imaging. Appl. Opt. **59**, 3376–3382 (2020)
33. Xue, Y., Cheng, S., Li, Y., Tian, L.: Reliable deep-learning-based phase imaging with uncertainty quantification. Optica **6**, 618–629 (2019)
34. Bostan, E., Heckel, R., Chen, M., Kellman, M., Waller, L.: Deep phase decoder: self-calibrating phase microscopy with an untrained deep neural network. Optica **7**, 559 562 (2020)
35. Wang, K., Li, Y., Qian, K., Di, J., Zhao, J.: One-step robust deep learning phase unwrapping. Opt. Express **27**, 15100–15115 (2019)
36. Dardikman-Yoffe, G., Roitshtain, D., Mirsky, S.K., Turko, N.A., Habaza, M., Shaked, N.T.: PhUn-Net: ready-to-use neural network for unwrapping quantitative phase images of biological cells. Biomed. Opt. Express **11**, 1107–1121 (2020)
37. Qin, Y., Wan, S., Wan, Y., Weng, J., Liu, W., Gong, Q.: Direct and accurate phase unwrapping with deep neural network. Appl. Opt. **59**, 7258–7267 (2020)
38. Zhang, Y.B., Koydemir, H.C., Shimogawa, M.M., Yalcin, S., Guziak, A., Liu, T., Oguz, I., Huang, Y., Bai, B., Luo, Y., Luo, Y., Wei, Z., Wang, H., Bianco, V., Zhang, B., Nadkarni, R., Hill, K., Ozcan, A.: Motility-based label-free detection of parasites in bodily fluids using holographic speckle analysis and deep learning. Light-Sci. Appl. **7**, 108 (2018)
39. Wu, Y., Calis, A., Luo, Y., Chen, C., Lutton, M., Rivenson, Y., Lin, X., Koydemir, H.C., Zhang, Y., Wang, H., Gorocs, Z., Ozcan, A.: Label-free bioaerosol sensing using mobile microscopy and deep learning. ACS Photonics **5**, 4617–4627 (2018)
40. Zhang, Y., Ouyang, M., Ray, A., Liu, T., Kong, J., Bai, B., Kim, D., Guziak, A., Luo, Y., Feizi, A., Tsai, K., Duan, Z., Liu, X., Kim, D., Cheung, C., Yalcin, S., Koydemir, H.C., Garner, O.B., Carlo, D.D., Ozcan, A.: Computational cytometer based on magnetically modulated coherent imaging and deep learning. Light-Sci. Appl. **8**, 91 (2019)
41. Wu, Y., Ray, A., Wei, Q., Feizi, A., Tong, X., Chen, E., Luo, Y., Ozcan, A.: Deep learning enables high-throughput analysis of particle-aggregation-based biosensors imaged using holography. ACS Photonics **6**, 294–301 (2019)
42. Yi, F., Moon, I., Javidi, B.: Automated red blood cells extraction from holographic images using fully convolutional neural networks. Biomed. Opt. Express **8**, 4466–4479 (2017)
43. Ahmadzadeh, E., Jaferzadeh, K., Shin, S., Moon, I.: Automated single cardiomyocyte characterization by nucleus extraction from dynamic holographic images using a fully convolutional neural network. Biomed. Opt. Express **11**, 1501–1516 (2020)
44. Kim, S.J., Wang, C., Zhao, B., Im, H., Min, J., Choi, H.J., Tadros, J., Choi, N.R., Castro, C.M., Weissleder, R., Lee, H., Lee, K.: Deep transfer learning-based hologram classification for molecular diagnostics. Sci. Rep. **8**, 17003 (2018)
45. Li, Y., Di, J., Wang, K., Wang, S., Zhao, J.: Classification of cell morphology with quantitative phase microscopy and machine learning. Opt. Express **28**, 23916–23927 (2020)
46. Ayyappan, V., Chang, A., Zhang, C., Paidi, S.K., Bordett, R., Liang, T., Barman, I., Pandey, R.: Identification and staging of B-cell acute lymphoblastic leukemia using quantitative phase imaging and machine learning. ACS Sensors **5**, 3281–3289 (2020)
47. Kandel, M.E., Rubessa, M., He, Y.R., Schreiber, S., Meyers, S., Naves, L.M., Sermersheim, M.K., Sell, G.S., Szewczyk, M.J., Sobh, N., Wheeler, M.B., Popescu, G.: Reproductive outcomes predicted by phase imaging with computational specificity of spermatozoon ultrastructure. Proc. Natl. Acad. Sci. U.S.A. **117**, 18302–18309 (2020)
48. Kim, G., Jo, Y.J., Cho, H., Min, H., Park, Y.K.: Learning-based screening of hematologic disorders using quantitative phase imaging of individual red blood cells. Biosens. Bioelectron. **123**, 69–76 (2019)

49. O'Connor, T., Hawxhurst, C., Shor, L.M., Javidi, B.: Red blood cell classification in lensless single random phase encoding using convolutional neural networks. Opt. Express **28**, 33504–33515 (2020)
50. O'Connor, T., Anand, A., Andemariam, B., Javidi, B.: Deep learning-based cell identification and disease diagnosis using spatio-temporal cellular dynamics in compact digital holographic microscopy. Biomed. Opt. Express **11**, 4491–4508 (2020)
51. Lin, Y.H., Liao, K.Y.K., Sung, K.B.: Automatic detection and characterization of quantitative phase images of thalassemic red blood cells using a mask region-based convolutional neural network. J. Biomed. Opt. **25**, 116502 (2020)
52. Goy, A., Rughoobur, G., Li, S., Arthur, K., Akinwande, A.I., Barbastathis, G.: High-resolution limited-angle phase tomography of dense layered objects using deep neural networks. Proc. Natl. Acad. Sci. U.S.A. **116**, 19848–19856 (2019)
53. Choi, G., Ryu, D.H., Jo, Y.J., Kim, Y.S., Park, W., Min, H., Park, Y.K.: Cycle-consistent deep learning approach to coherent noise reduction in optical diffraction tomography. Opt. Express **27**, 4927–4943 (2019)
54. Goy, A., Arthur, K., Li, S., Barbastathis, G.: Low photon count phase retrieval using deep learning. Phys. Rev. Lett. **121**, 243902 (2018)
55. Deng, M., Li, S., Goy, A., Kang, I., Barbastathis, G.: Learning to synthesize: robust phase retrieval at low photon counts. Light-Sci. Appl. **9**, 36 (2020)
56. Deng, M., Goy, A., Li, S., Arthur, K., Barbastathis, G.: Probing shallower: perceptual loss trained Phase Extraction Neural Network (PLT-PhENN) for artifact-free reconstruction at low photon budget. Opt. Express **28**, 2511–2535 (2020)
57. Rivenson, Y., Liu, T., Wei, Z., Zhang, Y., de Haan, K., Ozcan, A.: PhaseStain: the digital staining of label-free quantitative phase microscopy images using deep learning. Light-Sci. Appl. **8**, 23 (2019)
58. Gorocs, Z., Tamamitsu, M., Bianco, V., Wolf, P., Roy, S., Shindo, K., Yanny, K., Wu, Y., Koydemir, H.C., Rivenson, Y., Ozcan, A.: A deep learning-enabled portable imaging flow cytometer for cost-effective, high-throughput, and label-free analysis of natural water samples. Light-Sci. Appl. **7**, 66 (2018)
59. Wu, Y., Luo, Y., Chaudhari, G., Rivenson, Y., Calis, A., de Haan, K., Ozcan, A.: Bright-field holography: cross-modality deep learning enables snapshot 3D imaging with bright-field contrast using a single hologram. Light-Sci. Appl. **8**, 25 (2019)
60. Liu, T., Wei, Z., Rivenson, Y., de Haan, K., Zhang, Y., Wu, Y., Ozcan, A.: Deep learning-based color holographic microscopy. J. Biophotonics **12**, e201900107 (2019)
61. Liu, T., de Haan, K., Bai, B., Rivenson, Y., Luo, Y., Wang, H., Karalli, D., Fu, H., Zhang, Y., FitzGerald, J., Ozcan, A.: Deep learning-based holographic polarization microscopy. ACS Photonics **7**, 3023–3034 (2020)
62. Nygate, Y.N., Levi, M., Mirsky, S.K., Turko, N.A., Rubin, M., Barnea, I., Dardikman-Yoffe, G., Haifler, M., Shalev, A., Shaked, N.T.: Holographic virtual staining of individual biological cells. Proc. Natl. Acad. Sci. U.S.A. **117**, 9223–9231 (2020)
63. Wang, R., Song, P., Jiang, S., Yan, C., Zhu, J., Guo, C., Bian, Z., Wang, T., Zheng, G.: Virtual brightfield and fluorescence staining for Fourier ptychography via unsupervised deep learning. Opt. Lett. **45**, 5405–5408 (2020)
64. Kandel, M.E., He, Y.R., Lee, Y.J., Chen, T.H.Y., Sullivan, K.M., Aydin, O., Saif, M.T.A., Kong, H., Sobh, N., Popescu, G.: Phase imaging with computational specificity (PICS) for measuring dry mass changes in sub-cellular compartments. Nat. Commun. **11**, 6265 (2020)
65. Barbastathis, G., Ozcan, A., Situ, G.: On the use of deep learning for computational imaging. Optica **6**, 921–943 (2019)
66. Jo, Y.J., Cho, H., Lee, S.Y., Choi, G., Kim, G., Min, H., Park, Y.K.: Quantitative phase imaging and artificial intelligence: a review. IEEE J. Sel. Top. Quantum Electron. **25**, 6800914 (2019)
67. de Haan, K., Rivenson, Y., Wu, Y., Ozcan, A.: Deep-learning-based image reconstruction and enhancement in optical microscopy. Proc. IEEE **108**, 30–50 (2020)
68. Wetzstein, G., Ozcan, A., Gigan, S., Fan, S., Englund, D., Soljacic, M., Denz, C., Miller, D.A.B., Psaltis, D.: Inference in artificial intelligence with deep optics and photonics. Nature **588**, 39–47 (2020)

69. Guo, C., Bian, Z., Jiang, S., Murphy, M., Zhu, J., Wang, R., Song, P., Shao, X., Zhang, Y., Zheng, G.: OpenWSI: a low-cost, high-throughput whole slide imaging system via single-frame autofocusing and open-source hardware. Opt. Lett. **45**, 260–263 (2020)
70. Collins, J.T., Knapper, J., Stirling, J., Mduda, J., Mkindi, C., Mayagaya, V., Mwakajinga, G.A., Nyakyi, P.T., Sanga, V.L., Carbery, D., White, L., Dale, S., Lim, Z.J., Baumberg, J.J., Cicuta, P., McDermott, S., Vodenicharski, B., Bowman, R.: Robotic microscopy for everyone: the OpenFlexure microscope. Biomed. Opt. Express **11**, 2447–2460 (2020)
71. Forcucci, A., Pawlowski, M.E., Majors, C., Richards-Kortum, R., Tkaczyk, T.S.: All-plastic, miniature, digital fluorescence microscope for three part white blood cell differential measurements at the point of care. Biomed. Opt. Express **6**, 4433–4446 (2015)
72. Wong, C., Pawlowski, M.E., Forcucci, A., Majors, C.E., Richards-Kortum, R., Tkaczyk, T.S.: Development of a universal, tunable, miniature fluorescence microscope for use at the point of care. Biomed. Opt. Express **9**, 1041–1056 (2018)
73. https://cellscope.berkeley.edu/
74. Breslauer, D.N., Maamari, R.N., Switz, N.A., Lam, W.A., Fletcher, D.A.: Mobile phone based clinical microscopy for global health applications. PLoS ONE **4**, e6320 (2009)
75. Skandarajah, A., Reber, C.D., Switz, N.A., Fletcher, D.A.: Quantitative imaging with a mobile phone microscope. PLoS ONE **9**, e96906 (2014)
76. Ming, K., Kim, J., Biondi, M.J., Syed, A., Chen, K., Lam, A., Ostrowski, M., Rebbapragada, A., Feld, J.J., Chan, W.C.W.: Integrated quantum dot barcode smartphone optical device for wireless multiplexed diagnosis of infected patients. ACS Nano **9**, 3060–3074 (2015)
77. Kang, W., Huang, H., Cai, M., Li, Y., Hou, W., Yun, F., Wu, X., Xue, L., Wang, S., Liu, F.: On-site cell concentration and viability detections using smartphone based field-portable cell counter. Anal. Chim. Acta **1077**, 216–224 (2019)
78. Shan, Y., Wang, B., Huang, H., Jian, D., Wu, X., Xue, L., Wang, S., Liu, F.: On-site quantitative Hg^{2+} measurements based on selective and sensitive fluorescence biosensor and miniaturized smartphone fluorescence microscope. Biosens. Bioelectron. **132**, 238–247 (2019)
79. Switz, N.A., D'Ambrosio, M.V., Fletcher, D.A.: Low-cost mobile phone microscopy with a reversed mobile phone camera lens. PLoS ONE **9**, e95330 (2014)
80. D'Ambrosio, M.V., Bakalar, M., Bennuru, S., Reber, C., Skandarajah, A., Nilsson, L., Switz, N., Kamgno, J., Pion, S., Boussinesq, M., Nutman, T.B., Fletcher, D.A.: Point-of-care quantification of blood-borne filarial parasites with a mobile phone microscope. Sci. Transl. Med. **7**, 286re4 (2015)
81. Zhu, H., Mavandadi, S., Coskun, A.F., Yaglidere, O., Ozcan, A.: Optofluidic fluorescent imaging cytometry on a cell phone. Anal. Chem. **83**, 6641–6647 (2013)
82. Zhu, H., Yaglidere, O., Su, T.W., Tseng, D., Ozcan, A.: Cost-effective and compact wide-field fluorescent imaging on a cell-phone. Lab Chip **11**, 315–322 (2011)
83. Koydemir, H.C., Gorocs, Z., Tseng, D., Cortazar, B., Feng, S., Chan, R.Y.L., Burbano, J., McLeod, E., Ozcan, A.: Rapid imaging, detection and quantification of Giardia lamblia cysts using mobile-phone based fluorescent microscopy and machine learning. Lab Chip **15**, 1284–1293 (2015)
84. Muller, V., Sousa, J.M., Koydemir, H.C., Veli, M., Tseng, D., Cerqueira, L., Ozcan, A., Azevedo, N.F., Westerlund, F.: Identification of pathogenic bacteria in complex samples using a smartphone based fluorescence microscope. RSC Adv. **8**, 36493–36502 (2018)
85. Snow, J.W., Koydemir, H.C., Karinca, D.K., Liang, K., Tseng, D., Ozcan, A.: Rapid imaging, detection, and quantification of Nosema ceranae spores in honey bees using mobile phone-based fluorescence microscopy. Lab Chip **19**, 789–797 (2019)
86. Wei, Q., Qi, H., Luo, W., Tseng, D., Ki, S.J., Wan, Z., Gorocs, Z., Bentolila, L.A., Wu, T.T., Sun, R., Ozcan, A.: Fluorescent imaging of single nanoparticles and viruses on a smart phone. ACS Nano **7**, 9147–9155 (2013)
87. Wei, Q., Luo, W., Chiang, S., Kappel, T., Mejia, C., Tseng, D., Chan, R.Y.L., Yan, E., Qi, H., Shabbir, F., Ozkan, H., Feng, S., Ozcan, A.: Imaging and sizing of single DNA molecules on a mobile phone. ACS Nano **8**, 12725–12733 (2014)

88. Kuhnemund, M., Wei, Q., Darai, E., Wang, Y., Hernandez-Neuta, I., Yang, Z., Tseng, D., Ahlford, A., Mathot, L., Sjoblom, T., Ozcan, A., Nilsson, M.: Targeted DNA sequencing and in situ mutation analysis using mobile phone microscopy. Nat. Commun. **8**, 13913 (2017)

89. Szydlowski, N.A., Jing, H., Alqashmi, M., Hu, Y.S.: Cell phone digital microscopy using an oil droplet. Biomed. Opt. Express **11**, 2328–2338 (2020)

90. Orth, A., Wilson, E.R., Thompson, J.G., Gibson, B.C.: A dual-mode mobile phone microscope using the onboard camera flash and ambient light. Sci. Rep. **8**, 3298 (2018)

91. Sung, Y., Campa, F., Shih, W.C.: Open-source do-it-yourself multi-color fluorescence smartphone microscopy. Biomed. Opt. Express **8**, 5075–5086 (2017)

92. Freeman, E.E., Semeere, A., Osman, H., Peterson, G., Rajadhyaksha, M., Gonzalez, S., Martin, J.N., Anderson, R.R., Tearney, G.J., Kang, D.: Smartphone confocal microscopy for imaging cellular structures in human skin in vivo. Biomed. Opt. Express **9**, 1906–1915 (2018)

93. Wei, Q., Acuna, G., Kim, S., Vietz, C., Tseng, D., Chae, J., Shir, D., Luo, W., Tinnefeld, P., Ozcan, A.: Plasmonics enhanced smartphone fluorescence microscopy. Sci. Rep. **7**, 2124 (2017)

94. Rivenson, Y., Koydemir, H.C., Wang, H., Wei, Z., Ren, Z., Gunaydın, H., Zhang, Y., Gorocs, Z., Liang, K., Tseng, D., Ozcan, A.: Deep learning enhanced mobile-phone microscopy. ACS Photonics **5**, 2354–2364 (2018)

95. Heng, X., Erickson, D., Baugh, L.R., Yaqoob, Z., Sternberg, P.W., Psaltis, D., Yang, C.: Optofluidic microscopy-a method for implementing a high resolution optical microscope on a chip. Lab Chip **6**, 1274–1276 (2006)

96. Cui, X., Lee, L.M., Heng, X., Zhong, W., Sternberg, P.W., Psaltis, D., Yang, C.: Lensless high-resolution on-chip optofluidic microscopes for Caenorhabditis elegans and cell imaging. Proc. Natl. Acad. Sci. U.S.A. **105**, 10670–10675 (2008)

97. Lee, L.M., Cui, X., Yang, C.: The application of on-chip optofluidic microscopy for imaging Giardia lamblia trophozoites and cysts. Biomed. Microdevices **11**, 951–958 (2009)

98. Pang, S., Cui, X., DeModena, J., Wang, Y.M., Sternberg, P., Yang, C.: Implementation of a color-capable optofluidic microscope on a RGB CMOS color sensor chip substrate. Lab Chip **10**, 411–414 (2010)

99. Pang, S., Han, C., Lee, L.M., Yang, C.: Fluorescence microscopy imaging with a Fresnel zone plate array based optofluidic microscope. Lab Chip **11**, 3698–3702 (2011)

100. Zheng, G., Lee, S.A., Antebi, Y., Elowitz, M.B., Yang, C.: The ePetri dish, an on-chip cell imaging platform based on subpixel perspective sweeping microscopy (SPSM). Proc. Natl. Acad. Sci. U.S.A. **108**, 16889–16894 (2011)

101. Farsiu, S., Robinson, M.D., Elad, M., Milanfar, P.: Fast and robust multiframe super resolution. IEEE T. Image Process **13**, 1327–1344 (2004)

102. Farsiu, S., Elad, M., Milanfar, P.: Multiframe demosaicing and super-resolution of color Images. IEEE T. Image Process **15**, 141–159 (2006)

103. Lee, S.A., Leitao, R., Zheng, G., Yang, S., Rodriguez, A., Yang, C.: Color capable sub-pixel resolving optofluidic microscope and its application to blood cell imaging for Malaria diagnosis. PLoS ONE **6**, e26127 (2011)

104. Lee, S.A., Zheng, G.A., Mukherjee, N., Yang, C.: On-chip continuous monitoring of motile microorganisms on an ePetri platform. Lab Chip **12**, 2385–2390 (2012)

105. Lee, S.A., Erath, J., Zheng, G., Ou, X., Willems, P., Eichinger, D., Rodriguez, A., Yang, C.: Imaging and identification of waterborne parasites using a chip-scale microscope. PLoS ONE **9**, e89712 (2014)

106. Lee, S.A., Yang, C.: A smartphone-based chip-scale microscope using ambient illumination. Lab Chip **14**, 3056–3063 (2014)

107. https://www.foldscope.com/

108. Cybulski, J.S., Clements, J., Prakash, M.: Foldscope: Origami based paper microscope. PLoS ONE **9**, e98781 (2014)

109. Shaked, N.T.: Quantitative phase microscopy of biological samples using a portable interferometer. Opt. Lett. **37**, 2016–2018 (2012)

110. Girshovitz, P., Shaked, N.T.: Compact and portable low-coherence interferometer with off-axis geometry for quantitative phase microscopy and nanoscopy. Opt. Express **21**, 5701–5714 (2013)
111. Baek, Y.S., Lee, K.R., Yoon, J., Kim, K., Park, Y.K.: White-light quantitative phase imaging unit. Opt. Express **24**, 9308–9315 (2016)
112. O'Connor, T., Anand, A., Javidi, B.: Field-portable microsphere-assisted high resolution digital holographic microscopy in compact and 3D-printed Mach-Zehnder Interferometer. OSA Continuum **3**, 1013–1020 (2020)
113. Singh, A.S.G., Anand, A., Leitgeb, R.A., Javidi, B.: Lateral shearing digital holographic imaging of small biological specimens. Opt. Express **21**, 23617–23622 (2012)
114. Javidi, B., Markman, A., Rawat, S., O'Connor, T., Anand, A., Andemariam, B.: Sickle cell disease diagnosis based on spatio-temporal cell dynamics analysis using 3D printed shearing digital holographic microscopy. Opt. Express **26**, 13614–13627 (2018)
115. Rawat, S., Komatsu, S., Markman, A., Anand, A., Javidi, B.: Compact and field-portable 3D printed shearing digital holographic microscope for automated cell identification. Appl. Opt. **56**, D127–D133 (2017)
116. Mudanyali, O., Oztoprak, C., Tseng, D., Erlinger, A., Ozcan, A.: Detection of waterborne parasites using field-portable and cost-effective lensfree microscopy. Lab Chip **10**, 2419–2423 (2010)
117. Su, T.W., Erlinger, A., Tseng, D., Ozcan, A.: Compact and light-weight automated semen analysis platform using lensfree on-chip microscopy. Anal. Chem. **82**, 8307–8312 (2010)
118. Tseng, D., Mudanyali, O., Oztoprak, C., Isikman, S.O., Sencan, I., Yaglidere, O., Ozcan, A.: Lensfree microscopy on a cellphone. Lab Chip **10**, 1787–1792 (2010)
119. Isikman, S.O., Bishara, W., Sikora, U., Yaglidere, O., Yeah, J., Ozcan, A.: Field-portable lensfree tomographic microscope. Lab Chip **11**, 2222–2230 (2011)
120. Greenbaum, A., Sikora, U., Ozcan, A.: Field-portable wide-field microscopy of dense samples using multi-height pixel super-resolution based lensfree imaging. Lab Chip **12**, 1242–1245 (2012)
121. Greenbaum, A., Akbari, N., Feizi, A., Luo, W., Ozcan, A.: Field-portable pixel super-resolution colour microscope. PLoS ONE **8**, e76475 (2013)
122. Bishara, W., Sikora, U., Mudanyali, O., Su, T.W., Yaglidere, O., Luckhart, S., Ozcan, A.: Holographic pixel super-resolution in portable lensless on-chip microscopy using a fiber-optic array. Lab Chip **11**, 1276–1279 (2011)
123. Ray, A., Khalid, M.A., Demcenko, A., Daloglu, M., Tseng, D., Reboud, J., Cooper, J.M., Ozcan, A.: Holographic detection of nanoparticles using acoustically actuated nanolenses. Nat. Commun. **11**, 171 (2020)
124. Pushkarsky, I., Lyb, Y., Weaver, W., Su, T.W., Mudanyali, O., Ozcan, A., Di Carlo, D.: Automated single-cell motility analysis on a chip using lensfree microscopy. Sci. Rep. **4**, 4717 (2014)
125. Feizi, A., Zhang, Y., Greenbaum, A., Guziak, A., Luong, M., Chan, R.Y.L., Berg, B., Ozkan, H., Luo, W., Wu, M., Wu, Y., Ozcan, A.: Rapid, portable and cost-effective yeast cell viability and concentration analysis using lensfree on-chip microscopy and machine learning. Lab Chip **16**, 4350–4358 (2016)
126. Veli, M., Ozcan, A.: Computational sensing of Staphylococcus aureus on contact lenses using 3D imaging of curved surfaces and machine learning. ACS Nano **12**, 2554–2559 (2018)
127. Im, H., Pathania, D., McFarland, P.J., Sohani, A.R., Degani, I., Allen, M., Coble, B., Kilcoyne, A., Hong, S., Rohrer, L., Abramson, J.S., Dryden-Peterson, S., Fexon, L., Pivovarov, M., Chabner, B., Lee, H., Castro, C.M., Weissleder, R.: Design and clinical validation of a point-of-care device for the diagnosis of lymphoma via contrastenhanced microholography and machine learning. Nat. Biomed. Eng. **2**, 666–674 (2018)
128. Im, H., Castro, C.M., Shao, H., Liong, M., Song, J., Pathania, D., Fexon, L., Min, C., Avila-Wallace, M., Zurkiya, O., Rho, J., Magaoay, B., Tambouret, R.H., Pivovarov, M., Weissleder, R., Lee, H.: Digital diffraction analysis enables low-cost molecular diagnostics on a smartphone. Proc. Natl. Acad. Sci. U.S.A. **112**, 5613–5618 (2015)

129. Kesavan, S.V., Momey, F., Cioni, O., David-Watine, B., Dubrulle, N., Shorte, S., Sulpice, E., Freida, D., Chalmond, B., Dinten, J.M., Gidrol, X., Allier, C.: High-throughput monitoring of major cell functions by means of lensfree video microscopy. Sci. Rep. **4**, 5942 (2014)

130. Kesavan, S.V., Navarro, F.P., Menneteau, M., Mittler, F., David-Watine, B., Dubrulle, N., Shorte, S.L., Chalmond, B., Dinten, J.M., Allier, C.P.: Real-time label-free detection of dividing cells by means of lensfree video-microscopy. J. Biomed. Opt. **19**, 036004 (2014)

131. Momey, F., Coutard, J.G., Bordy, T., Navarro, F., Menneteau, M., Dinten, J.M., Allier, C.: Dynamics of cell and tissue growth acquired by means of extended field of view lensfree microscopy. Biomed. Opt. Express **7**, 512–524 (2016)

132. Dong, S., Guo, K., Nanda, P., Shiradkar, R., Zheng, G.: FPscope: a field-portable high-resolution microscope using a cellphone lens. Biomed. Opt. Express **5**, 3305–3310 (2014)

133. Aidukas, T., Eckert, R., Harvey, A.R., Waller, L., Konda, P.C.: Low-cost, sub-micron resolution, wide-field computational microscopy using opensource hardware. Sci. Rep. **9**, 7457 (2019)

134. Kim, J., Henley, B.M., Kim, C.H., Lester, H.A., Yang, C.: Incubator embedded cell culture imaging system (EmSight) based on Fourier ptychographic microscopy. Biomed. Opt. Express **7**, 3097–3110 (2016)

135. Jiang, S., Zhu, J., Song, P., Guo, C., Bian, Z., Wang, R., Huang, Y., Wang, S., Zhang, H., Zheng, G.: Wide-field, high-resolution lensless on-chip microscopy via near-field blind ptychographic modulation. Lab Chip **17**, 1058–1065 (2020)

136. Zhang, H., Bian, Z., Jiang, S., Liu, J., Song, P., Zheng, G.: Field-portable quantitative lensless microscopy based on translated speckle illumination and sub-sampled ptychographic phase retrieval. Opt. Lett. **44**, 1976–1979 (2019)

137. Meng, X., Huang, H., Yan, K., Tian, X., Yu, W., Cui, H., Kong, Y., Xue, L., Liu, C., Wang, S.: Smartphone based hand-held quantitative phase microscope using the transport of intensity equation method. Lab Chip **17**, 104–109 (2017)

138. Yang, Z., Zhan, Q.: Single-shot smartphone-based quantitative phase imaging using a distorted grating. PLoS ONE **11**, e0159596 (2016)

139. Mehta, S.B., Sheppard, C.J.R.: Quantitative phase-gradient imaging at high resolution with asymmetric illumination-based differential phase contrast. Opt. Lett. **34**, 1924–1926 (2009)

140. Tian, L., Waller, L.: Quantitative differential phase contrast imaging in an LED array microscope. Opt. Express **23**, 11394–11403 (2015)

141. Chen, M., Phillips, Z.F., Waller, L.: Quantitative differential phase contrast (DPC) microscopy with computational aberration correction. Opt. Express **26**, 32888–32899 (2018)

142. Phillips, Z.F., D'Ambrosio, M.V., Tian, L., Rulison, J.J., Patel, H.S., Sadras, N., Gande, A.V., Switz, N.A., Fletcher, D.A., Waller, L.: Multi-contrast imaging and digital refocusing on a mobile microscope with a domed LED array. PLoS ONE **10**, e0124938 (2015)

143. Tian, L., Liu, Z., Yeh, L.H., Chen, M., Zhong, J., Waller, L.: Computational illumination for high-speed in vitro Fourier ptychographic microscopy. Optica **2**, 904–911 (2015)

Printed by Printforce, the Netherlands